Pipefitting

Level One

Trainee Guide

Boston Columbus Indianapolis New York San Francisco Amsterdam
Cape Town Dubai London Madrid Milan Munich Paris Montreal Toronto Delhi
Mexico City Sao Paulo Sydney Hong Kong Seoul Singapore Taipei Tokyo

NCCER

President and Cheif Executive Officer: Don Whyte
President: Boyd Worsham
Vice President: Steve Greene
Chief Operations Officer: Katrina Kersch
Pipefitting Curriculum Project Manager: Elizabeth Schlaupitz
Senior Development Manager: Mark Thomas
Senior Production Manager: Tim Davis
Primary Technical Writer: Heidi Saliba
Technical Writer: Gary Ferguson
Managing Editor: Natalie Richoux
Desktop Publishing Manager: James McKay
Art Manager: Kelly Sadler
Digital Content Coordinator: Erin O'Nora
Desktop Publishing Specialists: Gene Page, Eric Caraballoso
Editors: Graham Hack, Jordan Hutchinson
Production Assistance: Rachel Downs

Pearson

Director of Alliance/Partnership Management: Kelly Trakalo
Content Producer: Alexandrina B. Wolf
Assistant Content Producer: Alma Dabral
Digital Content Producer: Jose Carchi
Senior Marketing Manager: Brian Hoehl
Composition: NCCER
Printer/Binder: LSC Communications
Cover Printer: LSC Communications
Text Fonts: Palatino and Univers

Credits and acknowledgments for content borrowed from other sources and reproduced, with permission, in this textbook appear at the end of each module.

ISBN-13: 978-0-13-580941-9
ISBN-10: 0-13-580941-X

25 2023

Preface

To the Trainee

There are some who may consider pipefitting synonymous with plumbing, but these are really two very distinct crafts. Plumbers install and repair the water, waste disposal, drainage, and gas systems in homes and in commercial and industrial buildings. Pipefitters, on the other hand, install and repair both high- and low-pressure pipe systems used in manufacturing, in the generation of electricity, and in the heating and cooling of buildings.

If you're trying to imagine a setting involving pipefitters, think of large power plants that create and distribute energy throughout the nation; think of manufacturing plants, chemical plants, and piping systems that carry all kinds of liquid, gaseous, and solid materials.

If you're trying to imagine a job in pipefitting, picture yourself in a job that won't go away for a long time. As the US government reports, the demand for skilled pipefitters continues to outpace the supply of workers trained in this craft. And high demand typically means higher pay, making pipefitters among the highest-paid construction workers in the nation.

While pipefitters and plumbers perform different tasks, the aptitudes involved in these crafts are comparable. Attention to detail, spatial and mechanical abilities, and the ability to work efficiently with the tools of their trade are key. If you think you might have what it takes to work in this high-demand occupation, contact your local NCCER Training Sponsor to see if they offer a training program in this craft, or contact your local union or non-union training programs. You might be the perfect fit.

We wish you success as you embark on your first year of training in the pipefitting craft and hope that you'll continue your training beyond this textbook. There are more than a half-million people employed in this work in the United States, and as most of them can tell you, there are many opportunities awaiting those with the skills and desire to move forward in the construction industry.

We invite you to visit the NCCER website at **www.nccer.org** for the latest releases, training information, Cornerstone magazine, and much more. You can also reference the Pearson product catalog online at **www.crafttraining.com**. Your feedback is welcome. You may email your comments to **curriculum@nccer.org** or send general comments and inquiries to **info@nccer.org**.

New with *Pipefitting Level One*

NCCER is proud to release *Pipefitting* with our latest instructional systems design, linking learning objectives to each module's content. Each module has been updated to reflect feedback from the subject matter experts committee who contributed to the update of curriculum information and provided recommendations for prioritizing competencies required by pipefitters, with portions of the previous content edited for clarity and efficiency. There are additional study questions and updated graphics and photos. In module 08101, the content has been rewritten in a concise and straightforward manner in order to make the text easy to remember. In module 08105, performance tasks have been updated to create efficiency during the evaluation process. Finally, 29102 – Oxyfuel Cutting, from the Welding craft, has replaced the previous edition's module on oxyfuel cutting (08104-06).

We invite you to visit the NCCER website at **www.nccer.org** for information on the latest product releases and training, as well as online versions of the *Cornerstone* magazine and Pearson's NCCER product catalog.

Your feedback is welcome. You may email your comments to **curriculum@nccer.org** or send general comments and inquiries to **info@nccer.org**.

NCCER Standardized Curricula

NCCER is a not-for-profit 501(c)(3) education foundation established in 1996 by the world's largest and most progressive construction companies and national construction associations. It was founded to address the severe workforce shortage facing the industry and to develop a standardized training process and curricula. Today, NCCER is supported by hundreds of leading construction and maintenance companies, manufacturers, and national associations. The NCCER Standardized Curricula was developed by NCCER in partnership with Pearson, the world's largest educational publisher.

Some features of the NCCER Standardized Curricula are as follows:

- An industry-proven record of success
- Curricula developed by the industry, for the industry
- National standardization providing portability of learned job skills and educational credits
- Compliance with the Office of Apprenticeship requirements for related classroom training (*CFR 29:29*)
- Well-illustrated, up-to-date, and practical information

NCCER also maintains the NCCER Registry, which provides transcripts, certificates, and wallet cards to individuals who have successfully completed a level of training within a craft in NCCER's Curricula. *Training programs must be delivered by an NCCER Accredited Training Sponsor in order to receive these credentials.*

Special Features

In an effort to provide a comprehensive and user-friendly training resource, this curriculum showcases several informative features. Whether you are a visual or hands-on learner, these features are intended to enhance your knowledge of the construction industry as you progress in your training. Some of the features you may find in the curriculum are explained below.

Introduction

This introductory page, found at the beginning of each module, lists the module Objectives, Performance Tasks, and Trade Terms. The Objectives list the knowledge you will acquire after successfully completing the module. The Performance Tasks give you an opportunity to apply your knowledge to real-world tasks. The Trade Terms are industry-specific vocabulary that you will learn as you study this module.

00104
INTRODUCTION TO POWER TOOLS

Objectives

When you have completed this module, you will be able to do the following:

1. Identify and explain how to use various types of power drills and impact wrenches.
   a. Identify and explain how to use common power drills and bits.
   b. Identify and explain how to use a hammer drill.
   c. Identify and explain how to use pneumatic drills and impact wrenches.
2. Identify and explain how to use various types of power saws.
   a. Identify and explain how to use a circular saw.
   b. Identify and explain how to use saber and reciprocating saws.
   c. Identify and explain how to use a portable band saw.
   d. Identify and explain how to use miter and cutoff saws.
3. Identify and explain how to use various grinders and grinder attachments.
   a. Identify and explain how to use various types of grinders.
   b. Identify and explain how to use various grinder accessories and attachments.
4. Identify and explain how to use miscellaneous power tools.
   a. Identify and explain how to use pneumatic and powder-actuated fastening tools.
   b. Identify and explain how to use pavement breakers.
   c. Identify and explain the uses of hydraulic jacks.

Performance Tasks

Under supervision of your instructor, you should be able to do the following:

1. Safely and properly demonstrate the use of three of the following tools:
   - Electric drill
   - Hammer drill or rotary hammer
   - Circular saw
   - Reciprocating saw
   - Portable band saw
   - Miter or cutoff saw
   - Portable or bench grinder
   - Pneumatic nail gun
   - Pavement breaker

Trade Terms

Abrasive
Alternating current (AC)
Arbor
Auger bit
Carbide
Chuck
Chuck key
Countersink
Direct current (DC)
Forstner bit
Grit
Ground fault circuit interrupter (GFCI)
Ground fault protection
Kerf
Masonry bit
Reciprocating
Revolutions per minute (rpm)
Ring test
Shank
Trigger lock

Industry Recognized Credentials

If you are training through an NCCER-accredited sponsor, you may be eligible for credentials from NCCER's Registry. The ID number for this module is 00104. Note that this module may have been used in other NCCER curricula and may apply to other level completions. Contact NCCER's Registry at 888.622.3720 or go to **www.nccer.org** for more information.

Trade Features

Trade features present technical tips and professional practices based on real-life scenarios similar to those you might encounter on the job site.

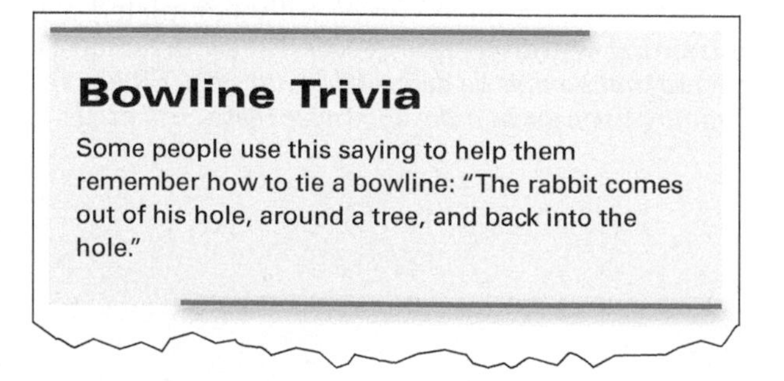
Bowline Trivia
Some people use this saying to help them remember how to tie a bowline: "The rabbit comes out of his hole, around a tree, and back into the hole."

Figures and Tables

Photographs, drawings, diagrams, and tables are used throughout each module to illustrate important concepts and provide clarity for complex instructions. Text references to figures and tables are emphasized with *italic* type.

Notes, Cautions, and Warnings

Safety features are set off from the main text in highlighted boxes and categorized according to the potential danger involved. Notes simply provide additional information. Cautions flag a hazardous issue that could cause damage to materials or equipment. Warnings stress a potentially dangerous situation that could result in injury or death to workers.

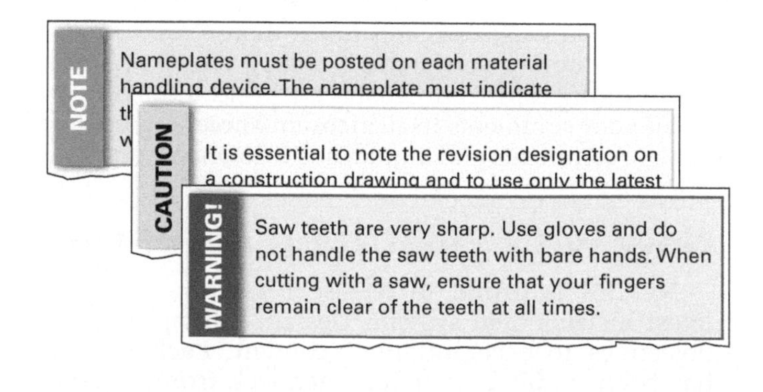
NOTE
Nameplates must be posted on each material
CAUTION
It is essential to note the revision designation on
WARNING!
Saw teeth are very sharp. Use gloves and do not handle the saw teeth with bare hands. When cutting with a saw, ensure that your fingers remain clear of the teeth at all times.

Case History

Case History features emphasize the importance of safety by citing examples of the costly (and often devastating) consequences of ignoring best practices or OSHA regulations.

Case History

Requesting an Outage

An electrical contractor requested an outage when asked to install two bolt-in, 240V breakers in panels in a data processing room. It was denied due to the 24/7 worldwide information processing hosted by the facility. The contractor agreed to proceed only if the client would sign a letter agreeing not to hold them responsible if an event occurred that damaged computers or resulted in loss of data. No member of upper management would accept liability for this possibility, and the outage was scheduled.

The Bottom Line: If you can communicate the liability associated with an electrical event, you can influence management's decision to work energized.

Going Green

Going Green features present steps being taken within the construction industry to protect the environment and save energy, emphasizing choices that can be made on the job to preserve the health of the planet.

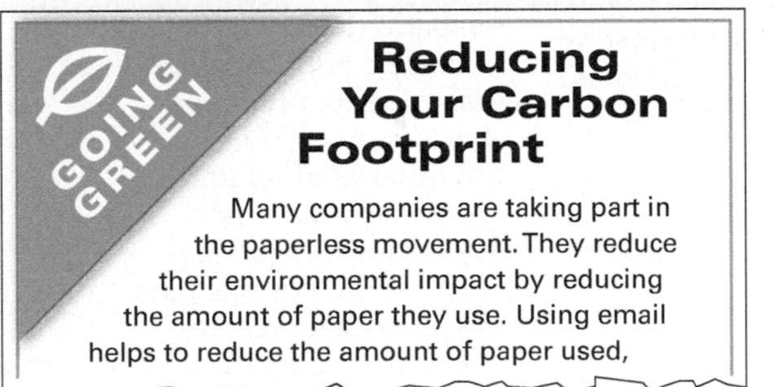

GOING GREEN
Reducing Your Carbon Footprint
Many companies are taking part in the paperless movement. They reduce their environmental impact by reducing the amount of paper they use. Using email helps to reduce the amount of paper used,

Did You Know

Did You Know features introduce historical tidbits or interesting and sometimes surprising facts about the trade.

Did You Know?
Safety First
Safety training is required for all activities. Never operate tools, machinery, or equipment without prior training. Always refer to the manufacturer's instructions.

Step-by-Step Instructions

Step-by-step instructions are used throughout to guide you through technical procedures and tasks from start to finish. These steps show you how to perform a task safely and efficiently.

Perform the following steps to erect this system area scaffold:

Step 1 Gather and inspect all scaffold equipment for the scaffold arrangement.

Step 2 Place appropriate mudsills in their approximate locations.

Step 3 Attach the screw jacks to the mudsills.

Trade Terms

Each module presents a list of Trade Terms that are discussed within the text and defined in the Glossary at the end of the module. These terms are presented in the text with **bold, blue** type upon their first occurrence. To make searches for key information easier, a comprehensive Glossary of Trade Terms from all modules is located at the back of this book.

During a rigging operation, the **load** being lifted or moved must be connected to the apparatus, such as a crane, that will provide the power for movement. The connector—the link between the load and the apparatus—is often a sling made of synthetic, chain, or **wire rope** materials. This section focuses on three types of slings:

Section Review

Each section of the module wraps up with a list of Additional Resources for further study and Section Review questions designed to test your knowledge of the Objectives for that section.

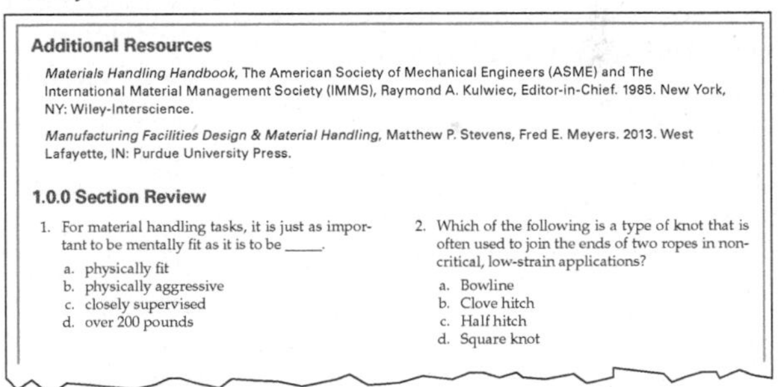

Additional Resources

Materials Handling Handbook, The American Society of Mechanical Engineers (ASME) and The International Material Management Society (IMMS), Raymond A. Kulwiec, Editor-in-Chief. 1985. New York, NY: Wiley-Interscience.

Manufacturing Facilities Design & Material Handling, Matthew P. Stevens, Fred E. Meyers. 2013. West Lafayette, IN: Purdue University Press.

1.0.0 Section Review

1. For material handling tasks, it is just as important to be mentally fit as it is to be _____.
   a. physically fit
   b. physically aggressive
   c. closely supervised
   d. over 200 pounds

2. Which of the following is a type of knot that is often used to join the ends of two ropes in non-critical, low-strain applications?
   a. Bowline
   b. Clove hitch
   c. Half hitch
   d. Square knot

Review Questions

The end-of-module Review Questions can be used to measure and reinforce your knowledge of the module's content.

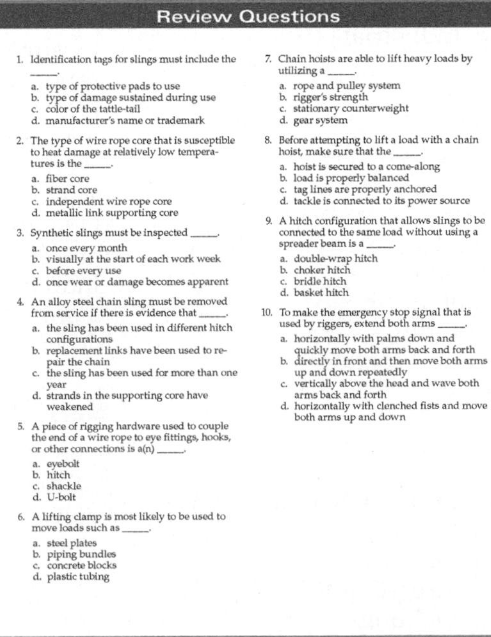

Review Questions

1. Identification tags for slings must include the _____.
   a. type of protective pads to use
   b. type of damage sustained during use
   c. color of the tattle-tail
   d. manufacturer's name or trademark

2. The type of wire rope core that is susceptible to heat damage at relatively low temperatures is the _____.
   a. fiber core
   b. strand core
   c. independent wire rope core
   d. metallic link supporting core

3. Synthetic slings must be inspected _____.
   a. once every month
   b. visually at the start of each work week
   c. before every use
   d. once wear or damage becomes apparent

4. An alloy steel chain sling must be removed from service if there is evidence that _____.
   a. the sling has been used in different hitch configurations
   b. replacement links have been used to repair the chain
   c. the sling has been used for more than one year
   d. strands in the supporting core have weakened

5. A piece of rigging hardware used to couple the end of a wire rope to eye fittings, hooks, or other connections is a(n) _____.
   a. eyebolt
   b. hitch
   c. shackle
   d. U-bolt

6. A lifting clamp is most likely to be used to move loads such as _____.
   a. steel plates
   b. piping bundles
   c. concrete blocks
   d. plastic tubing

7. Chain hoists are able to lift heavy loads by utilizing a _____.
   a. rope and pulley system
   b. rigger's strength
   c. stationary counterweight
   d. gear system

8. Before attempting to lift a load with a chain hoist, make sure that the _____.
   a. hoist is secured to a come-along
   b. load is properly balanced
   c. tag lines are properly anchored
   d. tackle is connected to its power source

9. A hitch configuration that allows slings to be connected to the same load without using a spreader beam is a _____.
   a. double-wrap hitch
   b. choker hitch
   c. bridle hitch
   d. basket hitch

10. To make the emergency stop signal that is used by riggers, extend both arms _____.
    a. horizontally with palms down and quickly move both arms back and forth
    b. directly in front and then move both arms up and down repeatedly
    c. vertically above the head and wave both arms back and forth
    d. horizontally with clenched fists and move both arms up and down

Module 00106 Introduction to Basic Rigging 31

NCCER Standardized Curricula

NCCER's training programs comprise more than 80 construction, maintenance, pipeline, and utility areas and include skills assessments, safety training, and management education.

Boilermaking
Cabinetmaking
Carpentry
Concrete Finishing
Construction Craft Laborer
Construction Technology
Core Curriculum: Introductory Craft Skills
Drywall
Electrical
Electronic Systems Technician
Heating, Ventilating, and Air Conditioning
Heavy Equipment Operations
Heavy Highway Construction
Hydroblasting
Industrial Coating and Lining Application Specialist
Industrial Maintenance Electrical and Instrumentation Technician
Industrial Maintenance Mechanic
Instrumentation
Ironworking
Manufactured Construction Technology
Masonry
Mechanical Insulating
Millwright
Mobile Crane Operations
Painting
Painting, Industrial
Pipefitting
Pipelayer
Plumbing
Reinforcing Ironwork
Rigging
Scaffolding
Sheet Metal
Signal Person
Site Layout
Sprinkler Fitting
Tower Crane Operator
Welding

Maritime

Maritime Industry Fundamentals
Maritime Electrical
Maritime Pipefitting
Maritime Structural Fitter
Maritime Welding
Maritime Aluminum Welding

Green/Sustainable Construction

Building Auditor
Fundamentals of Weatherization
Introduction to Weatherization
Sustainable Construction Supervisor
Weatherization Crew Chief
Weatherization Technician
Your Role in the Green Environment

Energy

Alternative Energy
Introduction to the Power Industry
Introduction to Solar Photovoltaics
Power Generation Maintenance Electrician
Power Generation I&C Maintenance Technician
Power Generation Maintenance Mechanic
Power Line Worker
Power Line Worker: Distribution
Power Line Worker: Substation
Power Line Worker: Transmission
Solar Photovoltaic Systems Installer
Wind Energy
Wind Turbine Maintenance Technician

Pipeline

Abnormal Operating Conditions, Control Center
Abnormal Operating Conditions, Field and Gas
Corrosion Control
Electrical and Instrumentation
Field and Control Center Operations
Introduction to the Pipeline Industry
Maintenance
Mechanical

Safety

Field Safety
Safety Orientation
Safety Technology

Supplemental Titles

Applied Construction Math
Tools for Success

Management

Construction Workforce Development Professional
Fundamentals of Crew Leadership
Mentoring for Craft Professionals
Project Management
Project Supervision

Spanish Titles

Acabado de concreto: nivel uno (*Concrete Finishing Level One*)
Aislamiento: nivel uno (*Insulating Level One*)
Albañilería: nivel uno (*Masonry Level One*)
Andamios (*Scaffolding*)
Carpintería: Formas para carpintería, nivel tres (*Carpentry: Carpentry Forms, Level Three*)
Currículo básico: habilidades introductorias del oficio (*Core Curriculum: Introductory Craft Skills*)
Electricidad: nivel uno (*Electrical Level One*)
Herrería: nivel uno (*Ironworking Level One*)
Herrería de refuerzo: nivel uno (*Reinforcing Ironwork Level One*)
Instalación de rociadores: nivel uno (*Sprinkler Fitting Level One*)
Instalación de tuberías: nivel uno (*Pipefitting Level One*)
Instrumentación: nivel uno, nivel dos, nivel tres, nivel cuatro (*Instrumentation Levels One through Four*)
Orientación de seguridad (*Safety Orientation*)
Paneles de yeso: nivel uno (*Drywall Level One*)
Seguridad de campo (*Field Safety*)

Acknowledgments

This curriculum was revised as a result of the farsightedness and leadership of the following sponsors:

Bechtel
Cianbro Companies
Fluor
KBR Services, Inc.
McAllen Careers Institute
Sundt Construction, Inc.
TIC - The Industrial Company
Turner Industries Group, LLC

This curriculum would not exist were it not for the dedication and unselfish energy of those volunteers who served on the Authoring Team. A sincere thanks is extended to the following:

Arnold Adame, Jr.
Tony Ayotte
Jacob Guzman
Rodney Landry
Josue Ponce
Shawn Reid
Brian Robinson
Fernando Sanchez
Troy Smith
Jody Suchanek

NCCER Partners

American Council for Construction Education
American Fire Sprinkler Association
Associated Builders and Contractors, Inc.
Associated General Contractors of America
Association for Career and Technical Education
Association for Skilled and Technical Sciences
Construction Industry Institute
Construction Users Roundtable
Design Build Institute of America
GSSC – Gulf States Shipbuilders Consortium
ISN
Manufacturing Institute
Mason Contractors Association of America
Merit Contractors Association of Canada
NACE International
National Association of Women in Construction
National Insulation Association
National Technical Honor Society
National Utility Contractors Association
NAWIC Education Foundation
North American Crane Bureau
North American Technician Excellence
Pearson
Prov
SkillsUSA®
Steel Erectors Association of America
U.S. Army Corps of Engineers
University of Florida, M. E. Rinker Sr., School of Construction Management
Women Construction Owners & Executives, USA

Contents

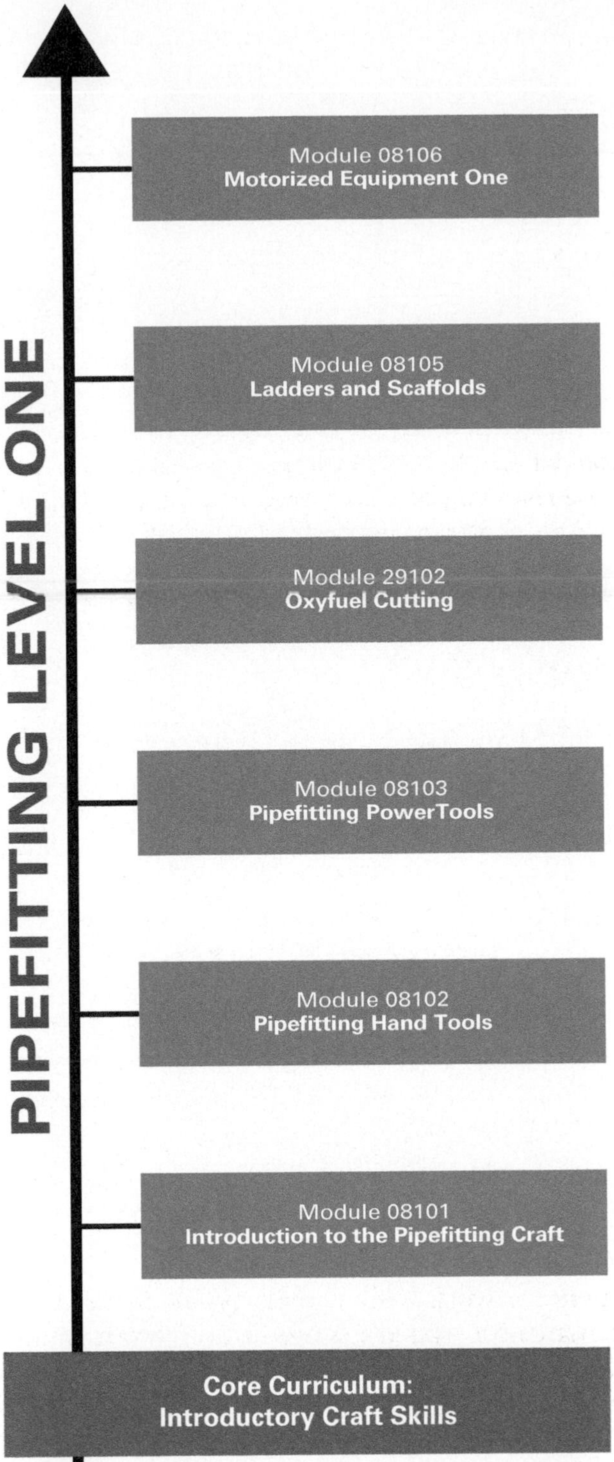

PIPEFITTING LEVEL ONE

Module 08106
Motorized Equipment One

Module 08105
Ladders and Scaffolds

Module 29102
Oxyfuel Cutting

Module 08103
Pipefitting PowerTools

Module 08102
Pipefitting Hand Tools

Module 08101
Introduction to the Pipefitting Craft

Core Curriculum:
Introductory Craft Skills

This course map shows all of the modules in *Pipefitting Level One*. The suggested training order begins at the bottom and proceeds up. Skill levels increase as you advance on the course map. The local Training Program Sponsor may adjust the training order.

This page is intentionally left blank.

Orientation to the Pipefitting Craft

Overview

Pipefitters work with many kinds of pipe, ranging from small, half-inch piping to that which is three or more feet in diameter. A pipefitter must know how to work with threaded, grooved, and welded piping systems and must be able to master a variety of tools and equipment.

Although both work with pipe, there is a major difference between pipefitters and plumbers. Pipefitters lay out and install piping systems primarily for industrial facilities such as chemical plants, oil refineries, food processing plants, and paper mills. Plumbers install and service water distribution and waste systems, primarily in residential and commercial applications. The successful pipefitter works well with all craft professionals, displaying a positive attitude and other qualities of professionalism, as well as an ongoing commitment to safety.

Module 08101

Trainees with successful module completions may be eligible for credentialing through the NCCER Registry. To learn more, go to www.nccer.org or contact us at 1.888.622.3720. Our website has information on the latest product releases and training, as well as online versions of our *Cornerstone* magazine and Pearson's product catalog.

Your feedback is welcome. You may email your comments to curriculum@nccer.org, send general comments and inquiries to info@nccer.org, or fill in the User Update form at the back of each module.

This information is general in nature and intended for training purposes only. Actual performance of activities described in this manual requires compliance with all applicable operating, service, maintenance, and safety procedures under the direction of qualified personnel. References in this manual to patented or proprietary devices do not constitute a recommendation of their use.

08101 V4.0

08101
Orientation to the Pipefitting Craft

Objectives

Successful completion of this module prepares trainees to:

1. Describe the pipefitting craft and the knowledge, skills, behaviors, and attitudes which contribute to a pipefitter's success.
   a. Describe the tools that pipefitters use and explain how they are chosen for a particular job.
   b. Explain the nature of pipefitting work and factors which influence it.
   c. State the basic safety obligations of construction-industry employees and employers.
2. Describe training pathways for the pipefitting professional and identify characteristics of successful crafters.
   a. List the components and requirements of an apprenticeship program and conditions which affect the length of time to complete the training.
   b. Describe the structure of NCCER craft-training programs and how they relate to pipefitting apprenticeships.
   c. Define professionalism and Identify characteristics common to successful craft workers.
   d. Explain the meaning and importance of human relations in the workplace.

Performance Tasks

This is a knowledge-based module; there are no Performance Tasks.

Trade Terms

Apprenticeship Training, Employer and Labor Services (ATELS)
Bevel
On-the-job training (OJT)
Occupational Safety and Health Administration (OSHA)
Sweat

Industry Recognized Credentials

If you are training through an NCCER-accredited sponsor, you may be eligible for credentials from NCCER's Registry. The ID number for this module is 08101. Note that this module may have been used in other NCCER curricula and may apply to other level completions. Contact NCCER's Registry at 888.622.3720 or go to www.nccer.org for more information.

Contents

Figures and Tables

SECTION ONE

1.0.0 THE PIPEFITTING CRAFT

Objective

Describe the pipefitting craft and the knowledge, skills, behaviors, and attitudes which contribute to a pipefitter's success.

a. Describe the tools that pipefitters use and explain how they are chosen for a particular job.
b. Explain the nature of pipefitting work and factors which influence it.
c. State the basic safety obligations of construction-industry employees and employers.

Trade Terms

Bevel: A cut made at an angle.

Sweat: A method of joining pipe in which solder is applied to the joint and heated until the solder flows into the joint.

Occupational Safety and Health Administration (OSHA): The federal government agency established to ensure a safe and healthy environment in the workplace.

Although pipefitters and plumbers both work with piping systems, these crafts are not the same. Plumbers work primarily with water distribution, drainage, and waste systems in residential and commercial environments. Pipefitters work with piping systems that carry water, gases, liquid chemicals, solids, and fuels in a variety of environments (*Figure 1*). Such environments include oil refineries, chemical plants, power plants, food processing plants, paper mills, ships, factories, and a host of other facilities. This training program covers the tools, materials, and techniques used by pipefitters to lay out, fabricate, and install piping systems.

1.1.0 Pipefitting Skills

A pipefitter must learn many skills, including working with a variety of specialized hand tools and power tools. The following are tasks performed by pipefitters:

- Reading and interpreting blueprints and specifications
- Planning and sketching piping systems
- Laying out and fabricating fittings
- Fitting up pipe for welding
- Measuring, cutting, threading, and assembling pipe
- Bending pipe
- Installing pipe, fittings, and valves using various joining techniques
- Inspecting and testing piping systems
- Performing routine maintenance and repairs on piping systems

Pipefitters must learn to work with several types of piping, including carbon steel, copper, plastic (a.k.a. polyvinyl chloride or PVC), stainless steel, and aluminum. In some environments, most of the pipe joints are welded, rather than threaded. This is especially true of aluminum and stainless steel pipe. Therefore, an important part of the pipefitter's work is to prepare and fit-up piping for welding.

Piping is likely to change direction several times between its origin, or source, and its ultimate destination. Because of this, a pipefitter must know how to select and apply fittings or bend the pipe to redirect the piping run. Some types of pipe, such as copper piping up to 2 inches in diameter, are fairly easy to bend. However, carbon steel piping is different. Because of its wall thickness, it is difficult to bend, so fittings such as tees and elbows are generally used to change its direction. It can, however, be bent using hydraulic benders. In some instances, such as high-pressure steam applications, the bending of carbon steel pipe is required by the job specifications.

Valves are used to control the flow of material in process control. As such, a pipefitter must understand the operation, application, and installation of the many types of manual and automatic valves used for this purpose (*Figure 2*).

A pipefitter may work on a wide variety of piping systems, such as:

- Cooling, heating, and refrigeration systems
- Compressed air systems
- High-pressure systems
- Steam systems
- Hydraulic systems
- Nuclear power systems
- Chemical storage and processing systems
- Oil, gasoline, and natural gas storage and processing systems
- Fire protection systems.

These systems all have one thing in common: they require properly aligned and correctly assembled piping using materials that are compatible with the product being transported. This means that attention to detail is an essential

(A) POWER STATION

(B) INDUSTRIAL PIPING SYSTEM

Figure 1 Piping system examples.

Figure 2 Piping section with valve.

characteristic of a successful pipefitter, as well as an understanding of the physical and chemical properties associated with each type of valve.

DID YOU KNOW?

Lead Pipes

Archaeologists and historians believe that the widespread use of lead pipe in ancient Roman cities may have contributed to the downfall of the Roman Empire. Lead is harmful to the brain, the nervous system, and vital organs such as the liver. Water carried through lead pipes may have gradually poisoned many people in the Roman Empire.

1.1.1 Safety on the Job

There are many safety issues to be considered on any construction site. In addition to the hazards normally found on a construction site, pipefitters encounter specific hazards because of the nature of their work, as described in *Table 1*. Precautions may be taken to mitigate these hazards, and it is the job of the pipefitting to identify dangers and work to control or avoid them.

A single large pipe or a bundle of smaller pipe is too heavy to lift by hand, so a machine such as a crane must be used to move the pipe. The pipefitter participates in the rigging process by helping to attach slings to the load. Tools used for flame cutting and beveling of pipe create a potential burn hazard, as well as hazards from flammable and explosive gases. When working with underground piping systems, the pipefitter may be exposed to trenching hazards and possibly confined space hazards. Pipefitters may also have to work at elevations in order to install and maintain piping systems. Therefore, fall protection is a key consideration. Power tools such as automatic pipe threaders also present cutting and electrical hazards.

Another hazard that concerns pipefitters involves the materials carried in piping systems. Some piping systems carry corrosive chemicals or hot liquids. Other processes operate under high pressure. It is important to know what material is being carried by the system and to take the appropriate precautions to avoid injury.

Table 1 Known Hazards in Pipefitting

Known Hazards in Pipefitting	
Hazards	**Introduced by**
Potential burn hazard	–Tools used for flame cutting and beveling of pipe; – Hazards from flammable and explosive gases in the pipefitting process
Trenching Hazards	–Working with underground piping systems
Confined Space	–Working with underground piping systems
Elevation Hazards	– Installation and maintenance of piping systems
Cutting and Electrical Hazards	– Power tools such as automatic pipe threaders
Corrosive Chemicals and Hot Liquids	– Materials carried in piping systems
Bursting and Explosions	– Piping processes operating under high pressure

1.1.2 Tools of the Trade

Pipefitters work with a variety of hand and power tools for cutting, threading, and joining pipe. Specialized pipe wrenches are used for joining pipe (*Figure 3*). The pipefitter must know when to use each of these wrenches, how to select the right size, and how to use them correctly. Power saws and other types of cutting tools are used for cutting pipe to the correct size. Reamers are used for removing burrs from pipe. Grinding tools are used to **bevel** pipe for welding. In some cases, an oxyacetylene cutting torch is used for cutting and beveling pipe prior to welding (*Figure 4*).

Threading tools (*Figure 5*) are used for applying threads to pipe so that sections of pipe can be joined with threaded fittings. Carbon steel pipe is usually threaded, but other types of pipe such as copper, plastic, and steel alloys can be threaded as well. Both manual and powered threading tools are used.

Copper piping is generally bent to the desired shape and connected using **sweat** joints. Different types of pipe benders are also used, depending on the size of the pipe.

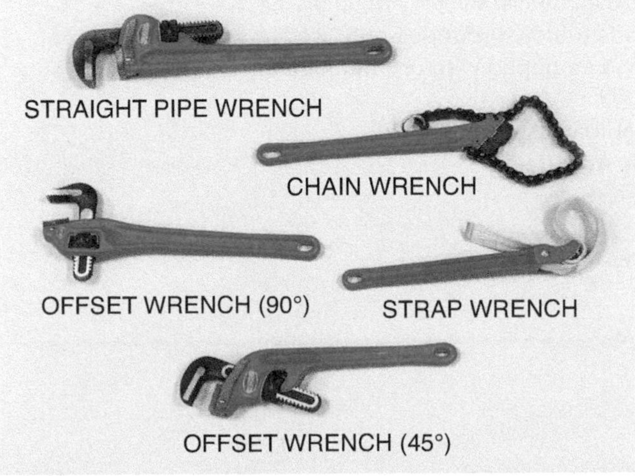

Figure 3 Common pipe wrenches.

Figure 4 Pipe cutting and beveling fixture.

Stainless steel piping is used in food processing and other specialized applications, such as medical facilities, where cleanliness is essential. Stainless steel is often joined by welding.

1.2.0 Opportunities in the Pipefitting Industry

Pipefitters are employed all over the United States and the world in refineries, power plants, and industrial facilities. This creates an opportunity to travel while earning an excellent income. A journey-level pipefitter will earn as much as, and

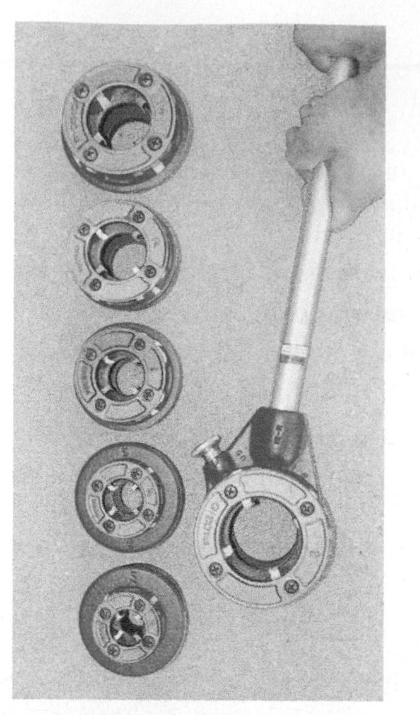

HAND THREADER

POWER THREADING MACHINE

Figure 5 Pipe threading tools.

Power Tool Safety Precautions

Power tools in any environment present their own set of concerns on top of those that already exist for the job at hand. When using power tools, even for a brief time, wear heavy protective clothing, a hard hat, goggles, and heavy work gloves to guard against injury from flying debris. The use of Personal Protective Equipment, or PPE, in combination with the tips below, will create a safer work environment:

- Use adequate hearing protection.
- Use appropriate respiratory protection when dust hazards are produced.
- The OSHA Respirable Crystalline Silica Standard for Construction applies to work involving common materials that produce tiny, airborne particles when manipulated. Employers are required to provide adequate safeguards for workers; these include the use of special masks, filters, and equipment adjustments to name a few. An overview of this standard may be found at **https://www.osha.gov/Publications/OSHA3681.pdf**. Detailed information is available at **https://www.osha.gov/dsg/topics/silicacrystalline/**.
- To avoid shock, ensure that electrically-powered tools are adequately grounded and used in a dry environment. For added protection, use tools that are double insulated and equipped with ground fault circuit interrupters (GFCIs).
- Do not use power tools in confined spaces where sparks could cause explosions.
- Use non-sparking tools around combustible and flammable materials.
- Do not exceed the rated speed of the tool.
- Use recommended tool guards.
- Inspect equipment regularly and repair or replace it as necessary.

sometimes more than, a college-educated office worker. Because of the demand, pipefitters can reasonably expect a solid level of job security.

A person starting in the pipefitting craft will first learn to properly handle and use the tools of the trade. They may be asked to bend, cut, and thread pipe and assist journey-level pipefitters in performing various other tasks. As their training progresses and their knowledge grows, apprentices will take on greater responsibilities.

Since pipes are joined by welding in many environments, pipefitters often work closely with welders. It is a natural progression then, for a pipefitter to learn welding. Pipe welding is very demanding and calls for a great deal of training and practice. The trade-off is promising though because once the pipefitter is certified as a pipe welder, the opportunities for jobs, income, and advancement increase significantly.

Career growth for a pipefitter can occur in many ways (*Figure 6*). If you have leadership qualities, you can become a foreman of a pipefitting crew. Some experienced pipefitters become instructors at contractor, union, or vocational training schools. Others become quality control inspectors, estimators, or piping designers. With further education and training, a pipefitter can become a piping system designer.

Within a company that does pipefitting work, there are typically opportunities to move into supervisory and management positions such as quality control manager, project superintendent, and plant superintendent. Each of these calls for a strong knowledge of the industry, time-test experience, and the demonstrated ability to teach, lead, mentor, and guide other workers to the on-time completion of a project.

Another avenue for growth is the inspection of pipe welds. Various technical methods, including x-ray techniques, are used to inspect and certify critical pipe welds. The inspector must be able to perform specialized tests and analyze the results.

1.3.0 Employer and Employee Safety Obligations

An obligation is like a promise or a contract. In exchange for the benefits of your employment and for your own well-being, you agree to work safely. In other words, you are obligated to be alert, aware, honest, respectful, responsible, and responsive to others. You agree to operate with good judgment and common sense, knowing that shortcuts, pranks, and any disregard for safety puts you and others at risk for injury and even death. You are obligated to make sure that anyone you supervise or work with is working safely. Your employer is obligated to maintain a safe workplace for all employees. **Safety is everyone's responsibility.**

Some employers will have safety committees. If you work for such an employer, you are then

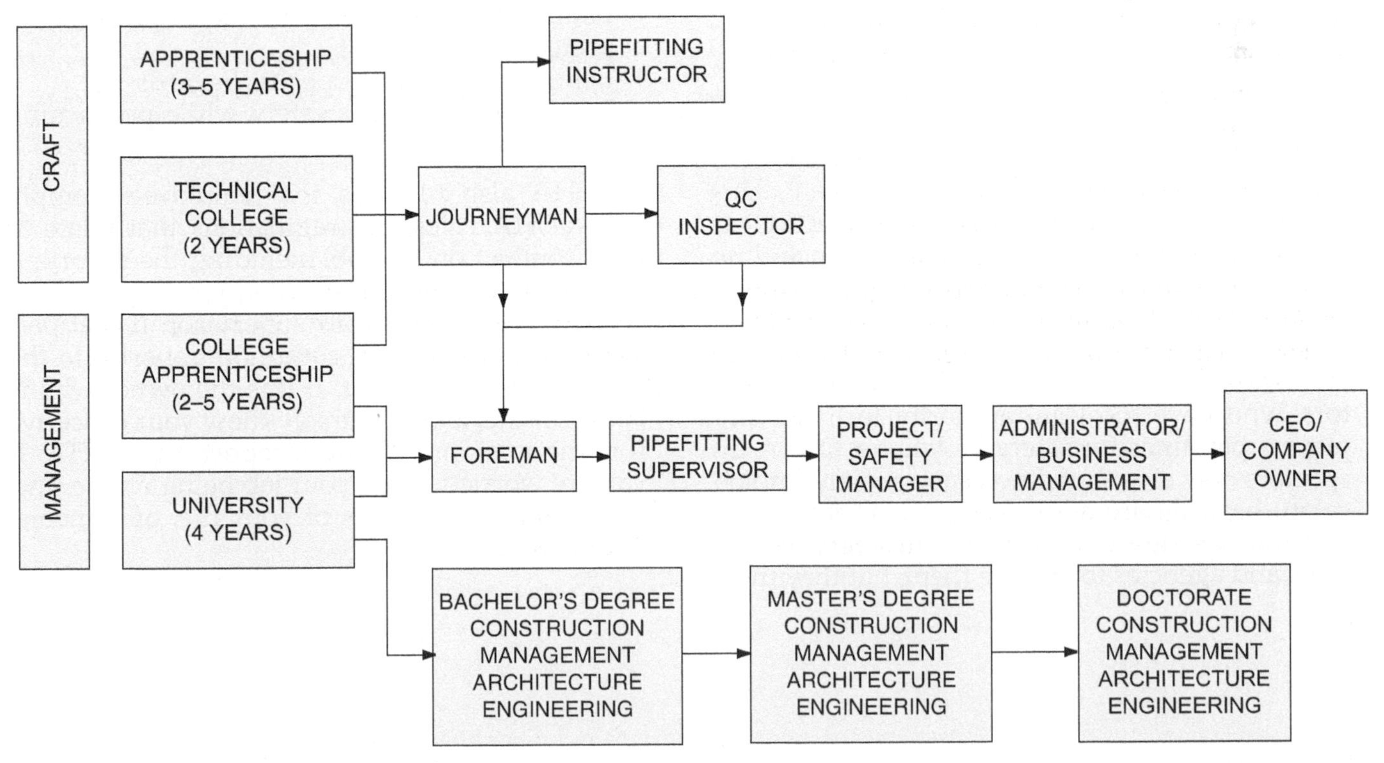

Figure 6 Career opportunities for pipefitters.

obligated to that committee to maintain a safe working environment. This means at least two things:

- Follow the safety committee's rules for proper working procedures and practices.
- Report any unsafe equipment and conditions directly to the committee or your supervisor.

On the job, if you see something that is not safe, report it! Do not ignore it. It will not correct itself. You have an obligation.

In the long run, even if you do not think an unsafe condition affects you, it does. Your employer is more likely to criticize you for not reporting a problem than to be irritated that you did. Your employer knows that the short time lost in making conditions safe again is nothing compared with shutting down the whole job because of a major disaster. If that happens, you are out of work anyway.

To address the need for creating safe workplaces, The US Congress passed the *Occupational Safety and Health Act* in 1970. It was adopted with the stated purpose "to assure as far as possible every working man and woman in the nation safe and healthful working conditions and to preserve our human resources."

This act also created the **Occupational Safety and Health Administration (OSHA)**, which is part of the US Department of Labor. The job of OSHA is to set occupational safety and health standards for all places of employment, enforce these standards, ensure that employers provide and maintain a safe workplace for all employees, and provide research and educational programs to support safe working practices.

OSHA requires each employer to provide a safe and hazard-free working environment. This applies to every industry (not just construction) as well as every type of business structure including contractors, subcontractors, private business owners, not-for-profit enterprises, and corporations of all sizes. The enforcement of this act of Congress is provided by the federal and state safety inspectors, who have the legal authority to make employers pay fines for safety violations. OSHA inspectors may also shut down a job site until noted safety hazards are addressed and resolved. The law allows states to have their own safety regulations and agencies to enforce them, but they must first be approved by the US Secretary of Labor. For states that do not develop such regulations and agencies, federal OSHA regulations must be obeyed.

These standards are listed in *OSHA Safety and Health Standards for the Construction Industry* (*29 CFR, Part 1926*), sometimes called *OSHA Standards 1926*. Other safety standards that apply to construction are published in *OSHA Safety and Health Standards for General Industry* (*29 CFR, Parts 1900 to 1910*).

The most important general requirements that OSHA places on employers in the construction industry are as follows:

- The employer must perform frequent and regular job site inspections of equipment.
- The employer must instruct all employees to recognize and avoid unsafe conditions, and to know the regulations that pertain to the job so they may control or eliminate any hazards.
- No one may use any tools, equipment, machines, or materials that do not comply with *29 CFR 1926*
- The employer must ensure that only qualified individuals operate tools, equipment, and machines.

According to OSHA, you are entitled to on-the-job safety training. As a new employee, you are entitled to the following:

- Being shown how to do your job safely
- Being provided with the required personal protective equipment
- Being warned about specific hazards
- Being supervised for safety while performing the work

OSHA also requires that employees comply with OSHA rules and regulations that relate to their conduct on the job, including the reporting of hazardous conditions. When you are aware of a safety hazard, tell your supervisor. If that person ignores the unsafe condition, report it to the next highest supervisor. If it is the owner who is being unsafe, let that person know your concerns. If nothing is done about it, report it to OSHA. If you are worried about your job being on the line, think about it in terms of your life, or someone else's, being at risk.

Additional Resources

In addition to the material in this section, the following resources are suggested for pipefitting trainees and practicing professionals.

The Pipe Fitters Blue Book, W. V. Graves. 1973. Webster, TX: Construction Trades Press.

Safety and Health Regulations for Construction. 2018. Occupational Safety and Health Administration. Available at: **https://www.osha.gov/pls/oshaweb/owastand.display_standard_group?p_part_number=1926&p_toc_level=1**.

Electronic Code of Federal Regulations (29 CFR 1926 and *29 CFR 1900–1910).* 2018. US Government Publishing Office. Available at: **www.ecfr.gov**.

OSHA's Respirable Crystalline Silica Standard for Construction. 2017. Occupational Safety and Health Administration. Available at: **https://www.osha.gov/Publications/OSHA3681.pdf**.

1.0.0 Section Review

1. True or false: Plumbers and pipefitters do essentially the same work because they both work with piping systems.
   a. True
   b. False

2. Which of the following is fairly easy to bend?
   a. copper piping up to 2 inches in diameter
   b. carbon steel piping less than 4 inches in diameter
   c. aluminum alloy piping that does not contain iron
   d. PVC piping systems, when thermal methods are used

3. Pipefitters often work closely with _____.
   a. plumbers
   b. diesel mechanics
   c. mobile crane operators
   d. welders

4. Safety is the responsibility of _____.
   a. the contractor
   b. the foreman
   c. the inspector
   d. everyone

SECTION TWO

2.0.0 TRAINING AND THE CHARACTERISTICS OF CRAFT PROFESSIONALS

Objective

Describe training pathways for the pipefitting professional and identify characteristics of successful crafters.

a. List the components and requirements of an apprenticeship program and conditions which affect the length of time to complete the training.
b. Describe the structure of NCCER craft-training programs and how they relate to pipefitting apprenticeships.
c. Define professionalism and Identify characteristics common to successful craft workers.

Trade Terms

Office of Apprenticeship: The US Department of Labor office that sets the minimum standards for training programs across the country.

On-the-job learning (OJL): Job-related learning acquired while working.

The US Department of Labor's (DOL) **Office of Apprenticeship** sets the minimum standards for training programs across the country. Office of Apprenticeship programs rely on mandatory classroom instruction, **on-the-job learning (OJL)**, and require 144 hours of classroom instruction per year and 2,000 hours of OJL per year. In a typical Office of Apprenticeship pipefitter apprentice program, trainees spend 576 hours in classroom instruction and 8,000 hours in OJL before receiving journey certificates.

2.1.0 Apprenticeship Program

Apprentice training goes back thousands of years, and its basic principles have stood the test of time. First, it offers a means for individuals entering a craft to learn from those who have mastered it. Second, it focuses on learning by doing and on the demonstration of real skills versus discussions of theory. Although some theories are presented in the construction classroom, they are generally presented in ways that help trainees understand the purposes behind the skills to be learned.

All apprenticeship standards prescribe certain types of OJL, in which tasks are broken down so the apprentice receives hands-on instruction. A specified number of hours is required for each task; the total number of hours for pipefitter apprenticeship programs is usually 8,000, which amounts to about four years of training.

In a traditional program, the training may be completed in increments of 2,000 hours per year, with the possibility of layoffs or illnesses affecting the duration. Classroom instruction and work-related training will not always run concurrently due to these reasons as well as the type of work that is needed in the field. In a competency-based program, it may be possible to shorten this time by testing out of specific tasks through a series of performance exams. Apprentices with special job experience or coursework may obtain credit toward their classroom requirements. These special cases will depend on the type of program and the regulations and standards under which it operates.

In any event, the apprentice must log all work time and turn it in to the apprenticeship committee so that accurate time control can be maintained. After each 1,000 hours of related work, the apprentice will likely receive a pay increase as indicated by the standards.

Informal on-the-job training provided by employers is usually less thorough than that provided through a formal apprenticeship program. The degree of training and supervision in this type of program often depends on the size of the employing firm. A small contractor may provide training in only one area, while a large company may be able to provide training in several areas.

For those entering an apprenticeship program, a high school or technical school education is desirable, as are courses in shop, mechanical drawing, and general mathematics. Manual dexterity, good physical condition, and quick reflexes are important. The ability to solve problems quickly and accurately and to work closely with others is essential. You must have a high concern for safety.

The prospective apprentice must submit certain information to the apprenticeship committee. This may include the following:

- Aptitude test (General Aptitude Test Battery [GATB] or GATB Form Test) results
  - Usually administered by the local Employment Security Commission
- Proof of educational background (candidate should have school transcripts sent to the committee)

Child Labor Laws

Federal law establishes the minimum standards for workers under the age of 18.

The *Child Labor Provisions of the Fair Labor Standards Act* forbid employers from using illegal child labor, and also forbid companies from doing business with any other business that does. DOL investigates alleged abuses of the law. In such cases, employers must provide proof of age for their employees.

In addition to the *Child Labor Provisions*, employers in the construction industry are required to follow DOL's *Child Labor Bulletin No. 101, Child Labor Requirements in Nonagricultural Occupations Under the Fair Labor Standards Act. Bulletin No. 101* does the following:

- Explains the coverage of the Child Labor Provisions
- Identifies minimum age standards
- Lists the exemptions from the Child Labor Provisions
- Sets out employment standards for 14- and 15-year-old workers
- Defines the work that can be performed in hazardous occupations
- Provides penalties for violations of the Child Labor Provisions
- Recommends the use of age certificates for employees

Some state and local governments may enforce stricter regulations. Employers are required to abide by all the laws that apply to them.

- Letters of reference from past employers and friends
- Results of a physical examination
- Proof of age
- If the candidate is a veteran, a copy of Form DD214
- A record of technical training received that relates to the construction industry and/or a record of any pre-apprenticeship training and/or a
- High school diploma or General Equivalency Diploma (GED)

The apprentice must do the following:

- Wear proper safety equipment on the job
- Purchase and maintain appropriate tools as needed and required by the contractor
- Submit a monthly OJT report to the committee
- Report to the committee if a change in employment status occurs
- Participate in classroom-related instruction and adhere to all classroom regulations such as attendance requirements.

2.2.0 Your Training Program

NCCER uses the minimum Office of Apprenticeship standards as the foundation for its comprehensive pipefitting curriculum. This four-year training program offers industry-driven training and education developed specifically to meet the needs of trainees. It adopts a competency-based teaching philosophy, which means that trainees must demonstrate the understanding and skills necessary to perform hands-on tasks. These are covered in each module and must be mastered before advancing to the next stage.

When an instructor is satisfied that a trainee has successfully demonstrated the required knowledge and skills for a particular module, that information is sent to NCCER and kept in the Registry system. The Registry can then confirm training and skills for workers as they move from state to state, company to company, or even within a company (see *Appendix*).

Whether you enroll in an NCCER program or another Office of Apprenticeship-approved program, it is advised that you work for an employer or sponsor who supports a nationally standardized training program that includes credentials to confirm your skill development.

2.3.0 Characteristics of Successful Craft Professionals

Pipefitting professionals must be able to use current craft materials, tools, and equipment to finish tasks safely and efficiently. They must be adept at adjusting methods to meet each situation and continuously train to remain knowledgeable about technical advancements and gain the skills to use them. Professionals never take chances with personal safety or the safety of others.

2.3.1 Professionalism

The word *professionalism* is a broad term that describes the desired overall behavior and attitude expected in the workplace. It is a concept that includes honesty, productivity, safety, civility, cooperation, teamwork, clear and concise communication, being on time and prepared for work, and having regard for one's impact on one's co-workers. Professionals pride themselves on performing a job well and on being punctual and dependable. Each job is completed in a professional way, never by cutting corners or reducing materials.

A valued professional maintains work attitudes and ethics that protect all types of property from damage or theft. It can be demonstrated in a variety of ways every minute of every situation in the workplace. Unfortunately, however, professionalism is often absent from the construction job site. Most people would argue that it must start at the top; while this is true – that management support of professionalism is important to its success in the workplace – it is even more important that individuals recognize their personal responsibility for behaving professionally.

It is extremely important that unprofessional behavior is not tolerated. That is not to say that unprofessional workers should be shunned; instead, each person should work to demonstrate professionalism in action and its subsequent rewards.

Essentially:

- Professionalism is both a benefit to the employer and the employee.
- It is a personal responsibility.
- Our industry is what each individual chooses to make of it.

Choose professionalism and the industry image will follow.

2.3.2 Honesty

Honesty and integrity are important traits of the professional; they go hand-in-hand with success. Honesty is not simply a choice between good and bad, but between success and failure. Dishonesty will always catch up with someone. Whether a person is stealing materials, tools, or equipment from the job site or simply lying about time worked, it will not take long for the employer to find out. Employment opportunities will eventually run out for those who are dishonest.

Honesty and integrity also mean giving a fair day's work for a fair day's pay. It means looking out for the employer's interests, as much as your own, by protecting their investments of tools, materials, and a sought-after workforce. Together, they speak carrying out one's side of a bargain. It means that someone's words convey truth, meaning, and reality. Thoughts as well as actions should be honest; employers place a high value on an employee whose integrity is unwavering.

2.3.3 Loyalty

Employees expect employers to look out for their interests, to provide them with steady employment, and to promote them to better jobs as openings occur. Employers feel that they, too, have a right to expect their employees to be loyal to them, to keep their interests in mind, to speak well of them to others, to keep any minor troubles strictly within the plant or office, and to maintain confidentiality about all matters that pertain to the business. Both employers and employees should keep in mind that loyalty is not something to be demanded; rather, it is something to be earned.

2.3.4 Willingness to Learn, Take Responsibility, and Cooperate

Every office and plant has its own way of doing things. Sometimes, a change in safety regulations or the purchase of new equipment makes it necessary for even experienced employees to learn new methods and operations. Employees may resent having to accept improvements because of the retraining that is involved; however, employers have a right to expect employees to put forth the effort for training and retraining as it becomes necessary. Methods must be kept up-to-date to meet competition and show a profit. It is this profit that enables the owner to continue in business and provide jobs for the employees.

In addition to staying current with training, employees are expected to see what needs to be done, then go ahead and do it. This speaks to initiative and foresight, and crosses into the areas of taking responsibility and cooperating. To cooperate means to work together, and it is the key to getting things done. Learn to work as a member of a team with your employer, supervisor, and fellow workers in a common effort to get the work done efficiently, safely, and on time.

People can work together well only if there is some understanding about what work is to be done, when and how it will be done, and who will do it. After having been asked once to do something, employees should assume the responsibility from then on, and they should follow the rules in doing so. Rules and regulations are necessary in all work situations so they should be respected and followed by all employees.

2.3.5 Punctuality and Attendance

Punctuality means being on time. Tardiness means being late, while absenteeism refers to being off the job for one reason or another. Consistent tardiness and frequent absenteeism are indications of poor work habits, unprofessional conduct, and a lack of commitment. Employers sometimes resort to docking pay, demoting workers, and even dismissing them in an effort to control tardiness and absenteeism. No employer likes to impose restrictions of this kind. However, in fairness to those workers who do come on time and who do not stay away from the job, an employer is sometimes forced to discipline those who will not follow the rules.

Failure to get to work on time results in confusion, lost time, and resentment on the part of those who do come on time. In addition, it may lead to penalties including dismissal. Although it may be true that a few minutes out of a day are not very important, you must remember that a principle is involved. Supervisors cannot keep track of people if they come in any time they please. Failure to be on time may hold up the work of fellow workers. Arriving early, on the other hand, shows interest in and enthusiasm for your work, which is appreciated by employers.

It is sometimes necessary to take time off from work. No one should be expected to work when sick or when there is serious trouble at home. However, it's possible to get into the habit of letting unimportant and unnecessary matters keep us from the job. This results in lost production and hardship on those who try to carry on the work with less help. Again, there is a principle involved. The person who hires us has a right to expect us to be on the job unless there is some very good reason for staying away. We should not stay up at night until we are too tired to go to work the next day. If we are ill, we should use the time at home to do all we can to recover quickly. This, after all, is no more than most of us would expect of a person we had hired to work for us, and on whom we depended to do a certain job.

If it is necessary to stay home, phone the office early in the morning so the boss can find another worker for the day. Time and again, employees have remained home without sending any word to the employer. This is the worst possible way to handle the matter as it leaves those at work uncertain about what to expect. They have no way of knowing whether you have merely been held up and will be in later, or whether immediate steps should be taken to assign your work to someone else. Courtesy alone demands that you let the boss know if you cannot come to work. As with frequent tardiness, frequent absences will reflect unfavorably on a worker when promotions are being considered.

2.4.0 Human Relations

Many people underestimate the importance of working well with others, so there is a tendency to pass off human relations as nothing more than common sense. But what exactly is involved in human relations? One response would be to say that part of human relations is being friendly, pleasant, courteous, cooperative, adaptable, and sociable.

As important as these are, they are not enough. Human relations are much more than just getting

Ethical Principles for All Professionals

Honesty: Be truthful in all dealings. Conduct business according to the highest professional standards. Faithfully fulfill all contracts and commitments. Do not deliberately mislead, deceive, or manipulate others.

Integrity: Demonstrate personal integrity and the courage of your convictions by doing what is right even where there is pressure to do otherwise. Do not sacrifice your principles because it seems easier.

Loyalty: Be worthy of trust. Demonstrate fidelity and loyalty to companies, employers and sponsors, co-workers, and trade institutions and organizations.

Fairness: Be fair and just in all dealings. Do not take undue advantage of another's mistakes or difficulties. Fair people are open-minded and committed to justice, respect for diversity and the equal treatment of individuals.

Respect for others: Treat all people with courtesy and dignity.

Obedience: Abide by laws, rules, and regulations relating to all personal and business activities.

Commitment to excellence: Pursue excellence in performing your duties, be well informed and prepared, and constantly try to increase your proficiency by gaining new skills and knowledge.

Leadership: By your own conduct, seek to be a positive role model for others.

people to like you — it is also knowing how to handle difficult situations as they arise. It's knowing how to work with supervisors who are often demanding and sometimes unfair. Human relations are learning how to handle frustrations without hurting others.

It involves understanding your own personality traits and those of the people you work with. Building sound working relationships in various situations is important. If working relationships have deteriorated for one reason or another, restoring them is essential.

2.4.1 Human Relations and Productivity

Effective human relations are directly related to productivity, which is one of the keys to business success. Employees must do everything they can to build strong, professional working relationships co-workers and clients. All employees, both new and experienced, are measured by the amount of quality work they can safely turn out. Employers quickly lose interest, however, in an employee with a great attitude but low productivity.

The employer expects every employee to do his or her share of the workload. If you are to be productive, you must do your share (or more than your share) without antagonizing your fellow workers. You must perform your duties in a manner that encourages others to follow your example.

2.4.2 Attitude

A positive attitude is essential to a successful career and it contributes to the productivity of others. It's far more than just a smile; in fact, some people transmit a positive attitude even though they seldom smile. They do this by the way they treat others, the way they look at their responsibilities, and the approach they take when faced with problems.

Being positive means being energetic, highly motivated, attentive, and alert, which helps ensure both efficiency and safety on the job. A persistently negative attitude, on the other hand, can spoil the perspectives of others. It's difficult to work alongside someone with a negative attitude. People favor a person who is positive because it makes a person's job more interesting and fulfilling. Supervisors can determine a subordinate's attitude by their approach to the job, reactions to directives, and the way they handle problems. A positive attitude will likely contribute to an employee's future success with the company.

Tips for a Positive Attitude

Keep these points in mind to help you develop and maintain a positive attitude:

- Remember that your attitude follows you wherever you go. If someone makes a greater effort to be a more positive person in their social and personal lives, it will automatically help them on the job. The reverse is also true; one effort will complement the other.
- Negative comments are seldom welcomed by fellow workers on the job. Neither are they welcome on the social scene. The solution is to focus on positive things, offer sincere compliments when they are in order, and offer suggestions and feedback in a constructive manner. Constant complainers and those who are highly critical of others do not build healthy and fulfilling relationships.
- Look for the good in others, especially your supervisor. No one is perfect and everyone has worthwhile qualities. If you dwell on people's good features, it will be easier to work with them.
- Look for things to appreciate at your job site. What are the factors that make it a good place to work? Is it the hours, the physical environment, the people, and/or the actual work being done? It the atmosphere? Keep in mind that you cannot expect to like everything. If you concentrate on the good things, the negative will seem less important and bothersome.
- Look for the good in the company. Just as there are no perfect assignments, there are no perfect organizations. Nevertheless, all have good features. Is the company progressive? What about its promotional opportunities? Are there chances for self-improvement? What about the wage and benefit package? Is there a good training program? You cannot expect to have everything you would like, but there should be enough to keep you positive.
- You may not be able to change the negative attitude of another employee, but you can protect your own attitude from becoming negative. You can also work to influence their perspectives, by pointing out the good when it is appropriate to do so.

Additional Resource

The 7 Habits of Highly Effective People: Powerful Lessons in Personal Change. Stephen R. Covey. 2004. New York, NY: Simon & Schuster.

2.0.0 Section Review

1. True or false? Apprenticeship training is a modern convention, brought about by the needs of the real estate industry in greater New York.
   a. True
   b. False

2. NCCER uses the minimum _____ standards as the foundation for its pipefitting curriculum.
   a. OSHA
   b. DOJ
   c. DOL
   d. ATELS

3. Apprenticeships are focused on _____.
   a. well-tested theories and scholarly applications of those theories
   b. learning by doing and the demonstration of real skills
   c. the learner's preferences for mode of learning and nuances of the trade
   d. redefining the craft and contributing new ideas to the field

4. True or false? Professionalism is a concept that specifically describes behaviors, rather than attitudes.
   a. True
   b. False

5. Human relations is _____.
   a. basic common sense
   b. smiling even when you don't feel like it
   c. built on sound working relationships
   d. not tied to productivity

SUMMARY

The pipefitting industry is one in which job growth and stability are very promising. Learning the craft is often accomplished through apprenticeship programs, which require on-the-job training, classroom learning, and other activities over the course of several years. With an emphasis on actual performance, an apprentice must demonstrate respect for others and the tools they use; the ability to choose the right tool for the task at hand; and the aptitude to finish a series of tasks correctly and efficiently. Although a pipefitter's work is physically demanding and it requires strong math skills and analytical abilities, a trainee can expect to be successful with the right combination of commitment to the craft, maintaining a positive attitude, and operating with loyalty and integrity.

Review Questions

1. Which of the following is a task not generally performed by pipefitters?
   a. Installing residential plumbing systems
   b. Inspecting piping systems
   c. Installing valves
   d. Planning piping systems

2. Which of the following is a correct statement about pipefitters?
   a. Pipefitters must learn how to weld because welding is an important part of their work.
   b. Pipefitters only work with carbon steel pipe.
   c. Pipefitters sometimes work in food processing plants.
   d. Pipefitting work is limited to the use of hand tools.

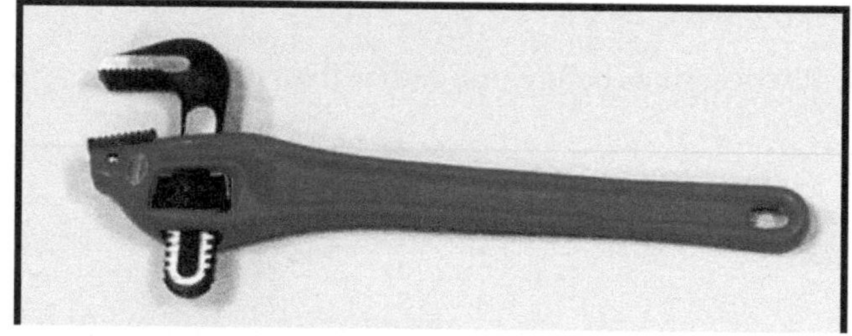

Figure RQ01N

3. The tool shown in *Figure RQ01N* is a(n) _____ wrench.
   a. straight pipe
   b. chain
   c. strap
   d. offset

4. Journey-level pipefitters often earn as much as, or more than, college-educated office workers.
   a. True
   b. False

5. Minimum standards for apprentice training programs are established by _____.
   a. OSHA
   b. Office of Apprenticeship
   c. NCCER
   d. your employer

6. If one of your co-workers does not act in a professional manner, the best approach is to _____.
   a. report the person to your supervisor
   b. advise other workers to shun the person
   c. demonstrate the benefits of professional behavior
   d. refuse to cooperate with the person

7. It is okay to be a little late for work as long as you make up the time.
   a. True
   b. False

8. The primary mission of OSHA is to _____.
   a. inspect job sites for safety violations
   b. fine companies that violate safety regulations
   c. distribute safety equipment to workers
   d. ensure that employers maintain a safe workplace

9. If you see a safety violation at your job site, you should _____.
   a. inform your supervisor
   b. ignore it unless it affects you directly
   c. report it to an OSHA inspector
   d. walk off the job until it is corrected

10. Failure to get to work on time can cause all of the following EXCEPT for _____.
    a. confusion
    b. safety violations
    c. resentment
    d. lost time

Trade Terms Quiz

Fill in the blank with the correct term that you learned from your study of this module.

1. You can solder copper pipe using the _______ method.
2. Nationwide training standards are set by _______.
3. Learning new skills while working is called _______.
4. A cut made at an angle is called a(n) _______.
5. Job-site safety and health rules are set by the _______.

Trade Terms

Office of Apprenticeship
Bevel
On-the-job learning (OJL)
Occupational Safety and Health Administration (OSHA)
Sweat

Cornerstone of Craftsmanship

Jacob Guzmán

KBR, Inc., Senior Human Resources Specialist,
Veteran Pipefitting Program Instructor

How did you choose a career in the industry?

I was a full-time student in Texas gearing up for nursing school and also teaching phlebotomy, CPR, and anatomy and physiology. In the break between finishing my prerequisite classes and starting nursing school, I found out I was going to be a father. I knew I had to find a full-time job.

Who inspired you to enter the industry?

The true inspiration was my newfound family. I had friends that worked in the field and after a few talks, I knew the industry could provide me the means to take care of my family.

What types of training have you been through?

I learned what I needed to do on the job from the journeymen in the field. Looking back, I wish I had some type of formal instruction before heading out.

How important are education and training in construction?

The importance of education and training in construction is truly a life or death matter. One must be trained to install, repair, and refit the pipe, valves, and other pieces of equipment that are saturated with all manner of dangerous products.

How important are NCCER credentials to your career?

The NCCER credentials are invaluable to my career. They opened doors closed tight to other applicants. After earning my Certified Plus credential in pipefitting, I was able to quickly rise to Pipe General Foreman for industry giants like Performance Contractors and KBR. Adding the Basic Rigger (Gold Card), Master Trainer, and Coordinator credentials did nothing but increase my earning potential.

How has training in construction impacted your life?

My life has been impacted in one of the most intimate ways. Without it, the quality of life for my wife and two kids would be very low. The industry allowed me to take my family from a one-bedroom apartment to a four-bedroom home with my name on the mortgage.

What kinds of work have you done in your career?

Starting out as a scaffold builder helper, I learned quickly that I was, in fact, not afraid of heights. Erecting a hanging scaffold 217 feet in the air isn't for the faint of heart. After switching to pipe, my experiences ranged from installing 16-inch, 300-pound blinds on a steam line in a heat-resistant suit to doing a first-line break on a 20-inch H2S line on breathing air. After being promoted to supervisor, I was able to lead a crew to safely perform all types of dangerous work.

Tell us about your present job.

I am currently serving at KBR as the sole instructor for its Veteran Pipefitting Program. I teach Industrial Pipefitting to active duty US Army soldiers who are on the verge of completing their contracts. Upon course completion, these service members will be entering the civilian workforce within 90 days. One success story is about an Infantry Sniper who was able to purchase a home for his wife and newborn baby boy within one year of separation from the military.

What do you enjoy most about your job?

What I truly enjoy most about my job is being able to help those that have fought and bled for this great nation, to transition into a field where they have the opportunity to continue their service in another way. They learn a skill set that allows them to not only go out and build their country, but build a career, build a life, and build a legacy.

What factors have contributed most to your success?

For me, it's the small things that have made the greatest difference. Integrity, respect, and discipline have had the most influence over my success. Owning up to mistakes, showing up ready to work, and speaking to both superiors and subordinates with the same level of professionalism is a true cornerstone of building any successful career.

Would you suggest construction as a career to others? Why?

Although this field isn't for everyone, I would suggest it to anyone who wants to secure their future. The need for skilled labor will never cease, and successfully navigating this industry all but guarantees a life free from financial worry.

What advice would you give to those new to the field?

The best advice I can give is to learn and develop your craft. The difference between a star quarterback in their final year of college and a first-round NFL quarterback draft pick in their rookie season is simple: Only one of them is a professional. A true craft professional is one that has spent the time to perfect their trade to the point where the only way they aren't working is by choice.

Tell us about an interesting career-related fact or accomplishment.

Before ever being considered for my current role, I was selected to be in G.I. Jobs magazine, featured as one of six veterans in uncommon but lucrative industries. Since then, Construction Citizen Magazine, BIC Magazine, Construction Superintendent, and several other newspapers and websites have featured me and/or the Veteran Pipefitting Program that I teach.

How do you define craftsmanship?

I suppose you can apply this definition to anything that is made in any industry. To me, craftsmanship is the pride of using your hands to build a product that is not only functional and high quality, but personal, necessary, and in a way, beautiful.

SAMPLES OF NCCER TRAINING CREDENTIALS

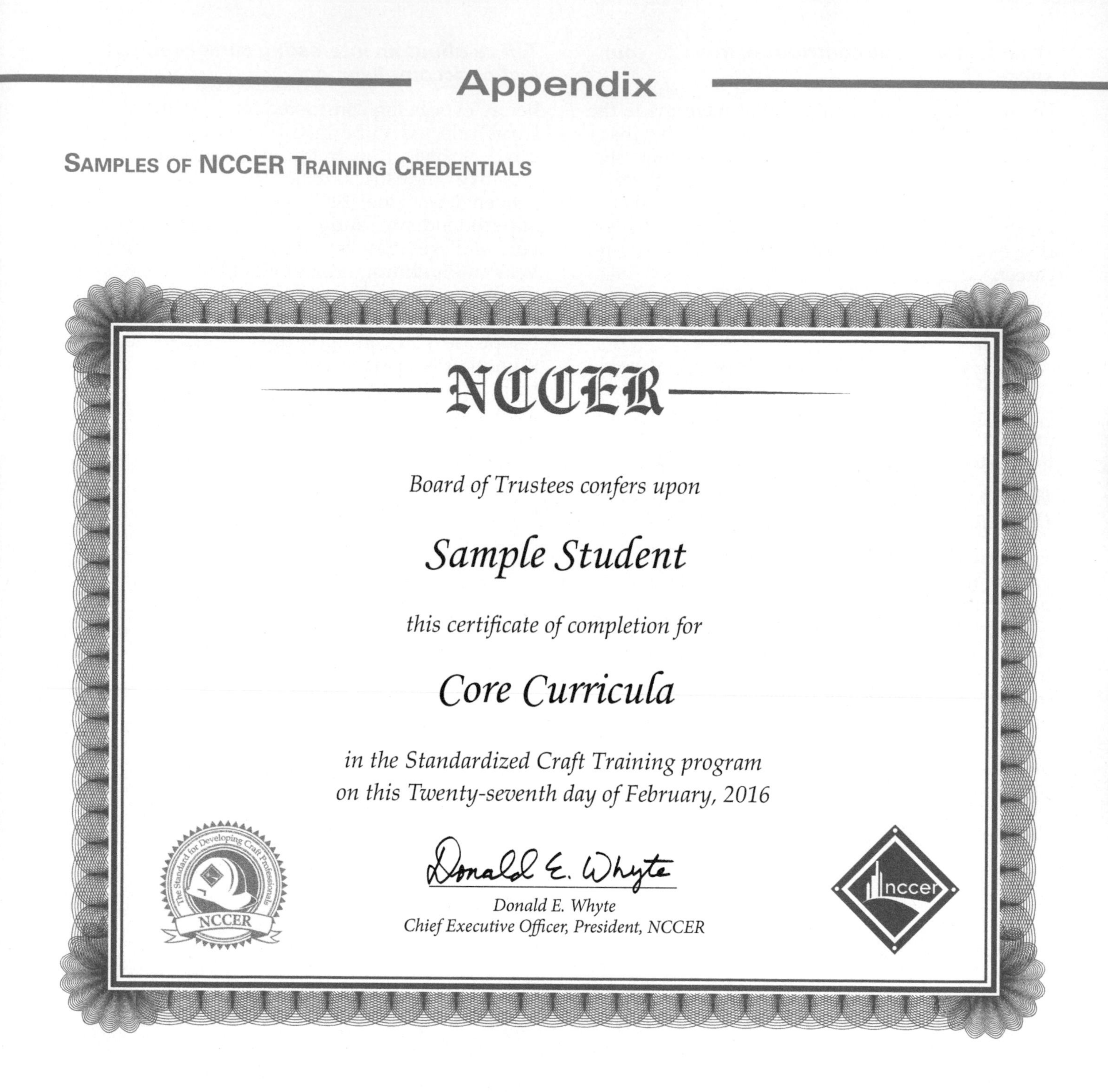
NCCER

Board of Trustees confers upon

Sample Student

this certificate of completion for

Core Curricula

in the Standardized Craft Training program
on this Twenty-seventh day of February, 2016

Donald E. Whyte
Chief Executive Officer, President, NCCER

nccer
Student 7 Samp
Certified Plus
4671784
nccer
SECURE SECURISE
Sample Student
2781481

Trade Terms Introduced in This Module

Office of Apprenticeship: The U.S. Department of Labor office that sets the minimum standards for training programs across the country.

Bevel: A cut made at an angle.

On-the-job learning (OJL): Job-related learning acquired while working.

Occupational Safety and Health Administration (OSHA): The federal government agency established to ensure a safe and healthy environment in the workplace.

Sweat: A method of joining pipe in which solder is applied to the joint and heated until the solder flows into the joint.

Additional Resources

This module presents thorough resources for task training. The following reference material is recommended for further study.

Apprenticeship. US Department of Labor: Office of Apprenticeship. 2018. Available at: **https://www.dol.gov/general/topic/training/apprenticeship**

Electronic Code of Federal Regulations (29 CFR 1926 and 29 CFR 1900-1910). 2018. US Government Publishing Office. Available at: **www.ecfr.gov**.

OSHA's Respirable Crystalline Silica Standard for Construction. 2017. Occupational Safety and Health Administration. Available at: **https://www.osha.gov/Publications/OSHA3681.pdf**.

*Safety and Health Regulations for Construction.*2018. Occupational Safety and Health Administration. Available at: **https://www.osha.gov/pls/oshaweb/owastand.display_standard_group?p_part_number=1926&p_toc_level=1**.

Silica, Crystalline: Overview. 2018. Occupational Safety and Health Administration. Available at: **https://www.osha.gov/dsg/topics/silicacrystalline/**.

The 7 Habits of Highly Effective People: Powerful Lessons in Personal Change., Stephen R. Covey. 2004. New York, NY: Simon & Schuester.

The Pipefitters Handbook, 3rd Edition. Forrest R. Lindsey. New York, NY: Industrial Press.

Section Review Answer Key

SECTION 1.0.0

Answer	Section Reference	Objective
1. b	1.0.0	1a
2. a	1.1.0	1a
3. d	1.2.0	1b
4. d	1.3.0	1c

SECTION 2.0.0

Answer	Section Reference	Objective
1. b	2.1.0	2a
2. d	2.2.0	2b
3. b	2.1.0	2a
4. b	2.3.1	2c
5. c	2.4.1	2d

NCCER CURRICULA — USER UPDATE

NCCER makes every effort to keep its textbooks up-to-date and free of technical errors. We appreciate your help in this process. If you find an error, a typographical mistake, or an inaccuracy in NCCER's curricula, please fill out this form (or a photocopy), or complete the online form at **www.nccer.org/olf**. Be sure to include the exact module ID number, page number, a detailed description, and your recommended correction. Your input will be brought to the attention of the Authoring Team. Thank you for your assistance.

Instructors – If you have an idea for improving this textbook, or have found that additional materials were necessary to teach this module effectively, please let us know so that we may present your suggestions to the Authoring Team.

NCCER Product Development and Revision

13614 Progress Blvd., Alachua, FL 32615

Email: curriculum@nccer.org
Online: www.nccer.org/olf

❑ Trainee Guide ❑ Lesson Plans ❑ Exam ❑ PowerPoints Other ____________________

Craft / Level: Copyright Date:

Module ID Number / Title:

Section Number(s):

Description:

Recommended Correction:

Your Name:

Address:

Email: Phone:

This page is intentionally left blank.

Pipefitting Hand Tools

OVERVIEW

Pipefitters use hand tools to grip, level, fabricate, cut, and bend pipe. Fabrication tools include squares, clamps, gauges, wraparounds, and pins. Vises and stands hold the work to free the crafter to work with both hands; levels are used to ensure that a pipe is level or plumb. Specialized tools are used to cut, thread, bend, and flare pipe.

Pipefitters must learn not only how and when to use their tools but also how to take care of them. Properly maintained tools are safer and more accurate. Using the correct tool – in a safe and appropriate manner – allows a crafter to skillfully complete the job.

Module 08102

Trainees with successful module completions may be eligible for credentialing through the NCCER Registry. To learn more, go to www.nccer.org or contact us at 1.888.622.3720. Our website has information on the latest product releases and training, as well as online versions of our *Cornerstone* magazine and Pearson's product catalog.

Your feedback is welcome. You may email your comments to curriculum@nccer.org, send general comments and inquiries to info@nccer.org, or fill in the User Update form at the back of this module.

This information is general in nature and intended for training purposes only. Actual performance of activities described in this manual requires compliance with all applicable operating, service, maintenance, and safety procedures under the direction of qualified personnel. References in this manual to patented or proprietary devices do not constitute a recommendation of their use.

08102 V4.0

08102
PIPEFITTING HAND TOOLS

Objective

Successful completion of this module prepares trainees to:

1. Identify and describe common pipefitting hand tools and their safe use.
   a. Identify general safety guidelines for hand tool use.
   b. Identify and describe the use of pipe vises and stands.
   c. Identify and describe the use of pipe wrenches and other common wrenches.
   d. Identify and describe the use of various levels.
   e. Identify and describe the use of various layout and fabrication tools.
   f. Identify and describe the use of common pipe clamps and alignment tools.
   g. Identify and describe the use of pipe cutting and reaming tools.
   h. Identify and describe the use of pipe threading tools.
   i. Identify and describe the use of bending and flaring tools.

Performance Tasks

Under the supervision of your instructor, you should be able to do the following:

1. Identify various pipefitting hand tools.
2. Secure a section of pipe in a vise and pipe stand.
3. Demonstrate the proper use of a wraparound.
4. Demonstrate the proper use of two of the following:
   - Straight pipe wrenches
   - Offset pipe wrenches
   - Strap wrenches
5. Demonstrate the proper use of two of the following:
   - Laser level
   - Torpedo and larger levels
   - Center finder
6. Check square and level:
   - Turn tongue 180 degrees from where it was
   - Flip level to ensure it is level
   - Using a square, check square from a fitting to a fitting, or a fitting to a pipe

Trade Terms

Burr
Conduit
Die
Female threads
Flare
Male threads
Outside diameter
Pipe fitting
Ratchet
Tack welds
Thread gauge

Industry Recognized Credentials

If you are training through an NCCER-accredited sponsor, you may be eligible for credentials from NCCER's Registry. The ID number for this module is 08102. Note that this module may have been used in other NCCER curricula and may apply to other level completions. Contact NCCER's Registry at 888.622.3720 or go to www.nccer.org for more information.

Contents

Figures and Tables

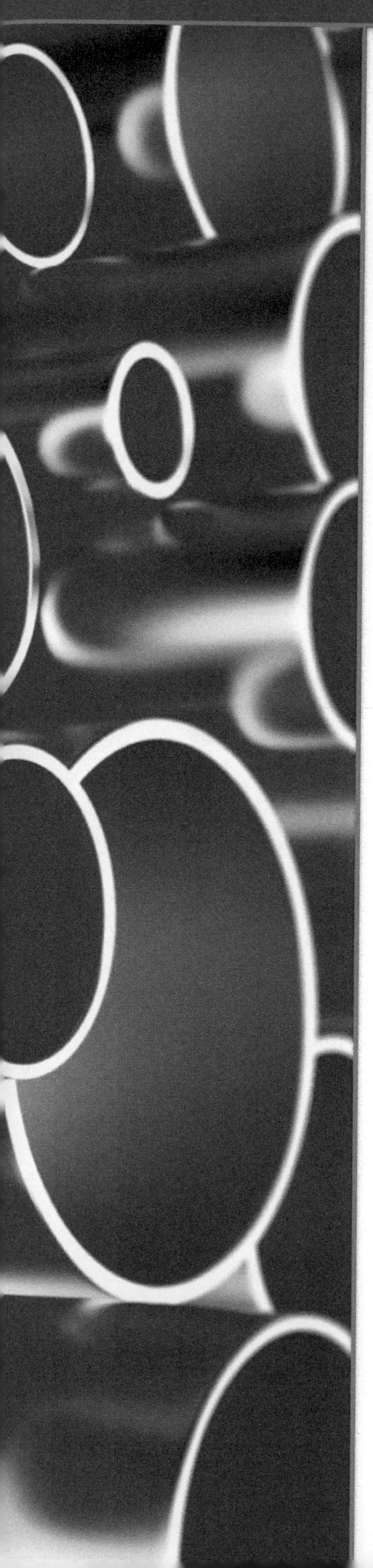

Figures and Tables (continued)

SECTION ONE

1.0.0 HAND TOOLS OF THE PIPEFITTING CRAFT

Objective

Identify and describe common pipefitting hand tools and their safe use.

a. Identify general safety guidelines for hand tool use.
b. Identify and describe the use of pipe vises and stands.
c. Identify and describe the use of pipe wrenches and other common wrenches.
d. Identify and describe the use of various levels.
e. Identify and describe the use of various layout and fabrication tools.
f. Identify and describe the use of common pipe clamps and alignment tools.
g. Identify and describe the use of pipe cutting and reaming tools.
h. Identify and describe the use of pipe threading tools.
i. Identify and describe the use of bending and flaring tools.

Performance Tasks

1. Identify various pipefitting hand tools.
2. Secure a section of pipe in a vise and pipe stand.
3. Demonstrate the proper use of a wraparound.
4. Demonstrate the proper use of two of the following:
   - Straight pipe wrenches
   - Offset pipe wrenches
   - Strap wrenches
5. Demonstrate the proper use of two of the following:
   - Laser level
   - Torpedo and larger levels
   - Center finder
6. Check square and level:
   - Turn tongue 180 degrees from where it was
   - Flip level to ensure it is level
   - Using a square, check square from a fitting to a fitting, or a fitting to a pipe

Trade Terms

Burr: A sharp, ragged edge of metal usually caused by cutting pipe.

Conduit: A round raceway, similar to pipe, that contains conductors.

Die: A tool used to make male threads on a pipe or bolt.

Female threads: Threads on the inside of a fitting.

Flare: A pipe end that has been forced open to make a joint with a fitting.

Male threads: Threads on the outside of a pipe.

Outside diameter: A measurement of the outside width of a pipe.

Pipe fitting: A unit attached to a pipe and used to change the direction of fluid flow, connect a branch line to a main line, close off the end of a line, or join two pipes of the same size or of different sizes.

Ratchet: A device that allows a tool to rotate in only one direction.

Tack welds: Short welds used to hold parts in place until the final weld is made.

Thread gauge: A tool used to determine how many threads per inch are cut in a tap, die, bolt, nut, or pipe. Also called a pitch gauge.

Pipefitting hand tools are generally categorized by their physical description and method of use. For example, a wraparound is a tool that wraps around a piece of pipe to lay out a straight line. A wide variety are available, with the more commonly-used tools offered in several sizes. In many cases, tools must be exactly the right size to prevent personal injury or equipment damage Understanding the use, maintenance, and potential hazards of nonpowered hand tools is essential to a safe and successful pipefitting career.

1.1.0 General Hand Tool Safety

Although hand tools do not present the immediate hazards of power tools, improper use can cause serious injury. Some hand tools are more dangerous than others and require additional precautions and care when being used. To safely use hand tools, follow these basic rules:

- Always choose the proper tool for the job.
- Never use a tool for anything other than its intended use. For example, do not use a screwdriver as a prybar. Do not modify any hand tool.

- Never use dull or broken cutting tools. Dull cutting tools require greater force to do the job. A sharp cutting tool is a safe tool.
- Keep your hands and fingers away from sharp edges of cutting tools. Wear gloves when using tools that require them.
- Work away from your body when using cutting tools.
- Ensure that a tool is in good condition and that handles are tightly fastened to the tools. Keep tools clean and free of rust.
- Wear eye protection when using chisels, punches, or other tools that may produce flying particles and debris.
- Use tools in a safe and proper manner. An improperly held wrench can slip, causing injury to the hand or knuckles.

1.2.0 Vises and Stands

Pipe vises and stands are used to temporarily support pipe or other materials. They are specially designed to securely hold pipe and other round objects.

1.2.1 Pipe Vises

Pipefitters use standard vises as well as yoke, chain, and strap vises. All vises are designed to hold or clamp, but the yoke, chain, and strap vises are designed for pipefitting jobs. The yoke vise is similar to a standard vise, except that it is hinged to open vertically so the pipe may be laid into the vise. Chain or strap vises also open so the pipe may be laid in easily, but chain or strap vises allow larger objects to be held. *Figure 1* shows a pipefitter's vises.

Follow these steps to use a pipefitter's vise:

Step 1 Obtain the pipe to be held.

Step 2 Identify where the work must be performed.

Step 3 Select a vise.

NOTE

The location of the work may determine the kind of vise needed. Standard bench vises may be used, but yoke, chain, or strap vises may be more appropriate depending on the task at hand.

Step 4 Inspect the selected vise for obvious damage, excessive jaw wear or screw play, worn-out chain links or strap, grease, rust, or excessive dirt. If any damage to the vise is found, fix or replace it. If grease or excessive dirt is found, clean and oil it as necessary.

Step 5 Check to see that the vise is securely mounted.

WARNING!

A vise must be securely mounted to prevent injury to the operator and damage to the vise, pipe, and other equipment in the area.

Step 6 Loosen and open the vise.

Step 7 Place the pipe into the vise.

CAUTION

To prevent marring of the pipe or other object that is being held, insert material that is softer than the object under the vise jaws or chain, or use a strap vise. For long objects, use a jack stand or similar support to hold the section of the object that is not in the vise.

Step 8 Close and secure the moveable side of the vise.

Step 9 Tighten the vise.

CAUTION

Tighten only enough to hold the object securely. Overtightening may crush the pipe in the vise.

Step 10 Perform the work on the pipe.

WARNING!

Wear safety equipment, such as glasses, gloves, apron, or face shield as appropriate.

CAUTION

When sawing an object while it is being held in a vise, saw as close to the vise as possible. Do not use a vise as an anvil. Never use an extension bar for extra tightening.

Step 11 Hold the pipe in the vise and loosen the vise grip until the moveable side of the vise can be fully opened.

Step 12 Remove the pipe from the vise.

Step 13 Inspect the vise for any obvious damage and correct any problems that are found.

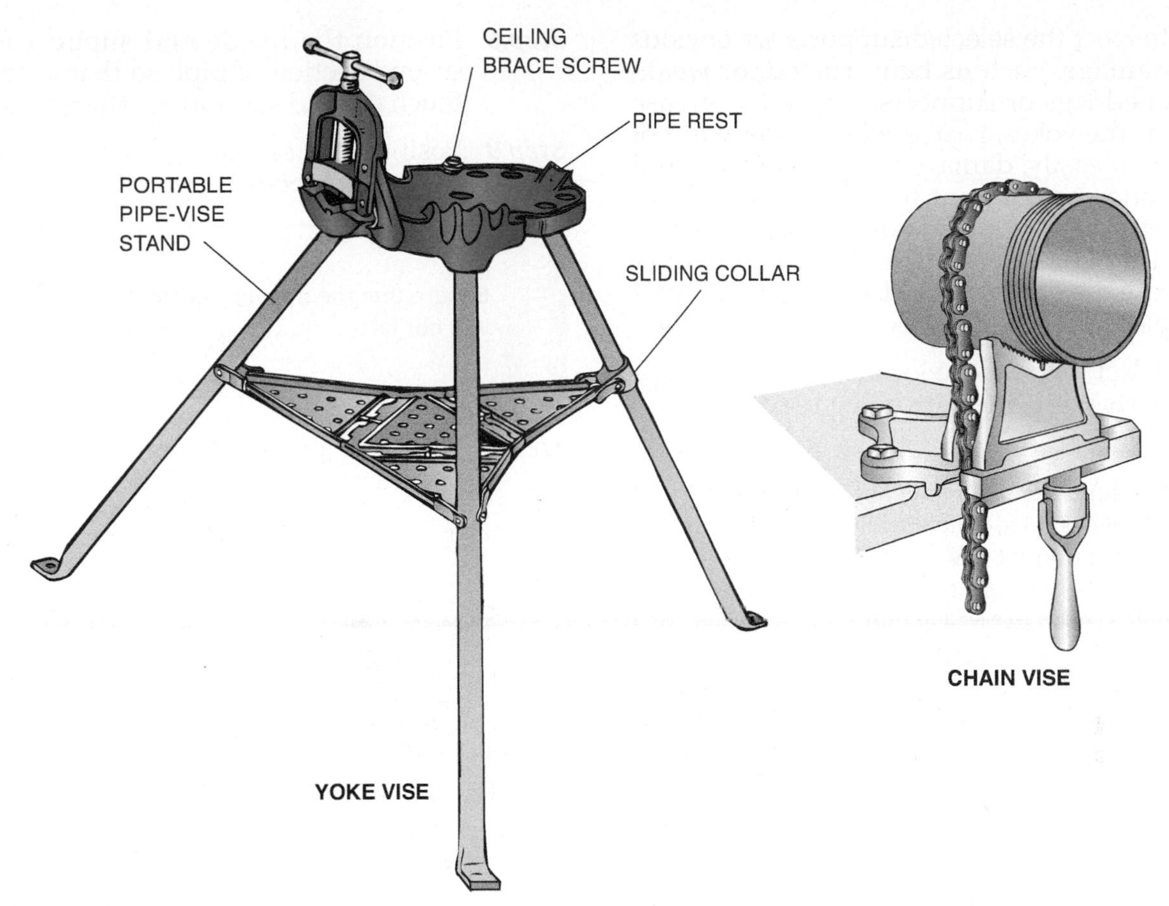

Figure 1 Yoke and chain vises.

Step 14 Clean the vise of any dust, dirt, grease, shavings, or chips, especially in the screw area and the chain links. Also, clean the general area around the vise.

Step 15 Close the moveable side of the vise and secure it.

Step 16 Store the vise if it is portable and no longer needed.

1.2.2 Pipe Jack Stands

Pipe jack stands are support devices (*Figure 2*) used in the repair or fabrication of pipe or other such objects. Some of these devices have rollers on the tops or yokes that allow the object supported to be easily maneuvered (rolled) and worked on safely. Each jack will have a specific load capacity; this must be stamped or written on the jack for easy reference. Do not exceed the noted rating of a jack, because this introduces safety hazards and violates the OSHA mandate that, "The operator shall make sure that the jack used has a rating sufficient to lift and sustain the load." (Source: **https://www.osha.gov/pls/oshaweb/owadisp.show_document?p_id=9851&p_table=standards)**

When a pipe is placed on the wrong stand, it introduces the possibility of incidents and accidents, in addition to significant financial penalties from OSHA. Be sure to follow manufacturers' recommendations for proper use of each stand, as well as company safety policies.

WARNING!

Jack stands and roller assemblies should not be field-fabricated. Load capacities and other safety considerations are built into these devices by their manufacturers, so only commercial jack stands and rollers should be used. Field-fabricated jack stands and rollers can fail and cause personal injury or even death.

Follow these steps to use jack stands:

Step 1 Select the kind of support device needed.

NOTE

For this exercise, assume that two, 6-foot sections of 4-inch steel pipe must be set up to be welded into one, 12-foot section. Since you will have to adjust the pipes with each other, four adjustable jacks are needed.

Step 2 Inspect the selected supports for obvious damage, such as bent, rusted, or weakened legs or supports. Check for: grease on the yokes, rusted wheels that will not turn easily, damaged screw threads, and collar adjustment handles that will not turn. If a support is damaged, replace it. If it needs cleaning and oiling, do it before the device is used. Oil should only be placed on the threads.

Step 3 Select a level, safe location to set the supports while they are being used.

> **WARNING!**
>
> These supports often hold heavy objects and must have safe footing. Position the supports out of the normal flow of work if possible. If using these supports on the ground, they should be set up on plywood.

Step 4 Position the supports in a straight line and space them so that two will support each section of pipe.

Step 5 Turn the collar adjustment on each support so that the yokes are fully lowered.

Step 6 Ensure that all the yokes are roughly level with each other.

> **WARNING!**
>
> Beware of pinch points between the collar adjust screws and the safety lock rings.

> **NOTE**
>
> Unlock the vertical lock screws and the safety lock ring on each support and adjust the main body (tube) of the support. When they are aligned, tighten the vertical lock screws so the tubes will not compress later.

Step 7 Position the first section of pipe onto one set of supports.

> **WARNING!**
>
> Check to see that the support is firmly under the pipe before the pipe is released. An object this heavy could injure a worker or damage equipment if it falls. Also ensure that the supports are located toward the ends of the pipe.

Step 8 Position the inside end support for the second section of pipe so that it does not touch the end support for the first section.

Step 9 Position the second section of pipe onto the second set of supports.

> **WARNING!**
>
> Ensure that the pipe is well seated and that it will not fall.

Step 10 Carefully push the two sections of pipe together until the ends meet.

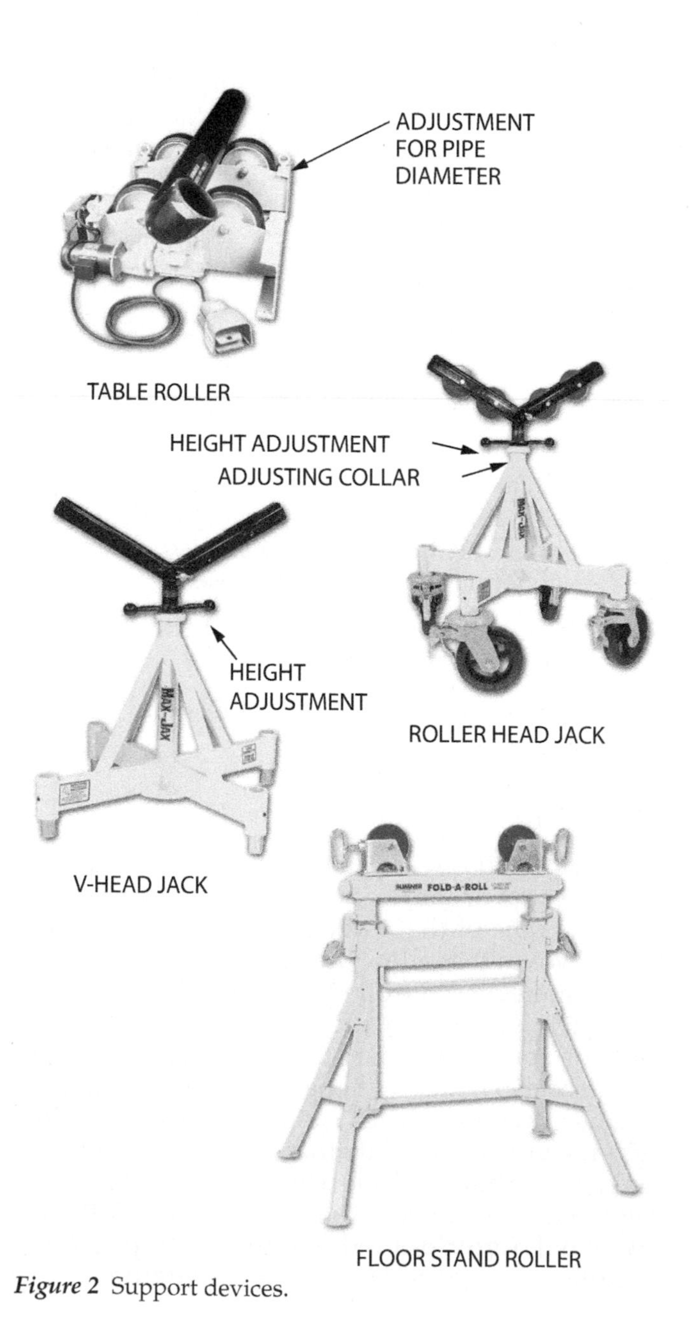

Figure 2 Support devices.

Step 11 Adjust the main body tubes and the collar adjustments of the supports to make the two sections of pipe roughly level with each other.

1.3.0 Pipe Wrenches

Many types and sizes of pipe wrenches are used in the field. Pipe wrenches are considered heavy-duty because they are made of durable cast iron or aluminum. They are used to grip and turn pipe or to grip and hold it in a stationary position. It is critical to select the right size pipe wrench for a given job. A pipe wrench that is too small will not hold the pipe firmly, and a handle that is too short will not provide enough leverage. A pipe wrench that is too large can strip the threads, break the pipe or fitting, or cause excessive marring or scratching of the pipe.

1.3.1 Pipe Wrench

The pipe wrench is used to grip and turn round pipes and tubing. Types include straight pipe wrenches, heavy-duty pipe wrenches, and 45-degree and 90-degree offset pipe wrenches (*Figure 3*). Wrenches come in a number of lengths, such as 6, 12, 14, 18, 24, 36, and 48 inches. The wrench's length determines the size of pipe that it can handle.

The straight pipe wrench is the most common type. When working in tight quarters, use offset wrenches; their angles (45- or 90-degrees) make it easier to reach pipe in cramped areas. The jaws of pipe wrenches always leave marks on the pipe; therefore, do not use them on pieces where appearance is important. When applying force to a wrench, always direct it toward the open side of the jaws. This technique will give you the best grip and leverage.

A variation on the straight pipe wrench is the compound-leverage pipe wrench (*Figure 4*). This design increases the leverage that can be applied on a pipe. This wrench is generally used on pipe joints that are frozen or locked together.

Nonferrous Alignment Tools

In some industries, fabricated alignment tools (wedges, strongbacks, and yokes) must be constructed of nonferrous materials, such as brass, to prevent sparking and to prevent stainless steel contamination. This is common in some segments of industries that use petroleum-based chemicals.

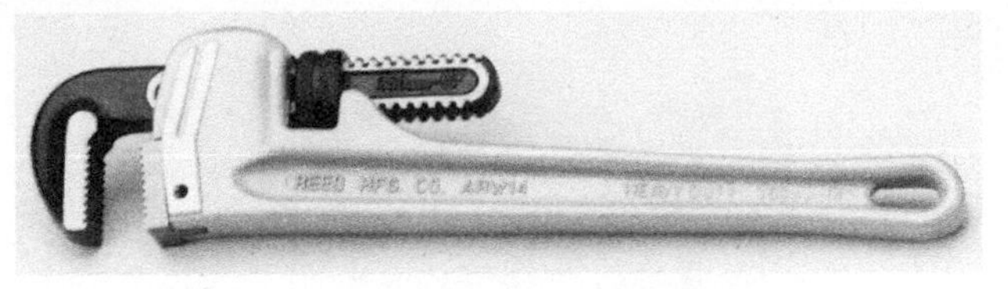

STRAIGHT PIPE WRENCH

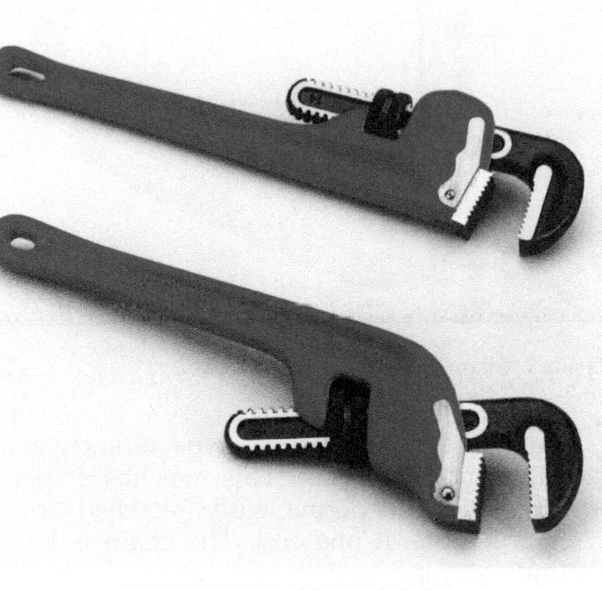

HEAVY-DUTY PIPE WRENCH
45-DEGREE OFFSET WRENCH

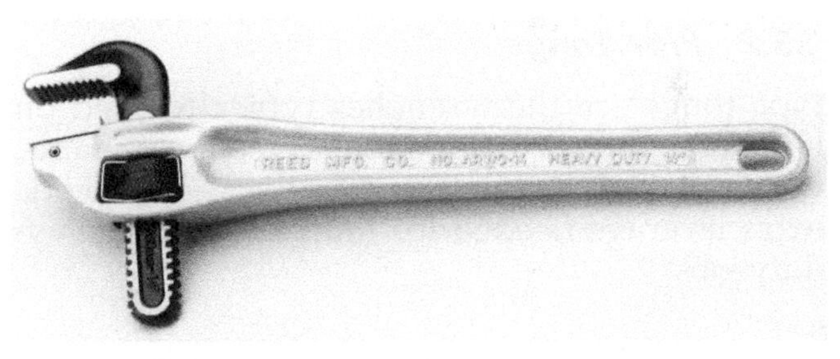

90-DEGREE OFFSET PIPE WRENCH

Figure 3 Pipe wrenches.

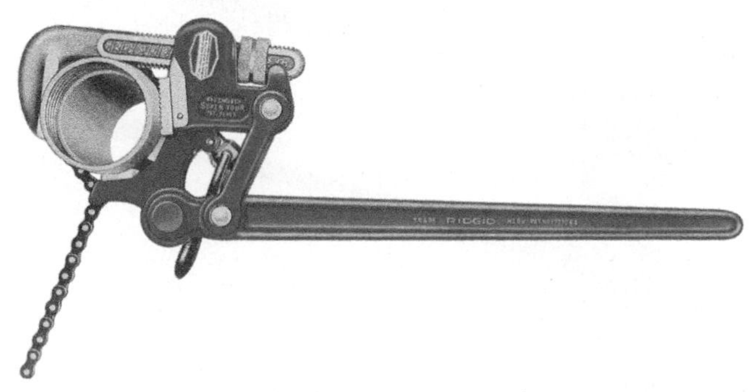

Figure 4 Compound-leverage pipe wrench.

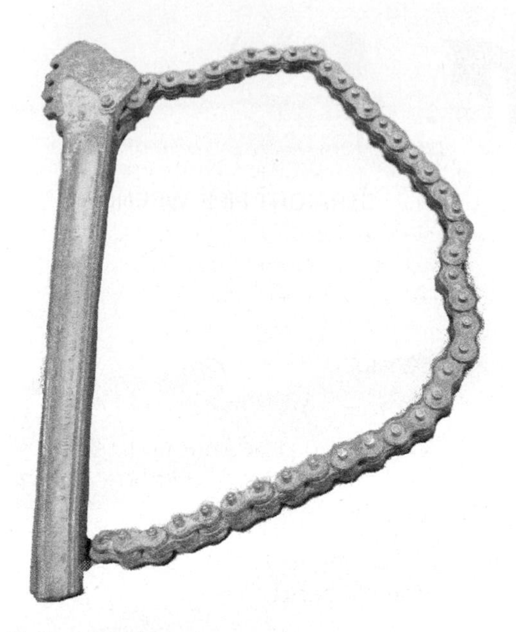

Figure 5 Chain wrench.

Another variation of the pipe wrench is the chain wrench (*Figure 5*). This tool has a specific length of chain permanently attached to the wrench handle at one end. The chain is looped around the pipe to grip and secure it while the other end of the chain can be secured to various positions on the wrench's handle. Oil the chain frequently to prevent it from becoming stiff or rusty.

1.3.2 Pipe Tongs

Pipe tongs are the wrenches typically used on large pipe (*Figure 6*). Pipe tongs also have chains, which need to be oiled frequently. They provide extra leverage needed for tougher, more heavy-duty jobs.

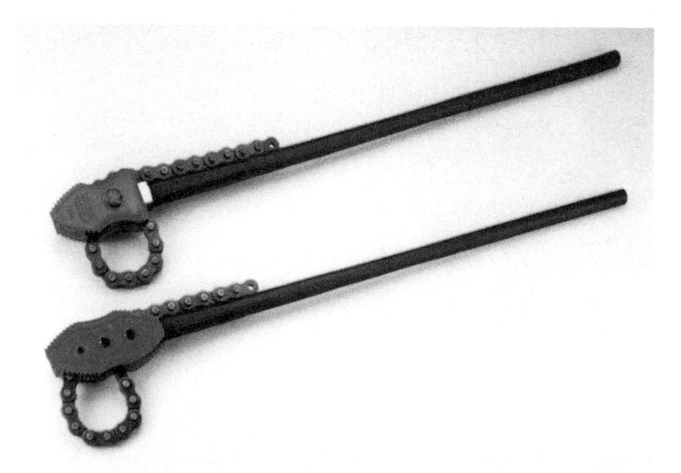

Figure 6 Pipe tongs.

1.3.3 Strap Wrench

The strap wrench (*Figure 7*) is used to hold chrome-plated, polished, or other types of finished pipe. The strap wrench does not leave jaw marks or scratches on the pipe because it grips with a strap rather than a jaw or chain. With some strap wrenches, rosin needs to be applied to the strap so that it does not slip. Other strap wrenches use vinyl straps that do not need rosin.

1.3.4 Other Common Wrenches

The open-end wrench (*Figure 8*) has two different-sized openings, each on one end of the tool. The size of each opening is stamped on the handle. This wrench is usually used for assembling fittings and fasteners that are less than 1 inch across.

The adjustable wrench, like the Crescent® wrench in *Figure 9*, is similar to the open-end wrench, except that it has an adjustable jaw. A roller at the stem of the jaw adjusts the size of the opening. Because it is so easy to adjust, this wrench is found in virtually every toolbox.

Figure 7 Strap wrench.

Figure 8 Open-end wrench.

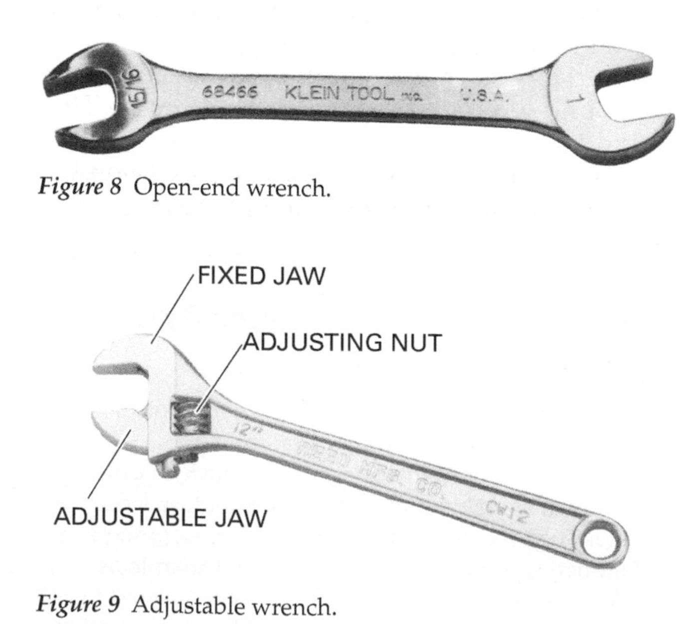

Figure 9 Adjustable wrench.

Did You Know?

The Pipe Wrench

The pipe wrench was invented by a steamboat firefighter named Daniel Stillson. Stillson had suggested to Walworth, a heating and piping firm, that it should design a wrench that could be used to join or separate pipes. The firm's owner told Stillson to take a crack at making a prototype that would twist off a stuck pipe. To the everlasting gratitude of future plumbers, Stillson was successful and his prototype twisted off the pipe. In 1870, Stillson's design was patented and Walworth began manufacturing the wrench. Even today, pipe wrenches are sometimes called Stillson wrenches.

Source: Letters Patent No. 95,744. Google Patent Library. **https://patents.google.com/patent/US95744**

Various sizes of adjustable wrenches are available from 6 to 18 inches in length.

Adjustable wrenches have smooth jaws, making these tools ideal for turning nuts, bolts, small pipe fittings, chrome-plated pipe fittings, and polished pipe fittings. Using an adjustable wrench can save considerable time because it eliminates the need to constantly switch wrench sizes as different sizes of nuts and bolts are being used.

1.3.5 Using and Caring for Pipe Wrenches

When using a pipe wrench, either place the wrench on the pipe or fitting with the jaw opening facing you and pull on the handle or place the wrench on the pipe or fitting with the jaw opening facing away from you and push on the handle.

WARNING!

Be careful if you are in a situation where you are pulling the handle toward you to keep the jaw from slipping off the pipe causing you to injure yourself. A safer practice is to push the wrench away from the body while bracing yourself so that you do not fall if the wrench slips off the pipe.

CAUTION

Never place the jaws of a pipe wrench onto the threaded end of a pipe because this will ruin the threads.

Follow these steps to tighten a fitting onto the threaded end of a pipe:

Step 1 Select a pipe wrench suitable for the size of pipe being used and the job being performed.

Step 2 Inspect the selected wrench for excessively worn jaw teeth and corners, as well as any noticeable damage to the handle. Replace the wrench if necessary.

Step 3 Secure the pipe in a chain vise, allowing the threaded end to extend from the vise about 8 inches.

Step 4 Tighten the fitting onto the end of the pipe by hand.

Step 5 Position yourself so that the fitting is on either your left or right side.

Step 6 Turn the knurled nut on the pipe wrench until the jaws are open slightly wider than the diameter of the fitting.

NOTE

Many straight pipe wrenches have pipe sizes marked on the hook jaw just above the housing. If your wrench has these marks, line up the correct mark with the top of the housing.

Step 7 Slide the jaws onto the fitting making sure they fit snugly. The fitting should be centered inside the jaws and there should be a gap between the fitting and the back of the hook jaw.

Step 8 Push or pull the handle as far as possible to tighten the fitting.

Step 9 Pull or push the handle in the opposite direction to release the jaws from the fitting.

Step 10 Regrip the fitting with the wrench and push or pull the wrench again.

Step 11 Repeat Steps 9 and 10 until the fitting is tight on the pipe.

CAUTION

Do not overtighten the fitting on the pipe or else the threads may strip. Three full threads should be exposed on the pipe at the end of the fitting.

Step 12 Remove the wrench from the pipe fitting.

Step 13 Clean the wrench and inspect it for any damage.

Step 14 Close the jaws and store the wrench.

It may be necessary to install or remove a fitting from a piece of pipe when a vise is not available or when the fitting is part of a piping system. In this case, use two pipe wrenches.

Follow these steps to remove a fitting from a pipe without using a vise:

Step 1 Position yourself with the fitting on your left or right side.

Step 2 Place one pipe wrench on the pipe.

Step 3 Place the other pipe wrench on the fitting with the jaw opening facing in the opposite direction (*Figure 10*).

Step 4 Pull the handle of the wrench on the pipe while pushing on the handle of the wrench on the fitting.

Step 5 Regrip the pipe and the fitting when you have turned the wrenches as far as you can.

Step 6 Repeat steps 2 through 5 until the fitting is loose enough to remove by hand.

Step 7 Remove the wrenches from the pipe and fitting.

Step 8 Clean the wrenches and inspect them for any damage.

Step 9 Close the jaws and store the wrenches.

Wrenches will only work well if they are used correctly and properly cared for. Follow these guidelines to use and care for pipe wrenches:

- Use pipe wrenches only to turn pipe and fittings. Do not use a pipe wrench to bend, raise, or lift a pipe.

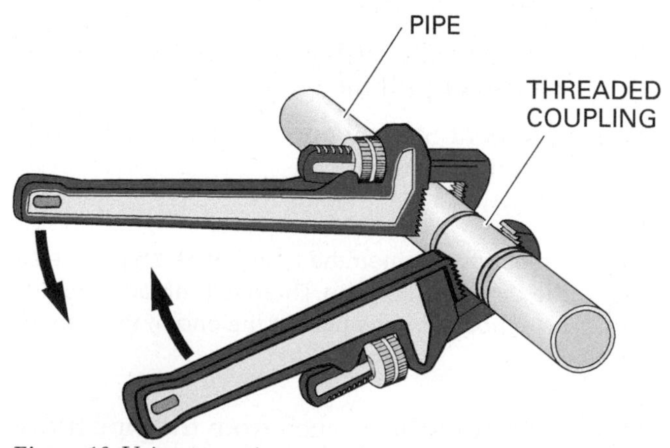

Figure 10 Using two pipe wrenches.

- Do not use the pipe wrench as a hammer. It is not designed for any sort of pounding.
- Do not drop or throw a pipe wrench. You may break or crack the wrench, which could cause injury to yourself or others.
- Check the teeth of the wrench often. They should be kept clean and sharp to keep the wrench from slipping on the pipe.
- Apply penetrating oil to the wrench threads to prevent them from sticking.

CAUTION

Never use cheaters (pipes used to extend wrenches) because they can cause equipment damage. Always use the right-sized wrench to avoid damaging the pipe.

1.4.0 Levels

Levels are used to determine the plumb or levelness of pipe. Plumb refers to vertical alignment, and level refers to horizontal alignment. Several types of levels are used by pipefitters, including framing levels, torpedo levels, and tubing water levels. Most levels used by pipefitters are made of tough, lightweight metals such as magnesium or aluminum. They generally have three vials: two are used to measure plumb and one is used to measure level. The amount of liquid each vial contains is not enough to fill the vial completely, this creates a bubble. Centering this bubble between the lines scribed on the outside of the vial produces plumb or level (*Figure 11*).

1.4.1 Framing and Torpedo Levels

Framing levels, also known as spirit levels, come in a variety of sizes. Longer levels are more accurate than shorter ones. The most common sizes for framing levels are 18, 24, 28, and 48 inches. Most pipefitters carry a 24-inch framing level. The frame of the framing level is milled and ground on both the top and bottom to be uniformly parallel and to ensure smoothness. Framing levels are available with or without a 45-degree vial. *Figure 12* shows a framing level.

A torpedo level (*Figure 13*) is approximately 9 inches long and tapered at both ends. Torpedo levels are best used in tight places or where accuracy is not critical. The top of the torpedo level has a groove running down the center from end to end that helps it sit on the round surface of a pipe. The bottom side of some torpedo levels has magnets encased in it so it will adhere to metal pipes. Torpedo levels have three vials and will measure level, plumb, and a true 45-degree angle. All pipefitters should carry a torpedo level.

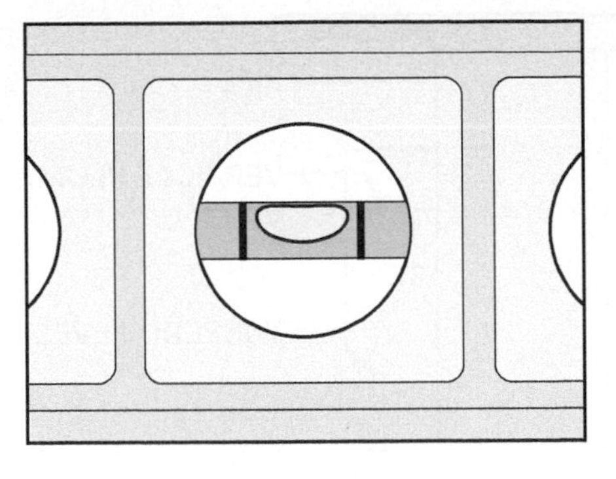

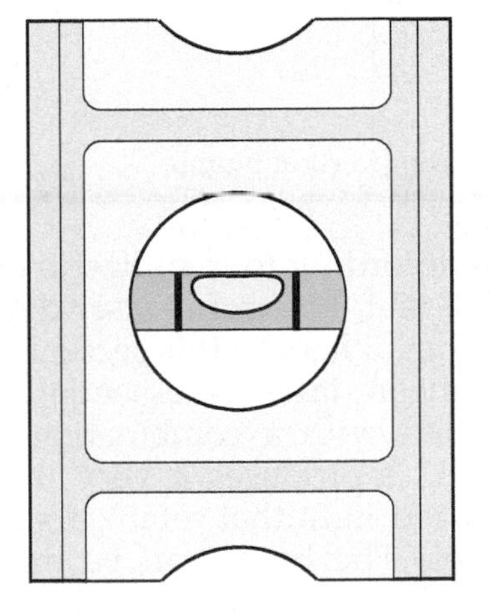

Figure 11 Correctly aligned spirit level vials.

1.4.2 Measuring Using Spirit Levels

Pipefitters use framing and torpedo levels in the same way to measure plumb and level. Framing levels can be used to measure true 45-degree angles if they have a 45-degree vial.

Follow these steps to use a framing or torpedo level to measure horizontal level:

Step 1 Make sure that the surface of the pipe (where the level is to be placed) is smooth and clean.

Step 2 Make sure that the sides of the level are clean.

Step 3 Place the level on the pipe to be evaluated.

Checking a Spirit Level

You should periodically check your spirit level to make sure it has not gotten out of calibration. Position the level on a flat surface so that the bubble registers in the center of the vial, then use the length of the level as a straightedge to draw a line. Flip the level over from side to side and lay it against the line. If the level is good, the bubble will be centered in the vial. Make sure the bubble reaches each side of the level to prevent potential errors.

NOTE

The bubble inside the vial will move to the high end of the pipe.

Step 4 Adjust the pipe so that the bubble lies between the two marks scribed on the vial.

Follow these steps to use a framing or torpedo level to measure vertical plumb:

Step 1 Make sure that the surface of the pipe or object is smooth and clean.

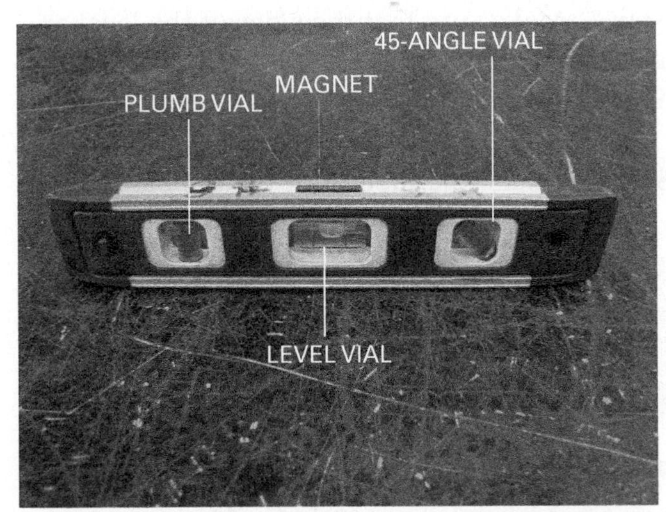

Figure 13 Torpedo level.

Figure 12 Framing level.

Step 2 Make sure that the sides of the level are clean.

Step 3 Hold the level vertically against the side of the pipe or object to measure vertical plumb from side to side. Hold the level so that the vertical plumb vial is toward the top of the pipe or object. *Figure 14* shows measuring vertical plumb.

Step 4 Adjust the pipe or object until the bubble lies between the scribe marks on the vial.

Step 5 Move the level 90 degrees around the pipe or object to measure vertical plumb from front to back.

Step 6 Adjust the pipe or object until the bubble lies between the scribe marks on the vial.

Step 7 Recheck the pipe or object for vertical plumb from side to side.

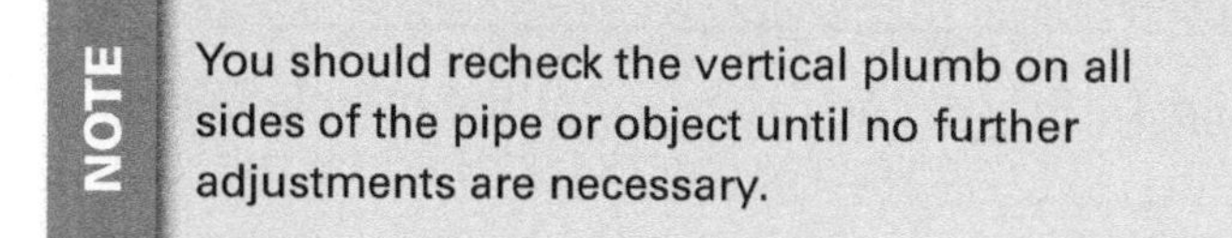

NOTE

You should recheck the vertical plumb on all sides of the pipe or object until no further adjustments are necessary.

Follow these steps to adjust a pipe to a true 45-degree angle, using a torpedo level:

Step 1 Make sure that the pipe is smooth and clean.

Step 2 Make sure that the sides of the level are clean.

Step 3 Place the torpedo level on the pipe so that the 45-degree vial is level.

Step 4 Adjust the pipe until the bubble lies between the marks scribed on the vial.

NOTE

As always, the bubble will move to the high side.

1.4.3 Laser Levels

Laser instruments are widely used today to determine level, plumb, and grade. Depending on the type of laser intensity, the beam produced may be invisible to the human eye. Visible laser beams have wavelengths in the visible light portion of the radiant energy spectrum. Invisible laser beams have wavelengths in the infrared portion of the radiant energy spectrum. Because these beams are invisible, they must be detected using electronic laser detectors. The accuracy of laser

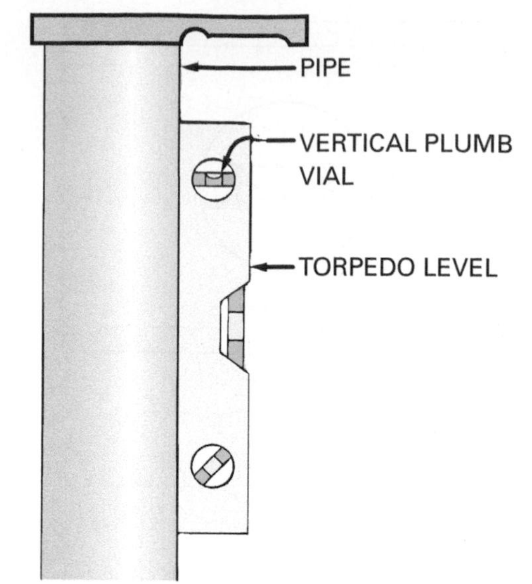

Figure 14 Measuring vertical plumb.

levels varies according to manufacturer, with $\frac{1}{8}$" to $\frac{1}{4}$" per 100 feet. Electronic laser detectors are described in detail later in this section.

A helium-neon laser is essentially an electron tube filled with a combination of helium and neon gas. It produces a very intense, narrow beam of red light that retains its shape over long distances. The beam can be manipulated, controlled, and detected quite easily. By expanding or focusing the beam with simple optical elements, the desired spot size can be achieved at almost any distance.

Modern laser instruments are typically battery-operated units that generate either a fixed laser beam, a rotating beam, or both. Fixed laser beam instruments can generate either single or multiple beams. One example of a fixed-beam device is the laser spirit level, which is a carpenter's level with a built-in visible beam laser (*Figure 15*). It can be used as an ordinary level or switched on to emit a laser beam, thereby extending the level's reference line typically between 200 and 300 feet. It can be mounted on a tripod or placed on a level surface. Another example of a single-beam device is a handheld distance meter (*Figure 16*) used to measure distances up to about 100 feet without a target, and even farther distances with a reflective target. The auto-leveling laser alignment tool is a good example of a multiple-use laser instrument. It can be used to provide simultaneous plumb, level, and square reference points. It is important to note that when performing calculations, the distance from the bottom of the laser level that rests on a pipe to the center of the beam may produce a difference of several inches.

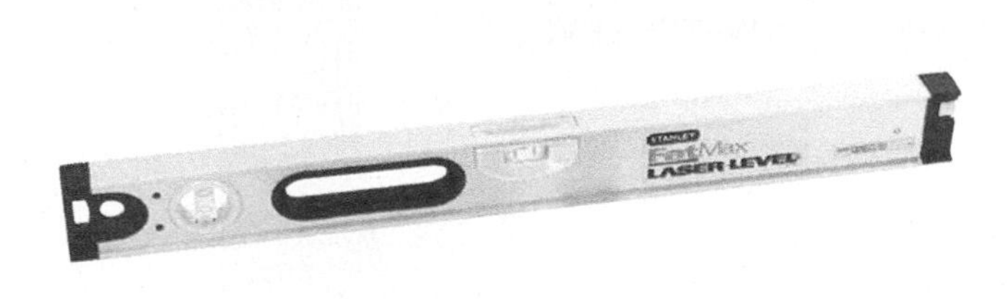

Figure 15 Laser input level.

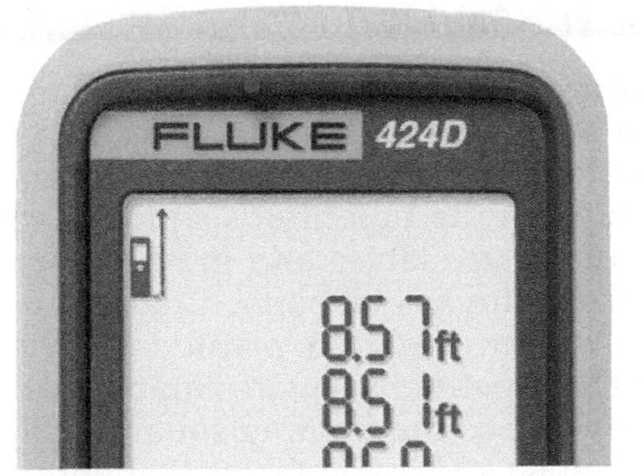

Figure 16 Fixed-beam laser instrument.

WARNING!

Never stare into the laser beam. It can cause blindness.

1.4.4 Tubing Water Level

The tubing water level (*Figure 17*) is a clear, plastic tube of a small diameter that is generally 50 feet long and partially filled with water. Pipefitters use the tubing water level to achieve an accurate reading of level between two distant points. It can also be used to transfer elevation bench marks between two locations. The tubing water level is preferred over the string line level when leveling a taut string or line because it is so accurate. It is more accurate, in fact, than a laser level at long distances.

When using a tubing water level, it is important to eliminate air bubbles in the line. The water lines have to be opened at the end so the lines can be stable. Two people are needed to use the level: One holds the end of the tube vertically near the reference point, or one end of the line, and adjusts the tube until the water level is even with the reference point. The other person takes the other end of the tube and holds it vertically at the other end of the line. The water level in the tube will rest exactly level with the water level at the other end of the tube. Then, a string is tied between these two points to make it perfectly level.

Laser Levels

In some situations, such as in bright sunlight, it may be difficult to see the laser beam. Enhancement goggles are available that allow the user to see the laser beam more easily. Some laser level instruments are specifically designed for outdoor use and have beams that are easily visible in sunlight.

1.5.0 Layout and Fabrication Tools

Pipe fabrication tools are used to lay out angles, find center lines of pipe, scribe straight lines around pipe, and plumb and square pipe sections together.

The most common fabrication tools include:

- Framing squares
- Pipefitter's squares
- Combination try squares
- Center finders
- Pipe line-up clamps
- Hi-Lo gauges
- Wraparounds
- Drift pins
- Two-hole pins
- Flange spreaders

1.5.1 Framing Squares

The framing square, also known as a carpenter's square or a steel square, can be used by pipefitters for a variety of tasks. The most common functions of a framing square are the following:

- Laying out guidelines for cutting steel and pipe, and then squaring up adjacent markings at right angles
- Using tables and scales on the square for measuring and fabricating
- Determining plumb or square fits in fabricating, with the use of a tape or rule
- Aligning two sections of pipe or a fitting to a pipe
- Finding and laying out the center lines of pipe
- Checking the squareness of the end of a pipe

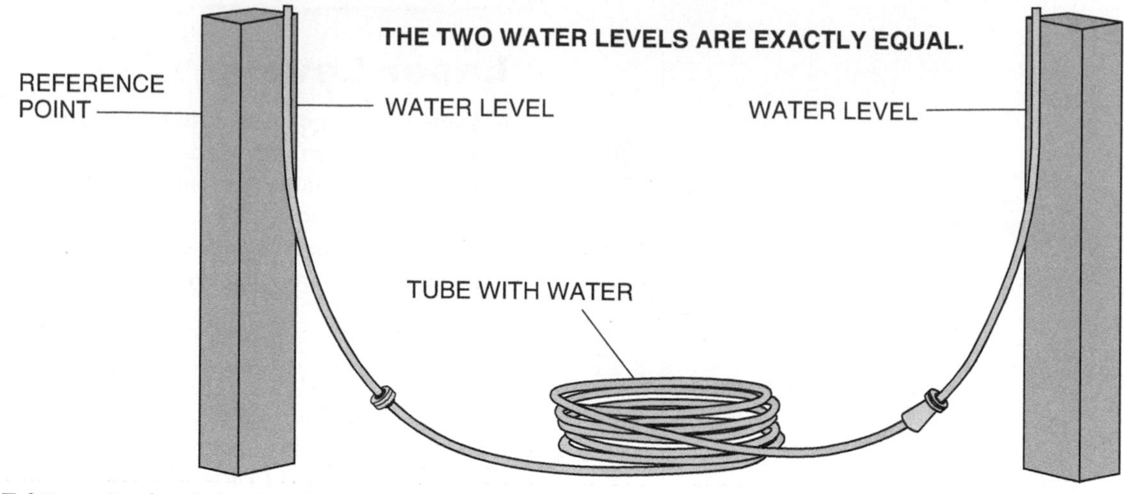

Figure 17 Tubing water level.

The framing square (*Figure 18*) consists of two parts: the tongue and the body, or blade. The tongue is the shorter, narrower part and the body is the longer, wider part. The point at which the tongue and the body meet on the outside edge is called the heel. The two most commonly used sizes are the 12 inch by 8 inch and the 24 inch by 16 inch.

Straight pipe alignment is one of the many applications of framing squares. To align two pieces of straight pipe, two framing squares are needed (see *Figure 19* and *Figure 20*). Check the framing squares for trueness by placing the blades and tongues side by side. Note that when aligning a 45-degree elbow to straight pipe, the squares must be perfectly aligned on top of the straight pipe or the end of the 45-degree elbow, otherwise the alignment will be off. More accurate tools for aligning are available; procedures for pipe alignment will be taught later.

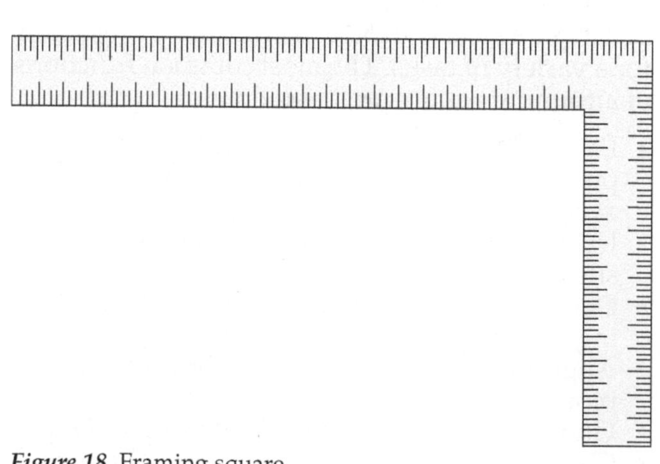

Figure 18 Framing square.

1.5.2 Pipefitter's Squares

Built on the same principle as the framing square, the pipefitter's square is designed specifically for pipefitting methods and techniques. The main differences between the pipefitter's square and the framing square are tables and graduations. The pipefitter's square has takeoffs, dimensions, and other useful information printed on the square itself. The pipefitter's square can be used in all the same ways as the framing square, with the added advantage of having tables that eliminate many calculations. Special care should be taken when using this tool because it can easily be bent out of square. In addition, it will rust when exposed to the elements and it is several times more expensive than a framing square.

1.5.3 Combination Try Squares

A combination try square (*Figure 21*) is a versatile tool composed of four parts, giving it a variety of uses in layout and fabrication. These four components are a grooved steel rule graduated into scales of $1/8$ inch, $1/16$ inch, $1/32$ inch, and $1/64$ inch; a locking square head with a level glass; a locking center head; and a protractor with a revolving turret marked with degrees. All of these parts are removable and can be used independently.

In addition to being used for many of the functions of a framing square, a combinatin try square can be used as a:

- Height and depth gauge
- Bevel or angle protractor
- Level
- Plumb
- Square
- Center finder
- Ruler

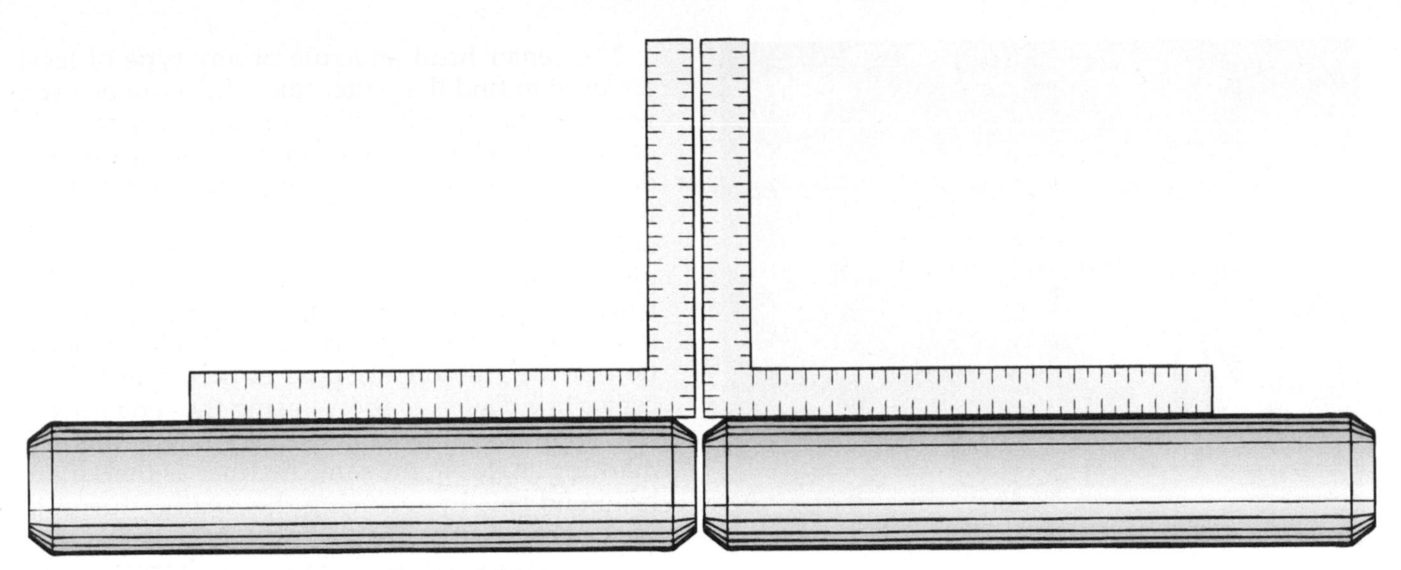

Figure 19 Positioning squares to align straight pipe.

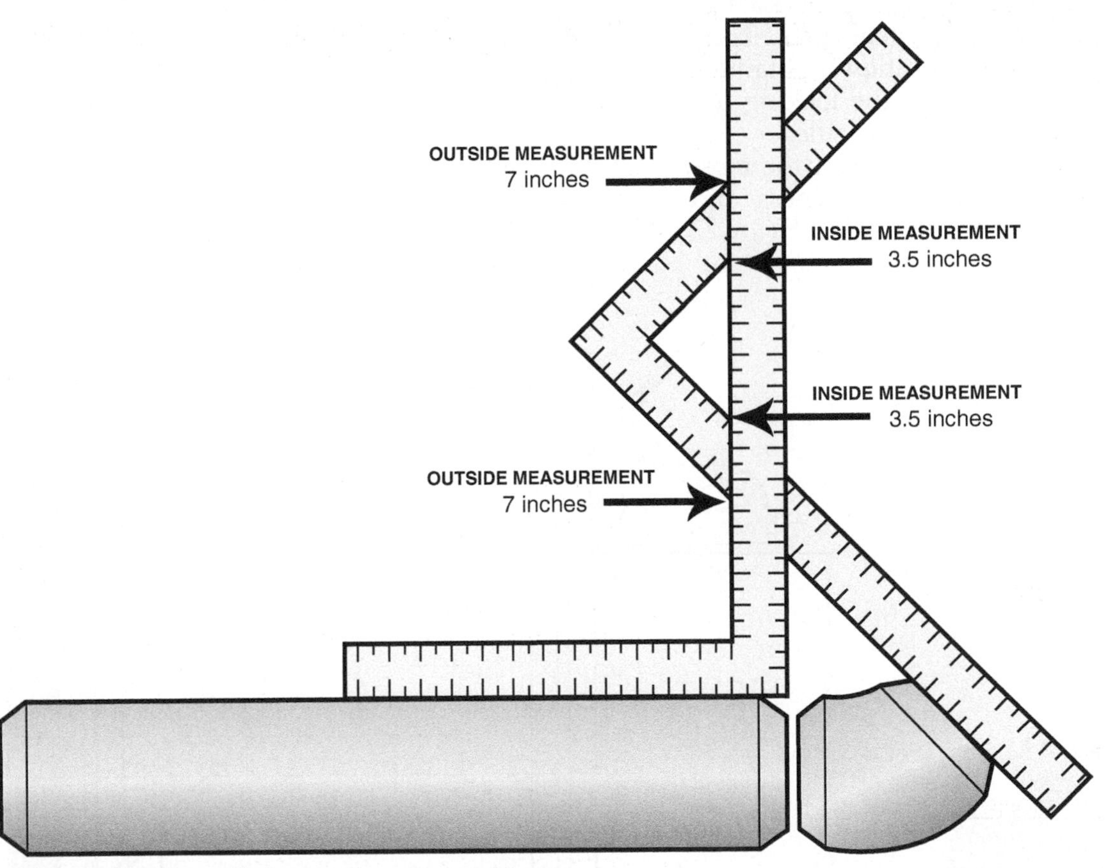

Figure 20 Aligning a 45-degree elbow to straight pipe.

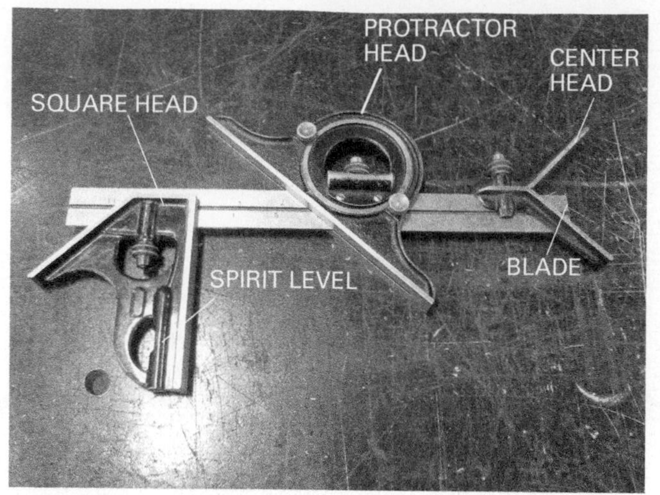

Figure 21 Combination try square.

To attach any of the components, loosen the spring-loaded locking nut, allowing the part to be fitted into the groove on the steel rule. Firmly lock the nut after the desired position is obtained. To remove a component, loosen the locking nut and slide the component off the steel rule.

The square head and rule are used for measuring height or depth, plumb or levelness, and squareness, as well as laying out angles. *Figure 22* shows uses of the square head.

The center head and rule of any type of level are used to find the center line which can be used when working with pipe or other cylindrical objects. The center line is found by adjusting the depth of the rule so that the protruding edge of the rule is approximately $1/16$ inch off the surface of the pipe when the center head is placed on the pipe. A level is placed against the long end of the rule, above the center head, and the try square is moved until the glass vial reads level. The location of the rule on the pipe is then marked for center line. *Figure 23* shows using the center head.

The protractor head is used for laying out and marking angles in the same manner as the square head and rule by revolving the rotating turret to the desired degree. By placing a level on the steel rule, the protractor can be used to determine the degree of slope of a pipe. It is not used on pipe, but rather plate steel or hangers. Place the edge of the protractor on the item, then rotate the turret until the level on the rule reads true. The angle on the protractor is the degree of slope.

1.5.4 Center Finders

Another tool used to find the center of a pipe is the center finder (*Figure 24*). The center finder, also known as a centering head, consists of a Y-

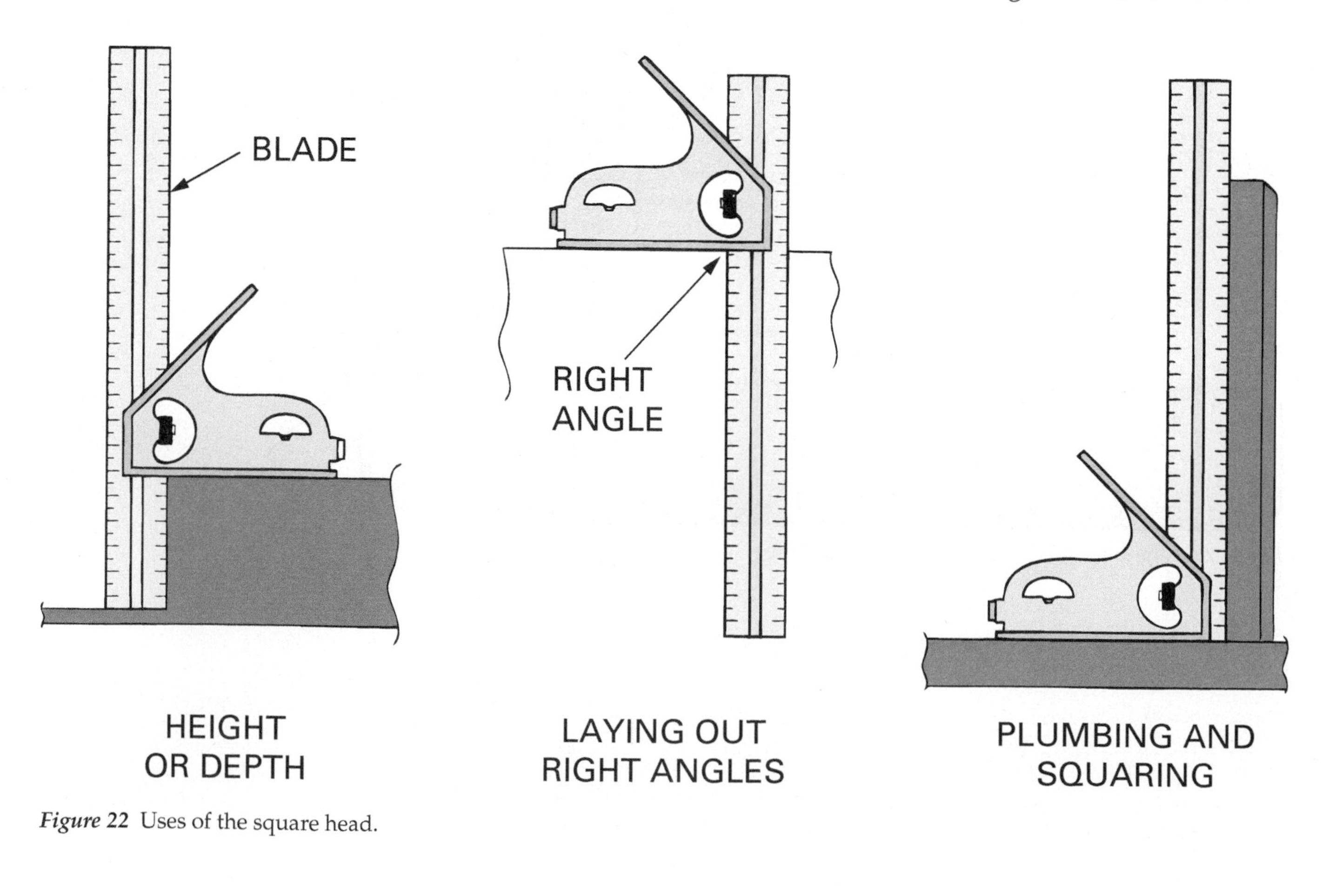

Figure 22 Uses of the square head.

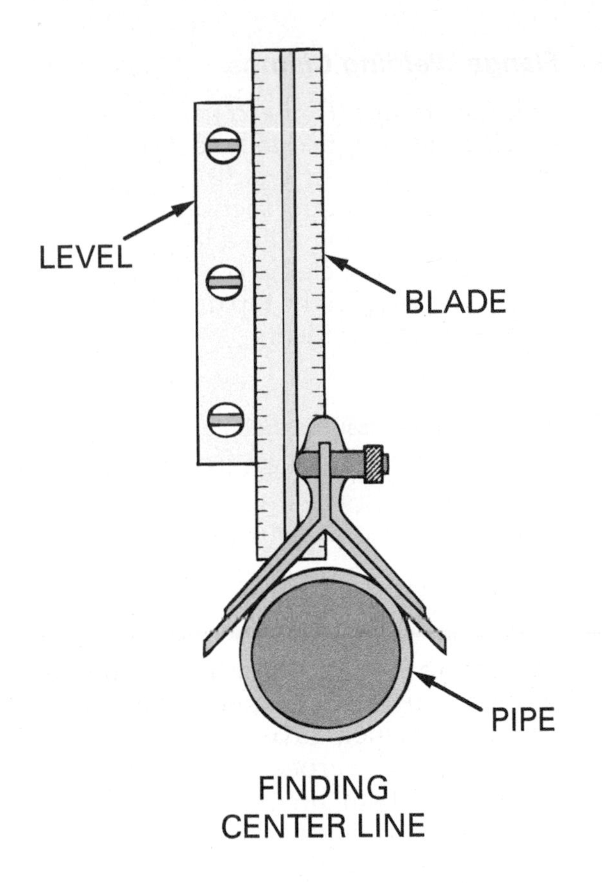

Figure 23 Using the center head.

Magnetic Protractor

The magnetic protractor on a conduit run can be used to measure the angle of a piping run. It has a grooved base that allows it to rest on the pipe.

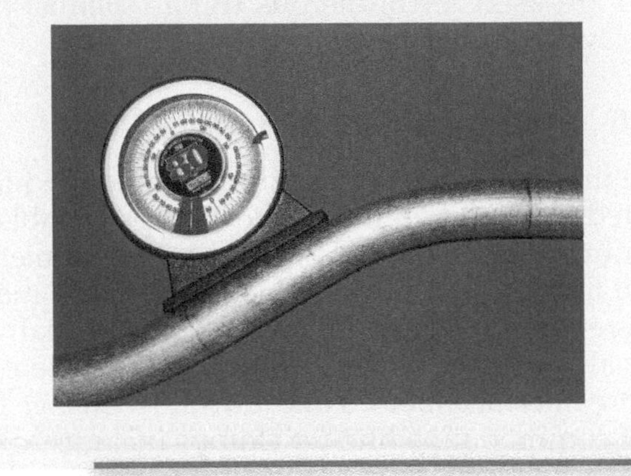

1.5.5 Wraparounds

A wraparound (*Figure 25*) is a piece of gasket material or leather belting that is used to draw a straight line around a pipe. A wraparound must have straight edges and must be long enough to wrap around the pipe one-and-a-half times.

Figure 24 Center finder.

type head that straddles the pipe, a centering pin, and an adjustable dial bubble protractor for indicating the center at some angle away from the top dead center. To mark the top dead center, set the center finder on top of the pipe and adjust it until the bubble indicates that it is centered. A light blow with a hammer to the centering pin makes an indentation on the surface to mark its center.

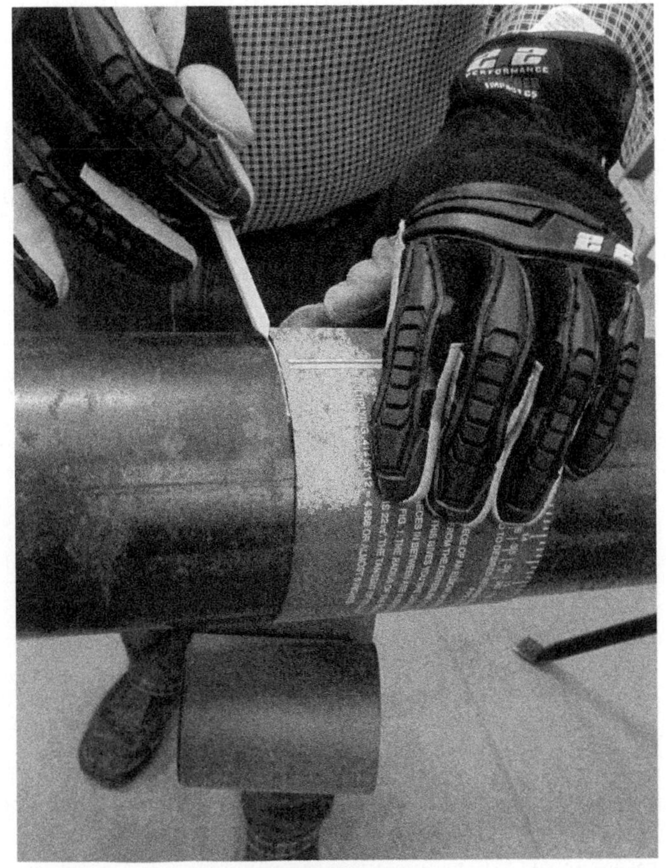

Figure 25 Using a wraparound.

When a wraparound is used correctly, the two overlapping edges should be aligned.

To use a wraparound, pull it tightly around the pipe until the edges overlap. Do not pull it tight enough to break it. Mark around the edge, using a scribe or soapstone. Markings on the wraparound itself are used for numerous things such as precise layouts and reference guides.

1.6.0 Pipe Line-Up Clamps

Pipe line-up clamps range from simple field-fabricated jigs to large hydraulic clamps capable of aligning pressure vessels with external diameters of 30 feet or more. Even though there are many differences in the types of pipe line-up clamps, they all serve the same purpose: to keep the pipe square and level before and during welding. The most common pipe line-up clamps include:

- Straight butt welding clamps
- Flange welding clamps
- Other chain-type welding clamps
- Shop-made aligning dogs

1.6.1 Straight Butt Welding Clamps

Straight butt clamps (*Figure 26*) are chain-type clamps used to secure two straight pipes together end-to-end. They consist of straight sections of steel approximately 14 to 18 inches long and two lengths of chain, one near each end, attached to screw locks that can tighten the chain once it has been wrapped around the pipe. This type of clamp is not used very often, however, newer clamps such as those by Walhonde take out the Hi-Lo and are preferred by pipefitters.

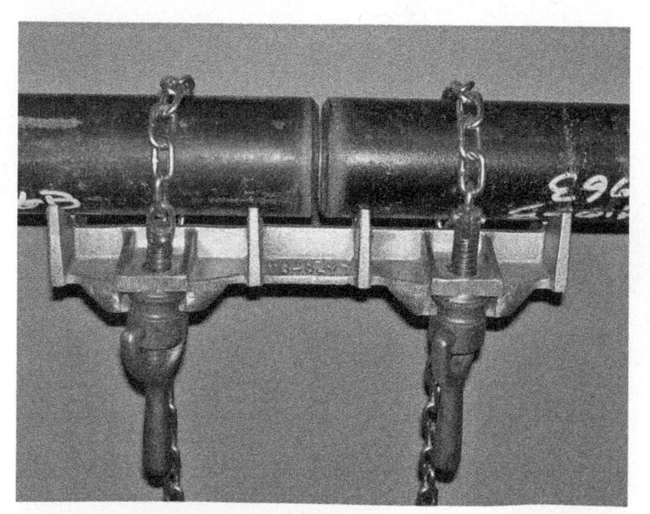

Figure 26 Straight butt welding clamps.

1.6.2 Flange Welding Clamps

Flange welding clamps (*Figure 27*) are chain-type clamps specifically designed to hold a pipe flange onto the end of a pipe for welding. The flange welding clamp is a straight section of steel approximately 12 inches long with a length of chain on one end and a screw clamp on the other. The clamp can be mounted on the top or bottom of the pipe and flange being welded.

> **NOTE**
> For this exercise, assume that a previously prepared 4-inch steel pipe will have a flange welded onto it.

Follow these steps to use a flange welding clamp:

Step 1 Inspect the flange welding clamp for any obvious damage, such as worn-out or broken chain links, bent or broken clamp feet, a warped clamp back, bent or broken chain hooks, damaged chain tighteners, a rusty or worn screw clamp, or grease, rust, or excessive dirt.

Step 2 Repair or replace any damaged part. Clean and oil any parts as needed.

Step 3 Set up two adjustable pipe stands to support the pipe.

Step 4 Secure the section of pipe onto the stands, with the end to be welded extending at least 12 inches beyond one of the stands.

Step 5 Loosen the screw clamp on the end of the flange welding clamp.

Figure 27 Flange welding clamp.

Step 6 Position the edge of the larger part (lip) of the flange into the screw clamp opening.

Step 7 Align the flange to the pipe.

Step 8 Tighten the screw clamp onto the flange lip.

Step 9 Position the clamp and the flange so that the smaller opening of the flange is aligned with the end of the pipe where the weld will be.

Step 10 Insert a spacer between the flange and the butt end of the pipe.

Step 11 Hold the flange and the clamp in place and wrap the clamp chain around the pipe.

Step 12 Attach the chain to the clamp.

Step 13 Turn the screw handle to tighten the chain.

Step 14 Check the flange and pipe joint again to see that they are still aligned with each other and ready for welding.

Step 15 After the welding has been completed, loosen the screw clamp from the flange lip.

WARNING!

Handle the pipe and clamps with care. Welded objects remain hot for several minutes after the welding is completed.

Step 16 Loosen the chain from the pipe.

Step 17 Remove the clamp from the flange and pipe.

Step 18 Clean the clamp if necessary and inspect it for damage. If it is damaged, replace it.

Step 19 Store the flange welding clamp.

1.6.3 Other Chain-Type Welding Clamps

Other chain-type welding clamps include the T-joint welding clamp and the elbow welding clamp. The T-joint welding clamp is used to hold a pipe section perpendicular to another pipe to be welded. The elbow welding clamp is used to hold a 90-degree elbow to the end of a straight pipe to be welded. The procedures for using these are the same as for the other chain-type clamps. *Figure 28* shows T-joint and elbow welding clamps.

1.6.4 Shop-Made Aligning Dogs

Many pipefitters fabricate their own aligning devices. These devices are known as aligning dogs in the pipefitting trade. *Figure 29* shows an example of an aligning dog.

The aligning dog shown in *Figure 29* is made from the same material as the pipe being welded, to which a nut has been welded. The aligning

Types of Welded Flanges

Some of the more common flanges installed using welded connections are the welding neck flange, lap joint flange, socket weld flange, and slip-on flange. Threaded flanges may also be back-welded as determined by the process. Welding neck flanges is preferred for use in severe service applications, such as those involving high pressure, sub-zero temperatures, or elevated temperatures.

When using a straight butt welding clamp, remember that it is used as an alignment device, not a rigging device. Instruction on how to use these types of clamps is typically provided on the job site, where each step is relevant to the specific task at hand.

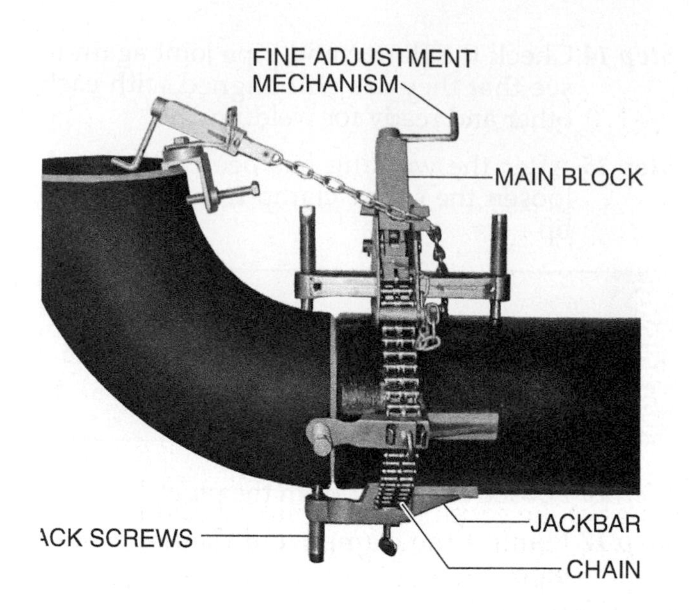

T-JOINT CLAMP

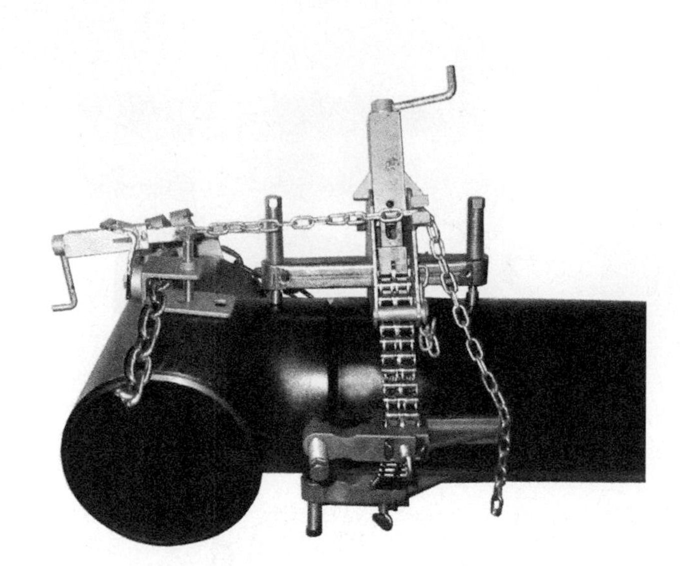

ELBOW WELDING CLAMP

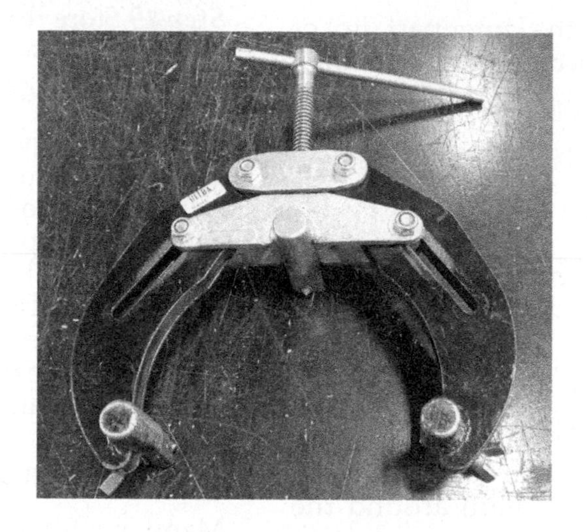

ULTRA WELDING CLAMP

Figure 28 Welding clamps.

devices use **tack welds** on one section of the pipe so that the other section can be moved by tightening the bolt. Usually two such dogs would be used to provide more control over the fit-up. Once the fit-up is complete, the aligning dogs are removed from the pipe. After the dog is removed from the pipe, the pipe must be ground to remove any welding deposits. Aligning dogs also require periodic grinding to remove welding deposits.

Welding codes in some areas prohibit anything being welded onto the pipe. Make sure you know the code before using any aids that must be welded to the pipe.

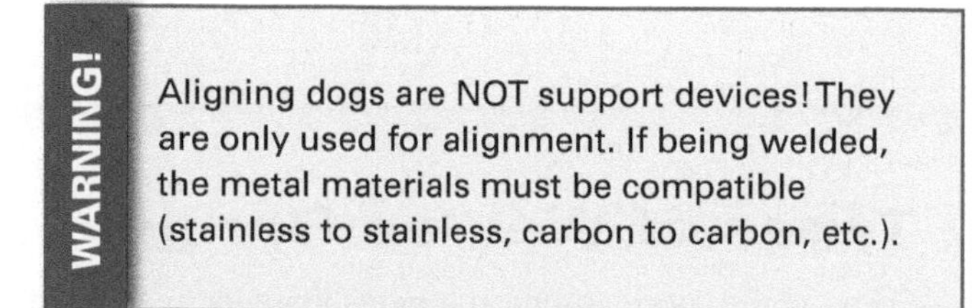
WARNING!

Aligning dogs are NOT support devices! They are only used for alignment. If being welded, the metal materials must be compatible (stainless to stainless, carbon to carbon, etc.).

1.6.5 Hi-Lo Gauges

The primary purpose of a Hi-Lo gauge is to check for internal pipe joint misalignment (*Figure 30*). The name of the gauge comes from the relationship of the alignment of one pipe to another pipe.

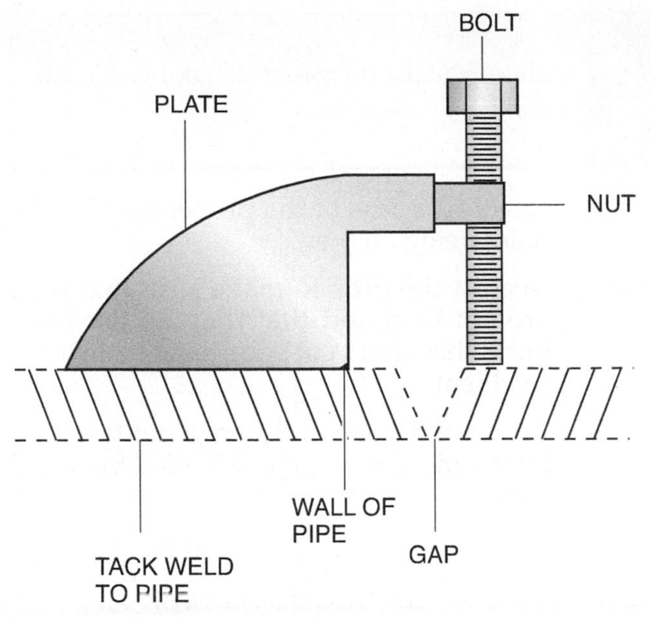

Figure 29 Aligning dog.

To check for internal misalignment, Hi-Lo gauges have two prongs, or alignment stops, that are pulled tightly against the inside diameter of the joint so that one stop is flush with each side of the joint. The variation between the two stops is read on a scale marked on the gauge. To measure misalignment using a Hi-Lo gauge, insert the prongs of the gauge into the joint gap. Pull up on the gauge until the prongs are snug against both inside surfaces and read the misalignment on the end of the scale.

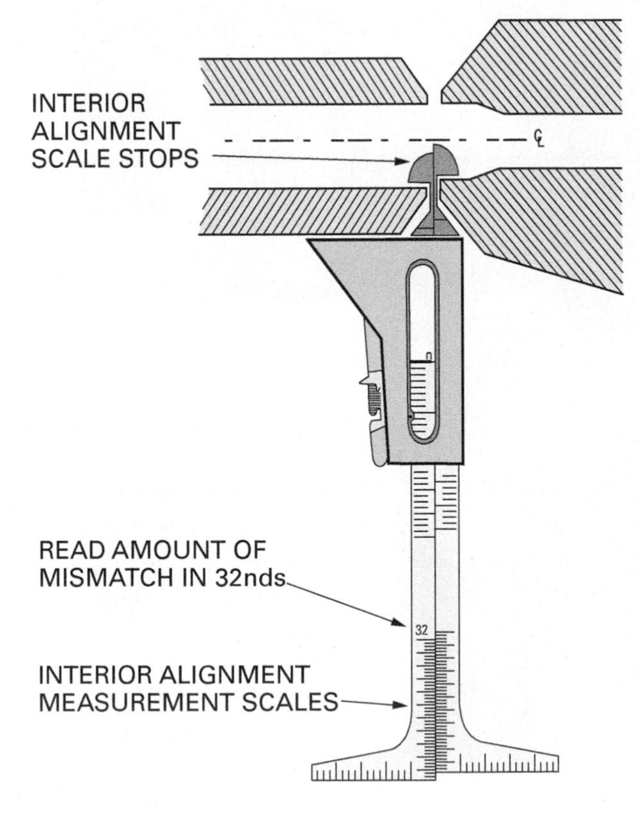

Figure 30 Checking internal misalignment using a Hi-Lo gauge.

Hi-Lo Application

When checking for internal misalignment, take two gauge readings: one before tacking together the two pieces of pipe and one after tack welding. The Hi-Lo gauge is thin enough to fit into the gap between the two pipes, even after tacking. It typically comes with both metric and English scales.

1.6.6 Drift Pins

A drift pin is a round piece of steel that is tapered on both ends and pipefitters use drift pins to align bolt holes when connecting flanges. Drift pins come in various sizes to fit different-sized bolt holes. *Figure 31* shows a common drift pin.

To use a drift pin:

Step 1 Select a pin of the proper size for the holes being aligned. The properly-sized pin will fit all the way through the bolt holes being aligned and will still be large enough and strong enough to pull the mating pieces together.

Step 2 Inspect the pin to make sure it is not bent and that it does not have any **burrs** that could cut your hands when using it.

Step 3 Insert the drift pin into one of the bolt holes of the mating flanges and manipulate the pin to align the flanges.

Step 4 Repeat Step 3 on the opposite side of the flanges.

Step 5 Insert bolts into the other bolt holes of the flanges and hand tighten the bolts.

Step 6 Remove the drift pin from the flanges.

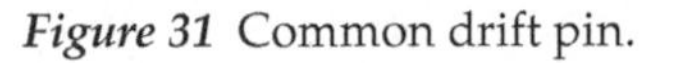

Figure 31 Common drift pin.

Step 7 Clean the drift pin using a rag and solvent if needed.

Step 8 Store the drift pin.

> **CAUTION**
>
> The drift pin needs to be tied off to prevent it from falling through the flange holes (this is especially important when working from heights, as it introduces safety concerns). Also, remember that drift pins are not wedges – they are strictly used for aligning the hole.

1.6.7 Two-Hole Pins

Two-hole pins are flange lineup pins used to orient a flange on the end of a pipe. These pins are available in a variety of sizes and are used to align the holes of a flange before it is welded. The pins are inserted into the top two holes on a flange. A level is then placed across them to allow for alignment of the flange. *Figure 32* shows two-hole pins used in conjunction with a flange lineup level.

To use two-hole pins:

> **NOTE**
>
> Make sure to put the torpedo on the side of the pin without the spring.

Step 1 Select two pins of the proper size for the holes being aligned.

Step 2 Inspect the pins to make sure that they are not bent and that they do not have burrs that could cut your hands when using them.

Step 3 Unscrew the knurled flange nut from the flange pin if using the threaded pins and nuts.

Step 4 Insert the flange pins into the two top bolt holes of the flange.

Step 5 Tighten the knurled flange nut onto the threaded end of the flange pin if using the threaded pins and nuts.

Step 6 Set a level over the two flange pins.

Step 7 Adjust the flange until it is level.

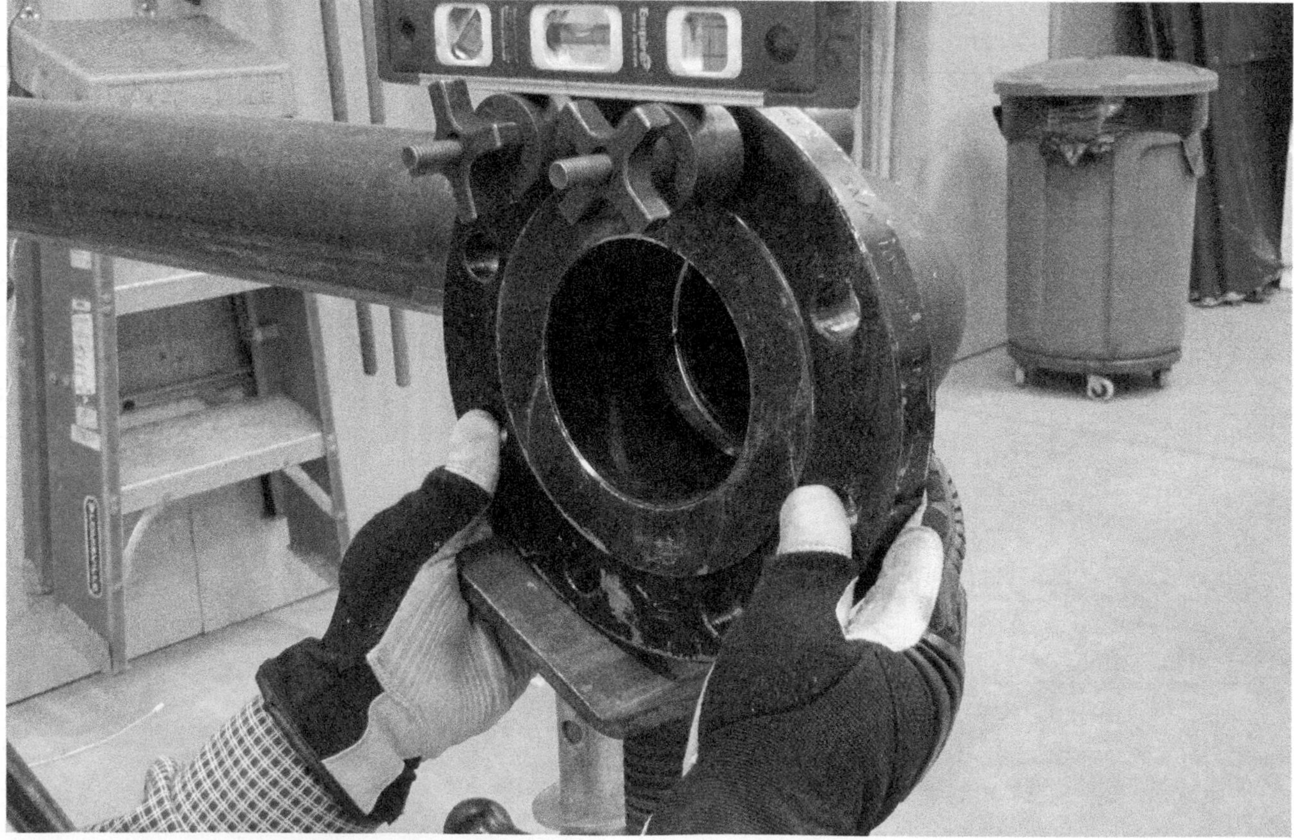

Figure 32 Two-hole flange lineup pins and level.

1.6.8 Flange Spreaders

Flange spreaders (*Figure 33*) are used to spread two mating flanges when gaskets need to be replaced without disassembling pipe lines. Flange spreaders exert tremendous pressure between the flanges smoothly and evenly, with no shock along the pipe and without disturbing the alignment of the flanges. Flange spreaders must be used in pairs, one on each side of the flange, for even distribution of pressure.

> **NOTE**
> Due to the variety of flange spreaders, be sure to follow manufacturer's instructions for use.

> **NOTE**
> Flange spreaders need to be tied off to prevent them from falling from heights, which is a safety issue.

1.7.0 Pipe Cutting and Reaming Tools

Pipe cutting tools are used to properly cut and prepare pipes for joining and connecting. A good connection depends on a clean, square cut and proper pipe reaming and threading. Common pipe cutting tools include:

- Hacksaws
- Soil pipe cutters
- Tube and pipe cutters
- Pipe reamers

Figure 33 Flange spreader.

1.7.1 Hacksaws

A hacksaw (*Figure 34*) is a specialty tool designed for cutting metals, plastics, and other synthetics. It has a pistol grip handle and uses a variety of

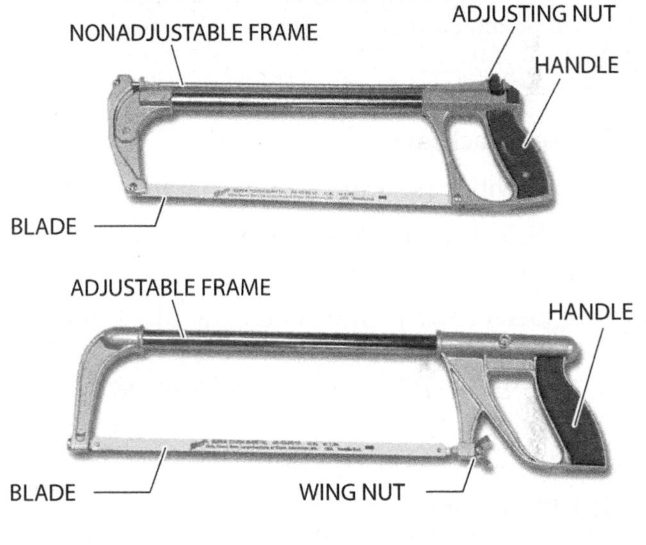

Figure 34 Hacksaws.

Case History

Field-Fabricated Jack Stands

A mechanic was crushed to death when a bus on which he was working fell off a set of jack stands. The stands were fabricated of plate steel by a local welding shop. The top plate of each stand was completely flat; neither of them had lips, which commercial stands always have. The front tires of the bus were not chocked and there was nothing to prevent it from falling off the stands.

The investigation of the accident revealed that the jack stands had not been tested or certified for their rated capacity, nor were they marked with such a capacity. They were not fabricated in accordance with commercial jack stand construction. The lips on commercial jack stands serve to cradle the area being supported. Also, commercial jack stands normally have three legs that help them compensate for irregular surfaces.

Sections of large-diameter pipe supported by jack stands may not be as large as buses, but they can still cause serious or fatal injuries if they fall from their supports. Always use commercial jack stands that have been tested and certified.

thin metal blades. Adjusting nuts on the frame secure the blade and allow for varying levels of tightness. Blades are selected by the number of teeth per inch and by their material composition. Blades with finer and more numerous teeth per inch are used for efficiently cutting harder, stronger materials. *Table 1* shows which blades to use for various materials.

The blade should always be installed in the saw with the teeth pointing away from the handle because the forward stroke is the cutting stroke. Because of their short length and brittleness, hacksaws require oiling to prevent rust and excessive wear.

Hacksaws are useful when an object cannot be taken to a power saw or when the sawing job is rather small.

NOTE

For this exercise, assume that the blade needs to be replaced.

To use a hacksaw:

Step 1 Identify the kind of material to be cut.

Step 2 Select the hacksaw best suited for the job (refer to Table 1).

Step 3 Select the proper hacksaw blade for the metal to be cut. Certain types of metals require specific blades for cutting.

Step 4 Loosen the adjusting nut near the handle until the blade is loose on the holding posts.

Step 5 Remove the old hacksaw blade.

Step 6 Check the post, saw frame, and handle for excessive wear or damage.

Step 7 Install the new hacksaw blade onto the posts. Make sure the teeth of the new blade are pointed away from the handle.

Step 8 Tighten the adjusting nut until the blade is tight.

CAUTION

Do not overtighten the blade.

Step 9 Use a vise to secure the object that is going to be sawed.

Step 10 Measure and mark the area where the cutting will be performed.

Step 11 Position your body so that you are balanced and facing the work area.

CAUTION

Always wear gloves, hard hat, and safety glasses when using a hacksaw. Follow all company policies regarding personal protective equipment.

Table 1 Blades for Various Materials

Stock to Be Cut	Pitch of Blade (Teeth per Inch)	Explanation
Machine steel Cold rolled steel Structural steel	14	The coarse pitch makes the saw free and fast cutting.
Aluminum Babbitt Tool steel High-speed steel Cast iron	18	Recommended for general use.
Tubing Tin Brass Copper Channel iron Sheet metal (over 18 gauge)	24	Thin stock will tear and strip teeth on a blade of coarser pitch.
Small tubing Conduit Sheet metal (greater than 18 gauge)	32	Recommended to avoid tearing.

NOTE

When using a hacksaw, grasp the saw handle and face the cut area. If the saw handle is in your right hand, position your left foot forward and toward the work. Place your other hand on the top/front of the saw frame.

Step 12 Place the saw blade on the cutting line with most of the blade on the side of the mark away from you.

Step 13 Pull the saw back across the mark while applying a slight, downward pressure. This will start the cut and allow for any needed correction. Keep the saw in line with your forearm as you begin with light, short strokes. Cutting takes place on the forward stroke.

Step 14 Press downward enough to make the teeth cut and push forward.

Step 15 Lift and pull the blade back to start another stroke. Do not lift the blade completely out of the groove. When sawing, keep your eyes on the sawing line rather than on the saw blade. Corrections may be made by a slight twist of the handle (and blade). Blow any cuttings off the line so that you can see.

Step 16 Continue sawing until the job is completed. When you near the end of the cut, saw slowly. Hold the waste piece in your other hand so that the material will not fall as you make the last cut.

Step 17 Remove the saw from the work area and carefully clean any buildup off the blade or saw.

WARNING!

Be careful during cleanups. Freshly sawed objects and recently used blades may still be hot and can burn an unprotected hand. Also, metal shavings can puncture the skin.

Step 18 Oil the blade lightly to prevent rust.

Step 19 Clean the area.

Step 20 Store the hacksaw and any additional blades.

1.7.2 Soil Pipe Cutters

A soil pipe cutter (*Figure 35*) is used to cut cast iron soil pipe. It consists of a **ratchet** handle with a length of chain permanently attached to one side of the cutter. Each link of the chain contains a cutting wheel. The chain is wrapped around the pipe and locked into the side of the cutter. As the ratchet handle is pumped, the cutter tightens the chain around the pipe. As the chain tightens, the cutting wheels act as small chisels that penetrate the pipe until it is cut.

To use a soil pipe cutter:

Step 1 Select a length of cast iron soil pipe to be cut.

Step 2 Measure and mark the pipe to be cut.

SOIL PIPE RATCHET CUTTER

SOIL PIPE SNAP CUTTER

Figure 35 Soil pipe cutters.

> **NOTE**
> Lay the pipe on the ground to cut it.

Step 3 Lift the pipe and place the chain underneath the cut mark.

Step 4 Place the cutter on top of the cut mark and pull the chain around the pipe.

Step 5 Attach the end of the chain in the open side of the cutter.

Step 6 Turn the chain tension knob clockwise to tighten the chain on the cut mark. The chain should be pulled as tight as possible.

Step 7 Pull out the lock knob and turn it so that the arrow points toward the handle.

Step 8 Push the handle down to start cutting the pipe.

Step 9 Continue to pump the handle until the pipe is broken. The ratchet will allow for lifting the handle without affecting the chain tension.

Step 10 Inspect the soil pipe cutter for any obvious damage.

Step 11 Store the cutter.

1.7.3 Tube and Pipe Cutters

Tube cutters (*Figure 36*) are used to cut thin-walled metal tubing made of copper, brass, aluminum, or steel. They are usually small, single-wheeled cutters. Some models have rollers that are tightened to force the pipe against the cutting wheel, while other models have a sliding cutting wheel that is forced against the pipe. All tube cutters must be rotated completely around the tube to make the cut.

Pipe cutters (*Figure 37*) are larger and more durable than tube cutters, but they operate in the same manner. Manual pipe cutters are designed to cut pipe such as carbon steel, brass, copper, cast iron, stainless steel, and lead (up to 6 inches in diameter). There are several types of pipe cutters, with the one-wheel cutter being the most commonly used. The one-wheel cutter has one alloy cutting wheel and two guide rollers. The rollers help the wheel make a straight cut and prevent burrs from forming on the surface of the pipe. The pipe cutters' main advantage over the hacksaw is accuracy and smoothness of cut.

The single-wheeled pipe cutter is rotated entirely around the pipe and can only be used where there is enough space. The four-wheeled cutter can be used where it is not possible to rotate the cutter around the pipe more than one-third of a turn. Since the four-wheeled cutter has no rollers to guide the cutter wheels, extra care must be taken to ensure perfect tracking.

After it is cut, the pipe must be reamed using a manual hand reamer. The outer edges of the cut should be filed smooth. Many tube cutters have a reamer attached to the body of the cutter; these can be used to clean the burrs inside the pipe after the cut has been made.

To use a tube or pipe cutter:

Step 1 Select, measure, and mark the pipe to be cut.

Step 2 Identify the size and kind of material to be cut.

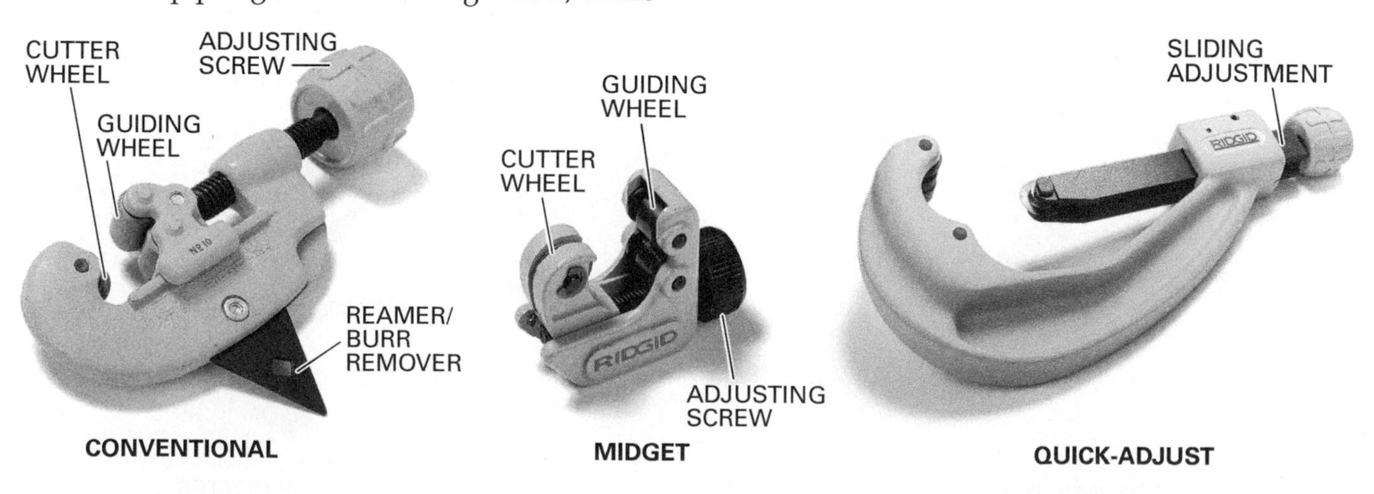

Figure 36 Types of tube cutters.

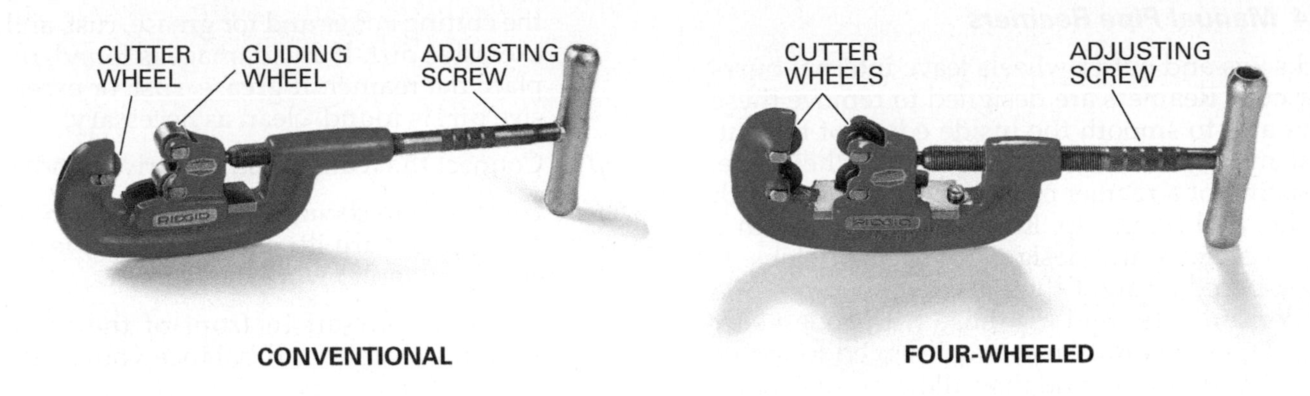

Figure 37 Pipe cutters.

Step 3 Select a cutting tool to match the material being cut.

CAUTION

Pipe cutters may be too strong or heavy for cutting thin-walled tubing or plastic pipe. Use a tubing cutter in these cases.

Step 4 Inspect the selected cutter for obvious damage or excessive wear on the cutting edges and for grease, rust, and excessive dirt. If any damage is found, replace the cutter. If grease, rust, or excessive dirt is found, clean as necessary and lightly oil the moving parts.

Step 5 Secure the pipe in a vise and install pipe supports if necessary. Secure the pipe near the cutting mark but leave enough room to work the cutter.

Step 6 Put on safety glasses.

Step 7 Place a catch pan on the floor underneath the cutting mark to catch the cutting oil coming off the object.

Step 8 Place the cutter around the pipe at the cutting mark. Position the cutter so that the opening is facing up underneath the object being cut.

Step 9 Line up the cutting wheel with the cutting mark and tighten the cutter handle until the cutting wheel is snug on the object.

CAUTION

Do not overtighten the handle. This can break the cutting wheel or crush plastic or light tubing.

Step 10 Rotate the cutter completely around the pipe. When using a four-wheeled cutter in a confined space where it is not possible to rotate the cutter around the pipe, move the cutter back and forth on the pipe.

Step 11 Tighten the cutter handle one-quarter of a turn to apply more pressure.

Step 12 Make another complete circle around the pipe.

NOTE

Occasionally apply a small amount of cutting oil to help ensure a smooth and easy cut. The oil also helps to prevent wear on the cutter.

Step 13 Continue tightening the cutter handle and circling the pipe until the cut is complete. When the cut appears to be almost complete, hold the excess piece to keep it from dropping when finished.

Step 14 Remove the cutter.

Step 15 Wipe the cutting oil off the pipe with a towel.

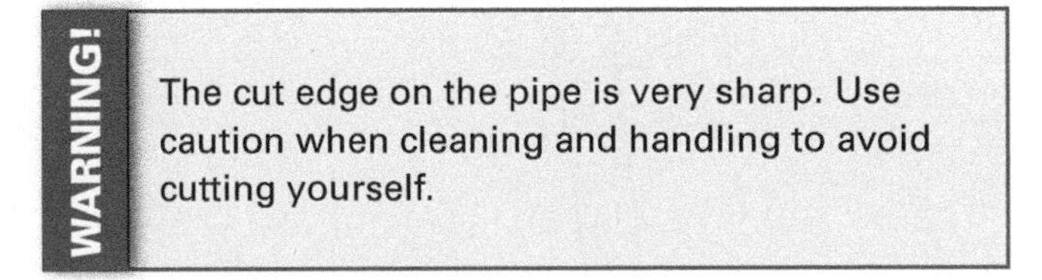

WARNING!

The cut edge on the pipe is very sharp. Use caution when cleaning and handling to avoid cutting yourself.

Step 16 Clean the cutter and check it for any obvious damage.

Step 17 Close the cutter before storing it.

1.7.4 Manual Pipe Reamers

Hacksaws and cutter wheels leave internal burrs after cuts. Reamers are designed to remove these burrs and to smooth the inside edges of the cut. Most manual reamers are of the ratchet type, consisting of a reamer cone, ratchet housing with handle, and handgrip. Reamer cones come in a variety of sizes and designs for specialty jobs. A long-tapered straight flute is used for pipes of a smaller diameter and for those with rough surfaces. Tapered spiral flutes are designed for pipes of a larger diameter, and they allow for easier and faster reaming. *Figure 38* shows ratchet-type manual pipe reamers.

Reamers should be kept lightly oiled and cleaned after each use. Always ream pipe before threading because reaming will sometimes stretch the pipe.

To use a ratchet-type pipe reamer:

Step 1 Identify the kind of reamer needed, then select the specific type and size appropriate for the job.

Step 2 Select a drive handle for the reamer according to the die head size.

Step 3 Inspect the selected reamer and handle for obvious damage or excessive wear on the cutting edges and for grease, rust, and excessive dirt. If any damage is found, replace the reamer. If grease, rust, or excessive dirt is found, clean as necessary.

Step 4 Connect the reamer and the drive handle.

Step 5 Use a vise to secure the item that is to be reamed. Secure it near its end but leave enough room to work the reamer.

Step 6 Position yourself in front of the work area, if possible, and balance your body as much as possible.

Step 7 Insert the tip of the reamer into the exposed end of the pipe. Guide it in with one hand and use the other hand for pressure on the rear of the reamer.

Step 8 Apply light forward pressure and start rotating the reamer clockwise. The reamer should bite as soon as the proper pressure is applied.

Step 9 Rotate the reamer a couple of times, withdraw it while continuing to rotate it, and check the progress. Continue if needed.

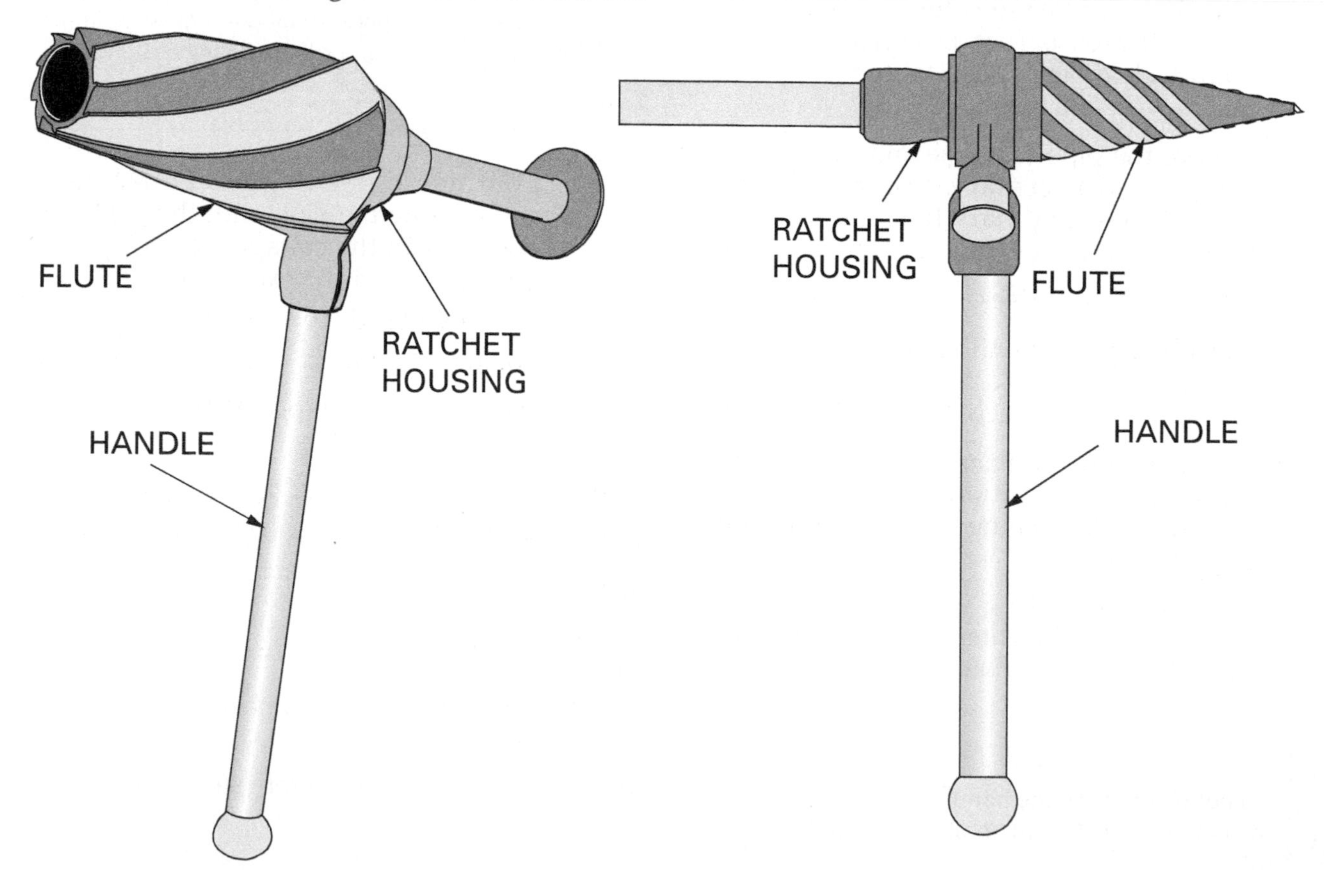

Figure 38 Ratchet-type manual pipe reamers.

CAUTION

The inside edge of the pipe should become very smooth. Do not over-ream. Stop when all burrs are gone.

Step 10 Remove the reamer.

Step 11 Clean the reamer with a shop cloth and check it for any obvious damage. If it is damaged, replace it.

Step 12 Disassemble the reamer and store all parts.

1.8.0 Pipe Threading Tools

Pipefitters often have to use hand pipe and bolt threaders. The ratchet-type hand pipe threader with removable, interchangeable die heads (*Figure 39*) is the most common. Die heads and dies are sized for specific diameters of pipe and are used to cut **male threads**. The size of the thread can be found on the face of the die heads. Special dies are available for threading brass, aluminum **conduit**, stainless steel, wrought iron, and cast iron; they are also available for left-handed threads. These dies should not be used on any material other than what they were designed for because they could be damaged or produce imperfect threads.

Most pipe threads are NPT threads, which designate National Pipe Threads. In some cases, metric threads are also available. The most common bolt threads are NC and NF, which designate National Coarse and National Fine. Always check the engineering specifications before selecting threading dies.

To cut threads correctly, the pipe threader must be kept in good condition.

When using hand pipe threaders:

- Inspect the tool for broken, worn, or damaged parts.
- Inspect dies to make sure they are sharp and not chipped.
- Use a good quality cutting oil to lubricate the cutting edges and remove chips and curls.
- Use good techniques when cutting. Sloppy or careless work will not produce good threads.

NOTE

NPT threads are not the same as bolt threads, so their dies will be different. Be sure to match metal materials (i.e., carbon to carbon).

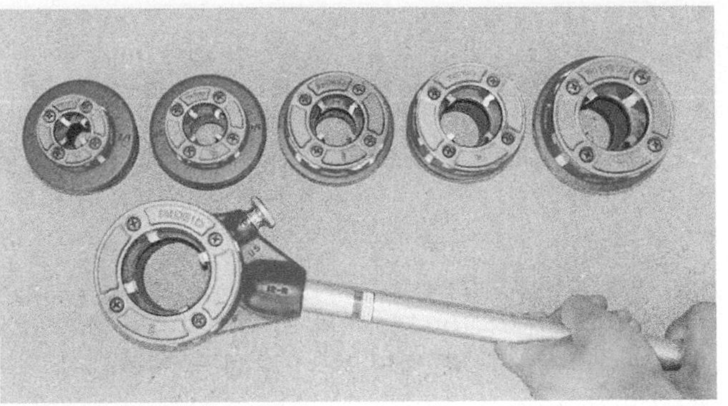

Figure 39 Hand threader.

1.8.1 Replacing Die Heads

Most handles will accept many different die heads for different jobs.

To replace the die head on a pipe threader handle:

Step 1 Pull out the lock knob and turn it so that the arrow points to the handle.

Step 2 Lift the die head straight out of the pipe threader handle.

Step 3 Insert a new die head into the pipe threader handle.

Step 4 Pull out and turn the lock knob so that the arrow is pointing away from the handle.

1.8.2 Replacing Dies

The die head holds the dies in place. When the dies become worn or damaged, they must be replaced.

To replace dies:

Step 1 Remove the die head from the handle.

Step 2 Remove the four screws in the top of the die head using a screwdriver.

CAUTION

When removing the screws, do NOT use any powered tools as this can strip the threads and make the die set useless.

Step 3 Remove the cover plate from the top of the die head.

Step 4 Slide the four dies out of the die head.

> **NOTE**
> Before replacing the dies, make sure the die head and the dies are clean, dry, and free of all metal chips. A small chip can raise a die and cause uneven pressure. This will tear the pipe thread and may break the die.

Step 5 Insert the new dies into the die head, making sure the number on each die matches the number on each die head slot.

> **NOTE**
> It is extremely important to keep the die sets together. Do not mix dies of one set with dies from another set. Usually, each die is marked with a serial number. The four dies that are made together have the same number.

Step 6 Replace the cover plate onto the die head.

Step 7 Tighten the four screws into the die head.

Step 8 Replace the die head into the pipe threader handle.

1.8.3 Threading Pipe

To thread pipe using a hand threader and vise:

Step 1 Identify the kind of material being threaded.

Step 2 Determine the diameter of the pipe.

Step 3 Select the die head best suited for the threading process.

Step 4 Insert the die head into the pipe threader handle.

Step 5 Inspect the die and die handle to ensure that all parts are clean and in good condition.

> **NOTE**
> Dies must be NPT dies.

Step 6 Secure the pipe in a vise.

Step 7 Make sure that the end of the pipe is square, clean, and deburred.

Step 8 Position your body so that you are balanced, facing the work area, and within easy reach of the work area.

Step 9 Position the die flush on the end of the pipe. Make sure that the arrow on the lock knob points in the direction that you are threading the pipe.

Step 10 Press the die against the end of the pipe and turn the die handle clockwise to start the threading process.

Step 11 Turn the die handle two full rotations around the pipe. Add cutting oil as necessary to keep a thin coat on the threads while cutting.

Step 12 Stop the work, pull out the lock knob, and turn the knob so that the arrow points in the opposite direction.

Step 13 Back the dies off the pipe. This allows you to make sure the threads are properly started and allows the threads to be cleaned from the cutting die.

Step 14 Clean the cutting threads from the die if necessary.

Step 15 Pull out the lock knob and turn it so that the arrow points in the direction in which you are threading.

Step 16 Rethread the die onto the pipe and continue the process until the threading is complete. The threading is complete when one full thread extends past the back edge of the dies.

> **CAUTION**
> Do not thread the pipe too far. This can cause an improper fit between the fitting and the pipe, which will result in leaks.

> **NOTE**
> Always follow client specifications; these are general starting guidelines for a manual threader. Every fitting is a little bit different. A threaded fitting should not be used more than once due to the changes that can occur for each time the pipe is inserted into the machine. The Pipe Fitter's Blue Book contains more information on thread engagement based on design specifications.

Step 17 Back the die off the pipe.

Step 18 Inspect the new threads and use a rag to remove any debris from the threads.

WARNING!

Freshly cut threads are very sharp. Use care in handling to avoid cuts.

1.8.4 Checking Threads

A **thread gauge** is used to check the threads on a bolt. *Figure 40* shows a thread gauge.

A thread gauge has many leaves. Each leaf measures 1 inch across and has a certain number of thread markers per inch as noted on the side of each leaf. This measures the number of threads per inch, also known as the pitch. The number of threads per inch depends on the diameter of the bolt. These are not commonly used on pipe threads.

1.8.5 Pipe Taps

Taps are used to cut female (or internal) threads of a pipe. *Figure 41* shows a pipe tap and a tap and die set. The tap shown in *Figure 41* has the first three threads ground off. This helps start the tap in a hole and evens the cutting pressure over several threads on the tap.

Pipe taps are used to cut threads that are larger than the stated tap size. A 1/2-inch pipe tap cuts threads that are almost 11/16-inch in diameter. This is because on a 1/2-inch pipe, the **outside diameter** is 11/16-inch (*Figure 42*).

The pipe wall must be considered when drilling a hole for **female threads**. A pipe tap marked 1/2-inch will tap threads for a 1/2-inch pipe, but the tap is actually larger than 1/2-inch to allow for the pipe walls. Use a 1/2-inch pipe tap and a 1/2-inch pipe will fit.

The hole that is drilled before making the pipe threads must be the proper size. If the hole is too small, the tap will have to be forced in and may break. If the hole is too large, the tap will be loose and will not cut proper threads. *Table 2* shows the drill size that should be used for each size of tap. Notice that the tap drill is always larger than the tap size. Again, this is because the tap is actually larger than its listed size.

To cut female threads in pipe using a pipe tap:

Step 1 Use a center punch to mark the spot where the hole is to be made.

Step 2 Refer to *Table 2* to determine the proper drill and tap to use.

Step 3 Obtain the drill and tap.

Step 4 Put on safety goggles and gloves.

Step 5 Drill the hole straight in toward the center of the pipe.

WARNING!

If the pipe has been in use, make sure it is empty and contains no pressure. If the pipe has been used to move acids, alkalis, or chemicals, purge it before drilling.

Step 6 Remove the drill and store it in a safe place.

Step 7 Fasten the tap to the head and insert the tap into the hole.

Step 8 Apply thread-cutting oil to the tap.

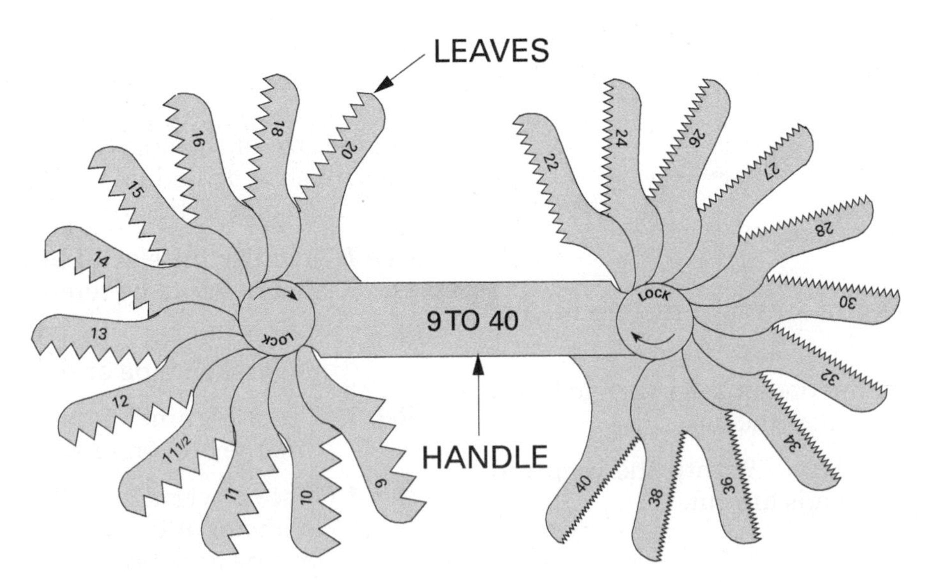

Figure 40 Thread gauge.

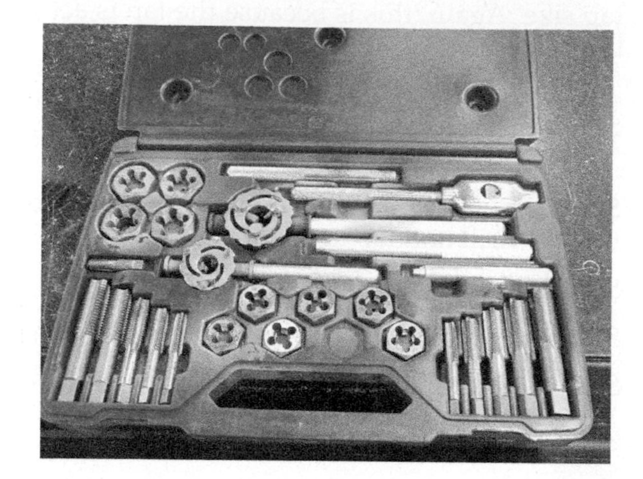

Figure 41 Pipe taps.

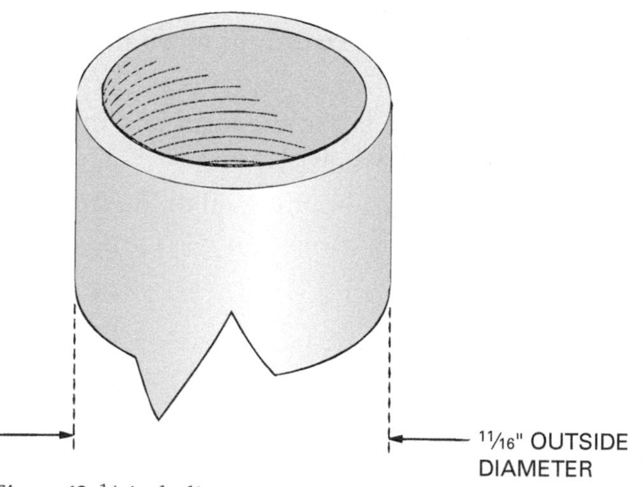

Figure 42 ½-inch diameter pipe.

Step 9 Apply pressure and turn the tap in one-half turn.

Step 10 Back off the tap one-quarter turn to break off chips and clear the tap.

Step 11 Repeat Steps 8 and 9 until the required number of threads are cut.

CAUTION

Do not allow chips to clog the tap. If the tap will not turn and has a springy feeling, stop and repeat Step 9 using a piece of wire to clear the tap.

1.8.6 Pipe Extractors

Pipe extractors (*Figure 43*), also referred to as easy-outs, are used to remove broken, threaded ends of pipes, plugs, and fittings. Each is plainly marked, showing the size of pipe it should be used with and the size of the hole that must be drilled if fittings must be drilled to receive the extractor.

To use a pipe extractor:

Step 1 Identify the size of broken pipe or fitting that needs to be removed.

Step 2 Select the correct size pipe extractor for the broken pipe or fitting.

Step 3 Use a hammer to drive the cutting end into the broken pipe or fitting.

Step 4 Place a wrench on the square end of the extractor and unscrew the broken fitting or pipe to remove it.

Table 2 Drill and Tap Sizes

Pipe Tap Size (Inches)	Drill Size (Inches)	Pipe Tap Size (Inches)	Drill Size (Inches)
1/8	11/32	2	2 3/16
1/4	7/16	2 1/2	2 5/8
3/8	16/32	3	3 1/4
1/2	23/32	3 1/2	3 3/4
3/4	15/16	4	4 1/4
1	1 5/32	4 1/2	4 3/4
1 1/4	1 1/2	5	5 5/16
1 1/2	1 23/32	6	6 3/8

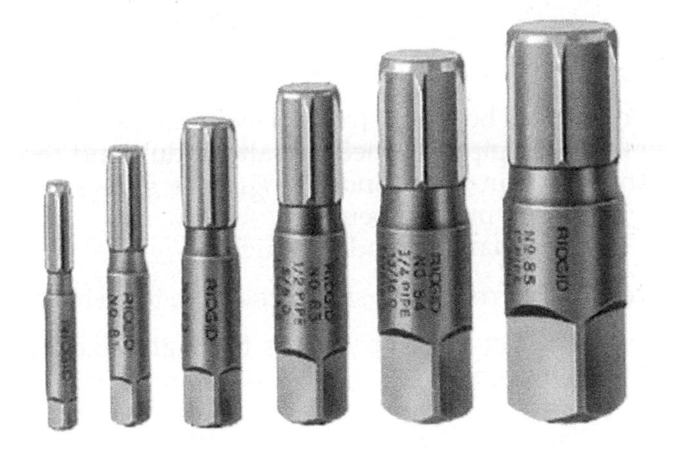

Figure 43 Pipe extractors.

Step 5 Remove the broken pipe or fitting from the extractor.

Step 6 Inspect the extractor for any damage and then store it.

1.9.0 Benders and Flaring Tools

Benders are the tools used to manually bend pipe and tubing. Spring benders, lever compression tube benders, and manual benders (hickeys) are the most common manual bending tools. It is important to get the desired bend without flattening, kinking, or wrinkling the tube or pipe. Pipe and tube bending require detailed calculations.

1.9.1 Spring Benders

The simplest hand bending tool is the spring bender (*Figure 44*). The spring bender should fit tightly over the tube and should be centered over the area to be bent. Spring benders are used to bend soft copper, aluminum, and other soft metals. They are available in five sizes for pipe between 1/4 inch and 5/8 inch in diameter. For this exercise, simply bend a piece of pipe or tubing without making prior calculations.

To bend tubing using a spring bender:

Step 1 Determine the size of the tubing to be bent.

Step 2 Choose an appropriately-sized spring bender for the tubing. The spring must fit tightly over the tubing to ensure a good bend.

Step 3 Position the spring over the center of the area to be bent.

Step 4 Grasp the spring between your fingers and thumbs and carefully bend the tubing until the proper radius and angle are formed.

Step 5 Slip the spring off the tube.

1.9.2 Lever Compression Tube Benders

Lever compression tube benders (*Figure 45*) are available in nine sizes for outside diameters of 3/16-inch to 7/8-inch. They can be used with soft and hard copper, brass, aluminum, steel, and stainless steel tubing. If the tubing is marked correctly prior to bending, bends are accurate to about 1/32 of an inch. A 90-degree bend in tubing adds approximately one tubing size diameter to the overall length of the tubing. For example, if you bend a 10-inch long, 1/2-inch diameter piece of tubing 90-degrees, the overall length of the tubing will be 10-1/2-inches after bending.

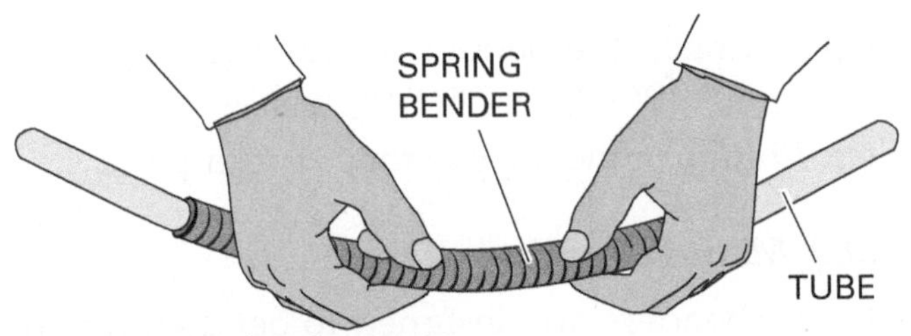

Figure 44 Tube-bending spring.

Figure 45 Lever compression tube benders.

To bend tubing using a lever compression bender:

Step 1 Hold the sheave block handle in your left hand and open the slide bar to its widest position.

Step 2 Slide the tube into the bender and align it so that the start of the bend will be on the zero mark on the bender.

Step 3 Snap the clip over the tube to hold the tube in place.

Step 4 Turn the slide bar around until the full length of the groove and slide bar are in contact with the tube.

Step 5 Pull the slide bar handle around in a smooth, continuous motion to bend the tube.

Step 6 Stop the bend at the required angle.

CAUTION

Be careful not to bend the tube past the required angle. It is difficult to straighten a tube without damaging it.

Step 7 Open the bender to its widest position.

Step 8 Raise the clip and remove the tube from the bender.

Step 9 Check the bender for the proper angle and for any signs of damage to the tube.

Step 10 Return the tools to their proper place.

1.9.3 Manual Benders

Manual benders are designed to bend electrical conduit but are sometimes used by pipefitters to bend pipe. There are two types of manual benders: one for pipe and heavy-wall conduit and the other for thin-wall conduit. *Figure 46* shows the two types of manual benders.

To use a manual bender or hickey:

Step 1 Determine the size of the pipe to be bent.

Step 2 Select a bender that is the right size for the job.

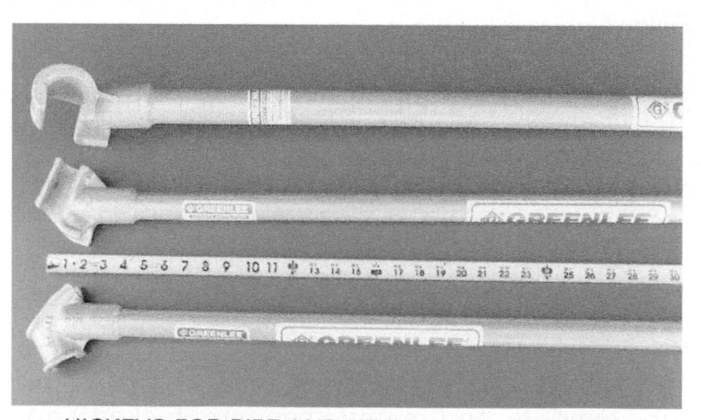

HICKEYS FOR PIPE AND HEAVY WALL CONDUIT

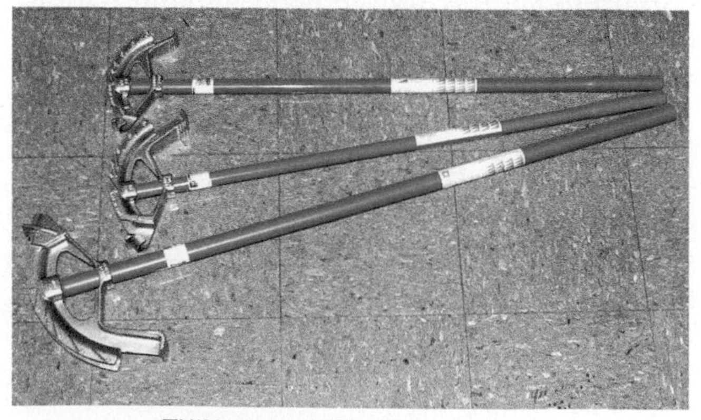

THIN WALL CONDUIT BENDERS

Figure 46 Manual benders.

Step 3 Mark the pipe at the beginning of the bend.

Step 4 Lay the pipe on the floor.

Step 5 Slide the bender onto the pipe at the start of the bend.

Step 6 Grip the handle and pull the bender down toward the floor to make the bend.

NOTE

The thin-wall conduit bender will make a smooth bend with one long pull of the handle; it has a heel that can be pressed with your foot to assist in bending the pipe. The pipe and heavy-wall conduit bender requires that a small bend be made, sliding the bender along the pipe a short distance and making another bend until the proper bend is achieved.

Step 7 Stop the bend when the required bend angle is achieved.

CAUTION

Be careful not to bend the pipe past the required angle. It is difficult to straighten pipe without damaging it.

Step 8 Remove the bender from the pipe.

Step 9 Check the pipe for cracking or kinking.

Step 10 Store the bender.

1.9.4 Flaring Tools

Flare joints may be required in situations where a torch (for welding, soldering, and brazing) is not allowed because of fire hazards. A flare joint is made by measuring, cutting, and reaming the pipes and using flare fittings. This kind of joint is commonly used to join soft copper tubing with diameters from ¼ to 2 inches. Soft copper fittings should be leakproof and easily dismantled with the right tools.

Two kinds of flare fittings are popular: the single-thickness flare and the double-thickness flare. For both types, use a special flaring tool (*Figure 47*) to expand the end of the tube outward into the shape of a cone or flare. The single-thickness flare forms a 45-degree cone that fits against the face of a flare fitting (*Figure 48*). In a single operation, the single-thickness flare is formed, and then the lip is folded back and compressed to make a double-thickness flare. The double-thickness flare (*Figure 49*) is preferable because of its larger-size tubing. A single-thickness flare

Figure 47 Flaring tool.

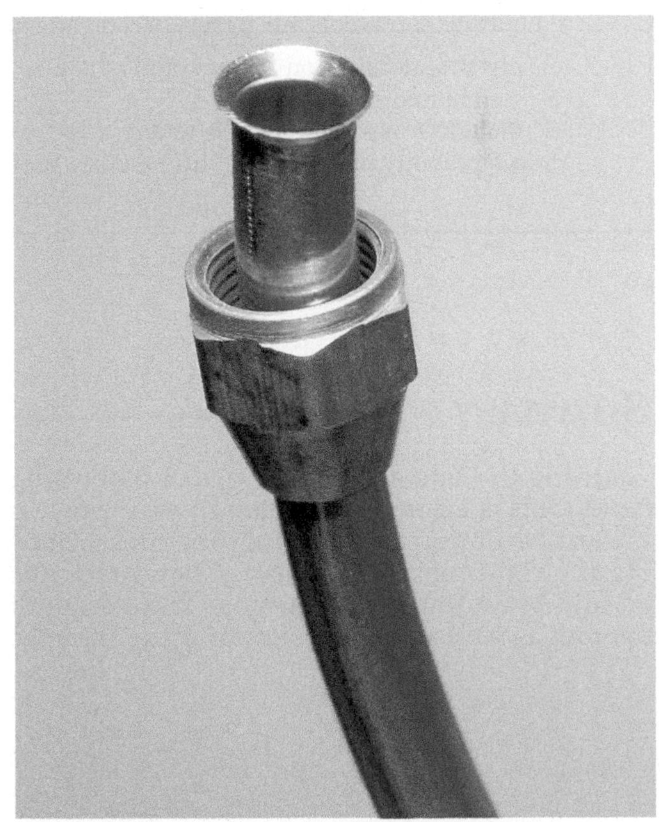

Figure 48 Single-thickness flare on copper tubing.

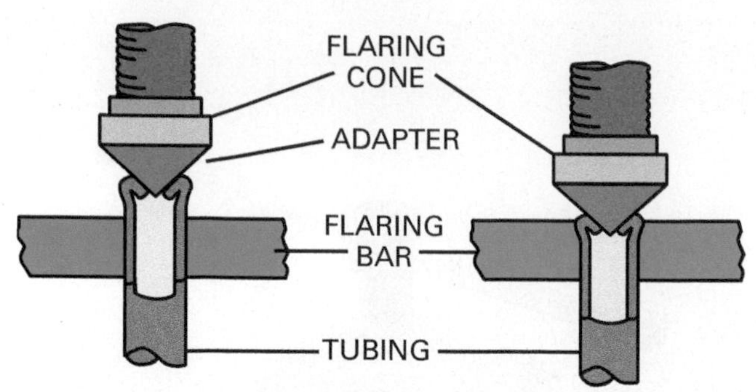

Figure 49 Double-thickness flare.

may be weak when used under excess pressure or where there is expansion. Double-thickness flare connections are easier to dismantle and reassemble without causing damage, as compared to single-thickness flare connections.

To make a flare joint:

Step 1 Measure, cut, and ream the tubing.

Step 2 Slip the flare nut over the tubing.

Step 3 Use the flaring tool to flare the tubing's ends until the fit is perfect.

Step 4 Slide the nuts onto each end of the flare tubing. Gently hand-bend or hand-shape the tubing over the male thread of each fitting.

Step 5 Use a smooth-jaw adjustable wrench to tighten until the fitting is snug.

Step 6 Test the joint for leaks.

Additional Resources

Hand and Power Tools. 2018. US Department of Labor: Occupational Safety and Health Administration. Available at: **https://www.osha.gov/SLTC/handpowertools/standards.html.**

Tools and Their Uses, Latest Edition. Naval Education and Training Program Development Center. Washington, DC: US Government Printing Office.

1.0.0 Section Review

1. Pipefitting hand tools are generally categorized by the _____.
   a. physical description and method of use
   b. weight and size
   c. materials used to make the tool
   d. adaptability of the tool in different situations

SUMMARY

Learning the safe and proper use of pipefitting hand tools is an important step in your process toward becoming a successful pipefitter. Selecting the right tool for the task at hand requires an understanding of the purpose and design of the tool as well as the logistics of using it safely.

When using hand tools, power tools, or any type of equipment on the job site, use the appropriate personal protective equipment and follow manufacturers' recommendations for each tool, job site rules and requirements, and OSHA regulations.

Review Questions

1. When inspecting pipe jack stands, what is something you do NOT look for?
   a. Bents legs or supports
   b. Weakened legs or supports
   c. Scratches
   d. Rust

2. The type of wrench that would be used on a chrome-plated pipe is the _____ wrench.
   a. strap
   b. chain
   c. Crescent®
   d. offset pipe

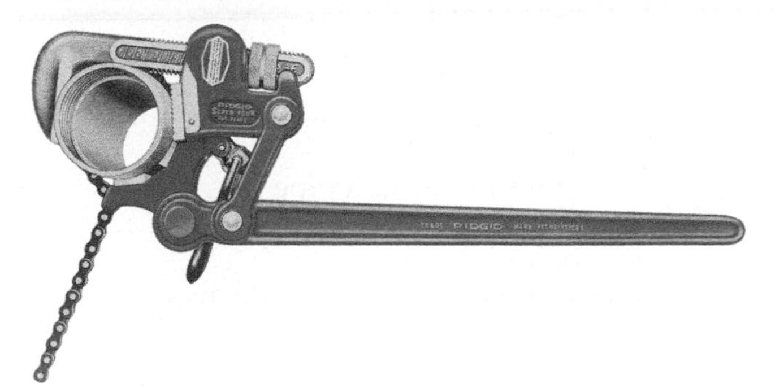

Figure RQ01

3. The tool shown in *Figure RQ01* is a(n) _____.
   a. pipe vise
   b. offset pipe wrench
   c. strap wrench
   d. compound-leverage pipe wrench

4. The term plumb relates to _____.
   a. horizontal alignment
   b. vertical alignment
   c. pipe height
   d. pipe location

5. A framing level can be used to measure _____.
   a. plumb only
   b. level only
   c. plumb or level
   d. d.none of the above

6. Which of the following is a task that cannot be performed with a framing square?
   a. Measuring the diameter of a pipe
   b. Aligning two sections of pipe
   c. Checking squareness of a pipe end
   d. Finding the center line of a pipe

Figure RQ02

7. *Figure RQ02* shows a tool used to _____.
   a. tap a hole in a pipe
   b. locate the center of a pipe
   c. measure the outside diameter of a pipe
   d. bevel a pipe

8. A Hi-Lo gauge is used to _____.
   a. check pressure in a piping system
   b. check for weld defects
   c. check pipe joint misalignment
   d. measure the inside diameter of a pipe

9. A wraparound is used to _____.
   a. align pipe joints for welding
   b. draw a straight line around a pipe
   c. measure the inside diameter of a pipe
   d. secure a pipe to a jack stand

10. Drift pins are used to _____.
    a. align bolt holes on pipe flanges
    b. secure pipes in jack stands
    c. keep pipe joints from moving while they are being welded
    d. check alignment of a pipe joint

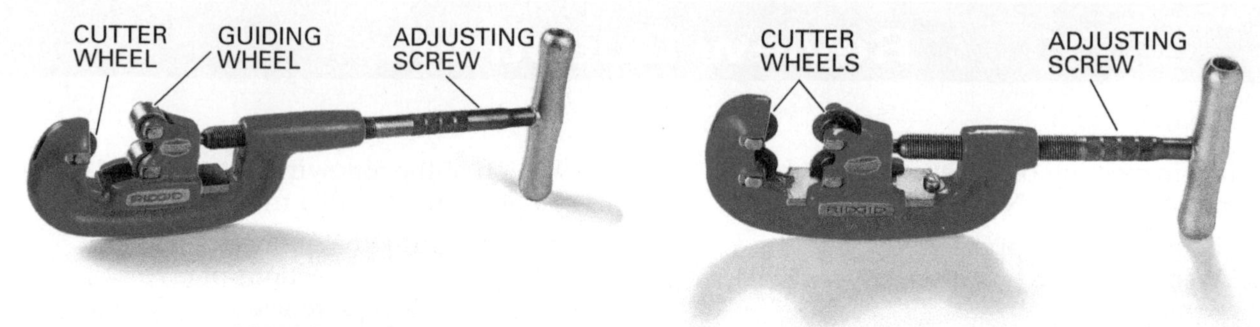

Figure RQ03

11. A hacksaw blade used to cut copper tubing should have _____ teeth per inch.
    a. 14
    b. 18
    c. 24
    d. 32

12. The tool shown in *Figure RQ03* is used to cut _____.
    a. cast iron soil pipe
    b. thin-wall copper tubing
    c. carbon steel pipe
    d. cement pipe

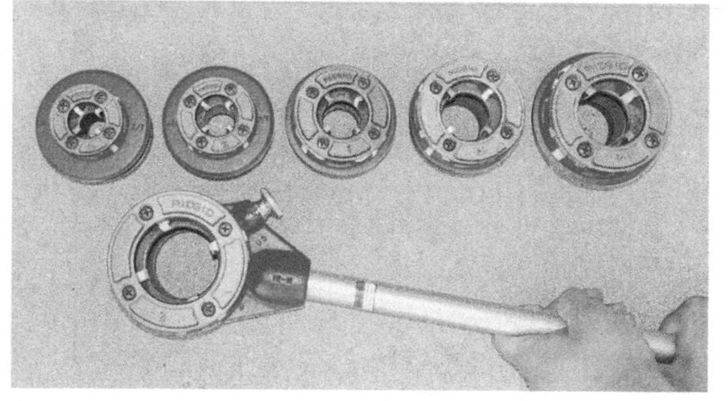

Figure RQ04

13. The tool shown in *Figure RQ04* is used to _____.
    a. thread pipe
    b. ream pipe
    c. bevel pipe ends
    d. position pipe for welding

14. Threading is complete when _____ full thread(s) extend(s) beyond the back edge of the die.
    a. one
    b. two
    c. three
    d. four

15. Pipe taps are used to _____.
    a. cut male threads of a pipe
    b. cut female threads of a pipe
    c. drill holes in pipe
    d. remove broken plugs and fittings

Trade Terms Quiz

Fill in the blank with the correct term that you learned from your study of this module.

1. To determine how many threads per inch are cut in a pipe, you would use a(n) ________.
2. Threads on the inside of a fitting are called ________.
3. Cutting pipe often leaves a ragged edge of metal called a(n) ________.
4. To cut male threads on a pipe, use a(n) ________.
5. A pipe end that has been forced open is called a(n) ________.
6. Before making the final weld, use ________ to hold the parts in place.
7. Threads on the outside of a pipe are called ________.
8. ________ is the outside width of a pipe.
9. A pipe-like raceway that contains electrical conductors is called a(n) ________.
10. A(n) ________ allows a tool to rotate in only one direction.
11. To change direction of a pipe or join two pipes, you would use a(n) ________.

Trade Terms

Burr
Conduit
Die
Female threads
Flare
Male threads
Outside diameter
Pipe fitting
Ratchet
Tack welds
Thread gauge

Trade Terms Introduced in This Module

Burr: A sharp, ragged edge of metal usually caused by cutting pipe.

Conduit: A round raceway, similar to pipe, that contains conductors.

Die: A tool used to make male threads on a pipe or bolt.

Female threads: Threads on the inside of a fitting.

Flare: A pipe end that has been forced open to make a joint with a fitting.

Male threads: Threads on the outside of a pipe.

Outside diameter: A measurement of the outside width of a pipe.

Pipe fitting: A unit attached to a pipe and used to change the direction of fluid flow, connect a branch line to a main line, close off the end of a line, or join two pipes of the same size or of different sizes.

Ratchet: A device that allows a tool to rotate in only one direction.

Tack welds: Short welds used to hold parts in place until the final weld is made.

Thread gauge: A tool used to determine how many threads per inch are cut in a tap, die, bolt, nut, or pipe. Also called a pitch gauge.

Additional Resources

This module presents thorough resources for task training. The following reference material is recommended for further study.

Hand and Power Tools. 2018. US Department of Labor. Occupational Safety and Health Administration. Available at: **https://www.osha.gov/SLTC/handpowertools/standards.html**.

Tools and Their Uses, Latest Edition. Naval Education and Training Program Development Center. Washington, DC: US Government Printing Office.

Figure Credits

Koike Aronson, Inc., Figure 2 (table roller)

Sumner Mfg. Co., Inc., Figure 2 (v-head jack, roller head jack, floor stand roller)

Reed Mfg. Co., Figure 3, 6, 7, 9

Klein Tools, Inc., Figure 8

Ridge Tool Company (RIDGID®) Figure 35,36, 37, 43, 45, RQ03

The Stanley Works, Figure 12, 13, 15

Pacific Laser Systems, Inc., Figure 16

RIDGID®, Figure 27

Section Review Answer Key

Section 1.0.0

Answer	Section Reference	Objective
1. a	1.0.0	1a

NCCER CURRICULA — USER UPDATE

NCCER makes every effort to keep its textbooks up-to-date and free of technical errors. We appreciate your help in this process. If you find an error, a typographical mistake, or an inaccuracy in NCCER's curricula, please fill out this form (or a photocopy), or complete the online form at **www.nccer.org/olf**. Be sure to include the exact module ID number, page number, a detailed description, and your recommended correction. Your input will be brought to the attention of the Authoring Team. Thank you for your assistance.

Instructors – If you have an idea for improving this textbook, or have found that additional materials were necessary to teach this module effectively, please let us know so that we may present your suggestions to the Authoring Team.

NCCER Product Development and Revision
13614 Progress Blvd., Alachua, FL 32615
Email: curriculum@nccer.org
Online: www.nccer.org/olf

❑ Trainee Guide ❑ Lesson Plans ❑ Exam ❑ PowerPoints Other ______________

Craft / Level: Copyright Date:

Module ID Number / Title:

Section Number(s):

Description:

Recommended Correction:

Your Name:

Address:

Email: Phone:

This page is intentionally left blank.

Pipefitting Power Tools

Overview

Pipefitters use power tools to cut, grind, thread, and shape all types of materials. It is very important to select the right tool for the job and make sure it is in good working order before use. Specialty tools, like threading machines and bevelers, should only be used for the specific jobs they were designed to perform. With power and pneumatic tools, it is especially important to follow all operating instructions and safety precautions because they pose safety hazards. When used correctly, these tools can greatly increase a pipefitter's productivity.

Module 08103

Trainees with successful module completions may be eligible for credentialing through the NCCER Registry. To learn more, go to www.nccer.org or contact us at 1.888.622.3720. Our website has information on the latest product releases and training, as well as online versions of our *Cornerstone* magazine and Pearson's product catalog.

Your feedback is welcome. You may email your comments to curriculum@nccer.org, send general comments and inquiries to info@nccer.org, or fill in the User Update form at the back of this module.

This information is general in nature and intended for training purposes only. Actual performance of activities described in this manual requires compliance with all applicable operating, service, maintenance, and safety procedures under the direction of qualified personnel. References in this manual to patented or proprietary devices do not constitute a recommendation of their use.

08103 V4.0

08103
Pipefitting Power Tools

Objectives

Successful completion of this module prepares trainees to:

1. State safety guidelines for power tool use.
   a. State general safety guidelines for power tool use.
   b. State electrical safety guidelines for power tool use.
2. Identify and describe the use of power tools for cutting pipe.
   a. Identify and describe the use of portable band saws.
   b. Identify and describe the use of abrasive saws.
3. Identify and describe the use of power tools for grinding and beveling pipe.
   a. Identify and describe the use of pipe grinding tools.
   b. Identify and describe the use of pipe beveling tools.
4. Identify and describe the use of power tools for pipe threading.
   a. Explain how to use a threading machine to load, cut, and ream pipe.
   b. Explain how to perform pipe threading operations.
   c. Identify and describe the use of portable power drives for pipe threading.

Performance Tasks

Under the supervision of your instructor, you should be able to do the following:

1. Cut pipe using a portable band saw (do not use a threading machine).
2. Use an end grinder/die grinder.
3. Operating a portable grinder, properly prep and bevel the end of a pipe.
4. Identify several types of pipe bevelers.
5. Replace dies in a threading machine.
6. Cut, ream, and thread pipe using a threading machine.
7. Cut and thread nipples using a nipple chuck.
8. Thread pipe using a portable power drive.

Trade Terms

Assured equipment grounding conductor program
Bevel
Chamfer
Chucks
Die
Fabrication
Ground fault circuit interrupter (GFCI)
Horsepower (hp)
Revolutions per minute (rpm)

Industry Recognized Credentials

If you are training through an NCCER-accredited sponsor, you may be eligible for credentials from NCCER's Registry. The ID number for this module is 08103. Note that this module may have been used in other NCCER curricula and may apply to other level completions. Contact NCCER's Registry at 888.622.3720 or go to **www.nccer.org** for more information.

Contents

Figures and Tables

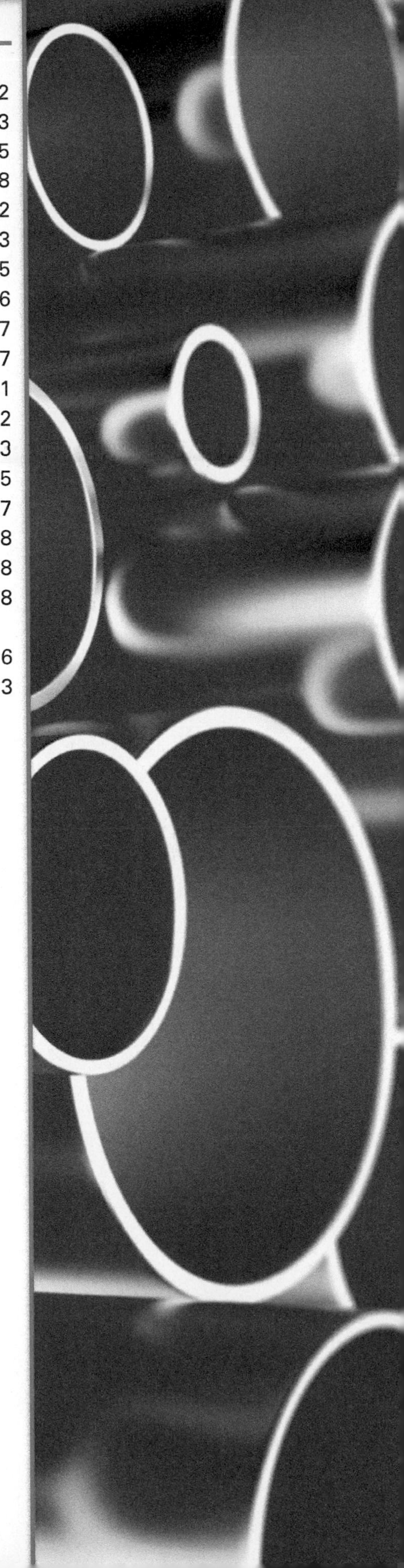

This page is intentionally left blank.

SECTION ONE

1.0.0 WORKING WITH POWER TOOLS

Objective

State safety guidelines for power tool use.

a. State general safety guidelines for power tool use.
b. State electrical safety guidelines for power tool use.

Trade Terms

Assured equipment grounding conductor program: A detailed plan specifying an employer's required equipment inspections and tests as well as a schedule for conducting those inspections and tests.

Ground fault circuit interrupter (GFCI): A fast-acting circuit breaker that senses small imbalances in the circuit caused by current leakage to ground and, in a fraction of a second, shuts off the electricity.

The widespread use of power tools in pipefitting has greatly increased productivity. Power tools are available for cutting, grinding, threading, and shaping all types of piping materials. Pipefitters must know how to safely operate these power tools for the specific jobs for which the tools are designed. Understanding how to work with electrical power tools and maintaining safety around electricity is essential to this operation.

1.1.0 Power Tool Safety

If used improperly, power tools can cause severe injuries and even death. The following safety precautions must be followed when using power tools:

- Wear the proper clothing according to company policies. Do not wear loose-fitting clothing when working with or around power tools. There is always the danger that loose clothing, especially shirt tails, will get caught in the moving parts of the tool.
  - Wear safety glasses when using power tools. Most power tools eject material which could cause severe eye damage. Types of safety glasses include foam-lined, Z87-plus rated, and mono-goggles.
- Wear properly fitting gloves according to job site procedures.
- Respect your tools and always follow the recommended operating and maintenance procedures. A well-maintained tool is a safe tool.
  - Inspect power tools before each use. Make sure there is no damage to the tool or power cord and that all the safety guards are in place.
  - Use tools only for their intended purposes, and use the proper tool for each job. Know the capabilities of each tool and work within those limits.
  - Avoid accidental starts. Make sure that the power switch is in the Off position before plugging in the power cord.
  - Be aware of the torque or kickback of the power tool being used. Maintain good balance and properly brace yourself when using power tools.
  - Do not overreach. Any fall can be dangerous, but a fall while using a power tool can be much worse.
  - Do not alter or modify a power tool in any way.
  - Never carry a power tool by its power cord or an air tool by its air line.
  - Never force a tool beyond its capabilities. Doing so wears out the tool prematurely and places you and others in unsafe situations.
- Disconnect the power source before performing maintenance on a tool or changing accessories.
- Allow only competent technicians to repair power tools.
- Store tools when they are not in use. Allowing power tools to lie around increases the possibility of incidents, including trip and fall hazards that could harm co-workers as well as damage the equipment itself.
- Keep the work area clean. Many accidents are caused by materials or tools carelessly spread about the work area.

1.1.1 Pneumatic Power Tool Safety

Pneumatic power tools can also be dangerous and can cause severe personal injury if misused. The following safety precautions apply to pneumatic power tools:

- Make sure all air hoses are properly drained or blown clear before the hose is attached to the tool.
- Check the hose pressure to make sure it is applicable to the tool.

- Ensure that the oiler reservoir is at the proper level before using the tool.
- Never direct the air flow from an air nozzle toward yourself or anyone else.
- Secure all hose connections to avoid accidental disconnection.
- Do not turn on the air until you trace the hose and ensure that the end is securely attached to an air tool. Use whip checks (*Figure 1*) and proper pins.
- Do not crimp the hose to turn off the air to the tool or to change a tool.
- Inspect the air hose for damage before using the tool.
- Do not attempt to repair a damaged air hose. Always replace it.

WARNING!

When connecting into an air hose, make sure that it is an air hose and not an inert gas line, such as nitrogen or argon.

- Make sure the air hose is rated for the air pressure with which it will be used. The air hose should have a pressure rating printed on the hose.
- Air hoses must be tied off at least 7 feet above the ground or floor to prevent tripping hazards.

1.2.0 Electric Power Tool Safety

Electric power tools are extremely dangerous and can cause severe injury and even death if misused. The following safety precautions apply to electric power tools:

- Do not use electric power tools in wet or damp locations. There is always a danger of electric shock, even if the tool is properly grounded.
- Respect power cords. Keep them out of situations where they can be frayed, cut, or damaged. Always disconnect a power cord by grasping the plug and pulling. Never yank on the cord to disconnect it.
- Ensure that the available current is the same current for which the tool is designed. This will be indicated on the identification plate attached to the tool.
- All power cords must be secured to prevent a tripping hazard. One way to do this is to tie them off overhead, at least 7 feet off the floor, or cover them with cord mats when they are running across on the ground.

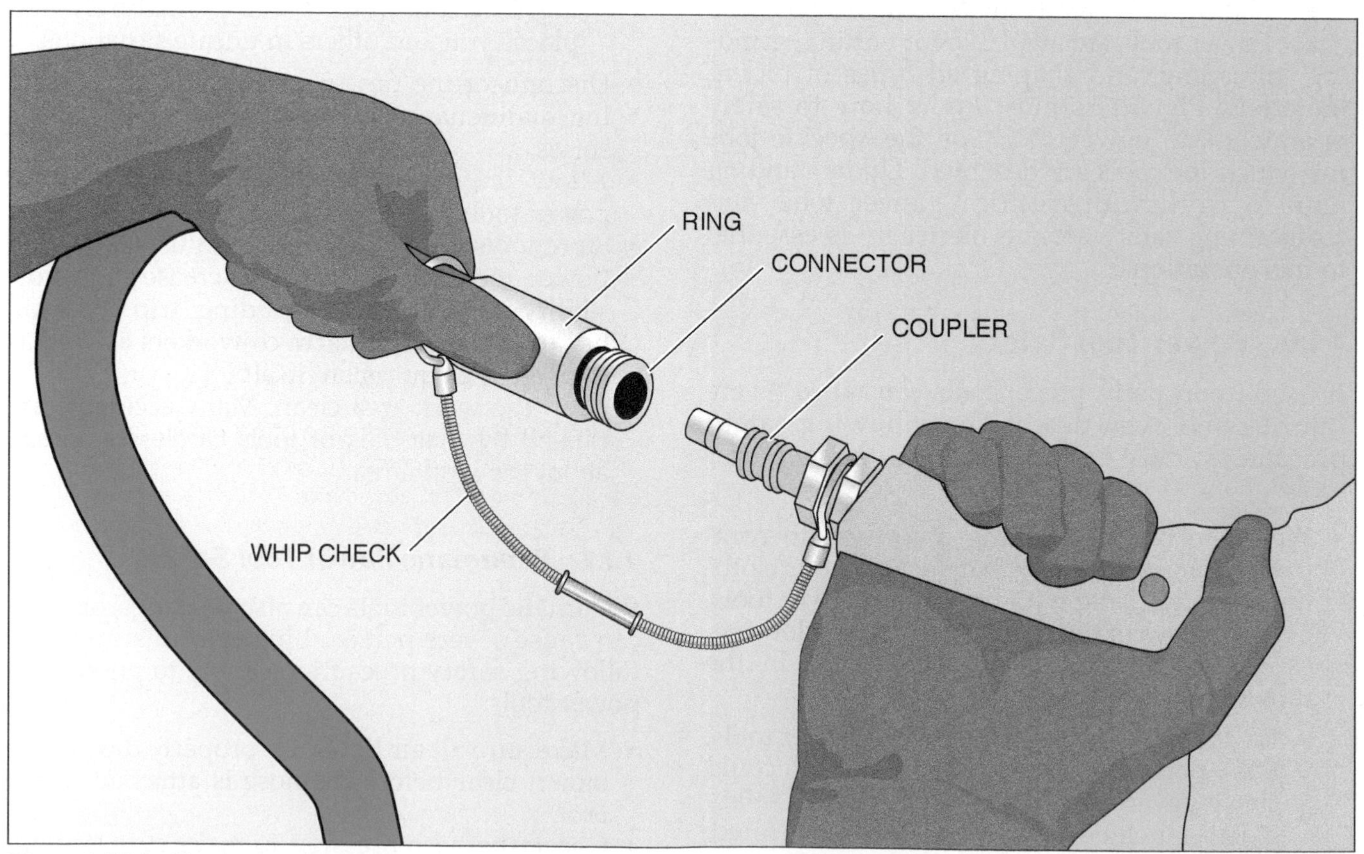

Figure 1 Whip Check

- Do not repair a frayed or damaged power cord. It must be replaced by a licensed electrician.
- Never remove the grounding pin from a three-wire power cord or extension cord.

> **NOTE**
> In some work situations, such as in mining and at shipyards, trigger locks must be removed from power tools.

1.2.1 Ground Fault Circuit Interrupters

A **ground fault circuit interrupter (GFCI)** is a fast-acting circuit breaker that senses small imbalances in the circuit caused by current leakage to ground. A GFCI continually matches the amount of current going to an electrical device against the amount of current returning from the device. A GFCI detects loss of power, not a surge in power, and it will interrupt the electric power within $\frac{1}{40}$ of a second. *Figure* 2 shows an extension cord with a GFCI.

A GFCI will not protect you from line-to-line contact hazards such as holding either two hot wires or a hot and a neutral wire in each hand. It does provide protection against the most common form of electrical shock, which is a ground fault. It also provides protection from fires, overheating, and wiring insulation deterioration.

GFCIs can be used successfully to reduce electrical hazards on construction sites. Tripping of GFCIs—interruption of circuit flow—is sometimes caused by wet connectors and tools. Limit the amount of water that tools and connectors come into contact with by using watertight or sealable connectors. Having more GFCIs or shorter circuits can prevent tripping caused by cumulative leakage from several tools or from extremely long circuits.

Figure 2 Extension cord with a GFCI.

1.2.2 GFCI Requirements

OSHA allows two methods of ensuring a proper grounding system for temporary wiring. The first is to provide GFCIs for all 15A and 20A circuits. All GFCIs must be tested every three months. The second method is known as an **assured equipment grounding conductor program**. This is a monthly inspection of all electric tools and extension cords, which are not protected by GFCI, to ensure that the ground wire is intact, and the equipment is safe to operate. A written program must be in place to establish the frequency and method of logging the inspections. Some companies use a color code system to indicate when the equipment has been inspected. The current edition of the *National Electrical Code®* has more stringent requirements for temporary wiring.

1.2.3 Electrical Cord Safety

Electrical cords are frequently seen on construction sites, yet they are often overlooked. Use the following safety guidelines to ensure your safety and the safety of other workers:

- Every electrical cord should have an Underwriters Laboratories (UL) label attached to it. Check the UL label for specific wattage. Do not plug more than the specified number of watts into an electrical cord.
- If the UL label indicates that a cord set is not for outdoor use, only use it indoors.
- Do not remove, bend, or modify any metal prongs or pins of an electrical cord.

Case History

Grounding and Insulation Saves Lives

One worker was climbing a metal ladder to hand an electric drill to the journeyman installer on a scaffold 5 feet above him. When the worker climbing the ladder reached the third rung from the bottom of the ladder, he received a fatal shock. He died because the cord of the drill he was carrying had an exposed wire that made contact with the conductive ladder.

The Bottom Line: Inspect all equipment before using it. Never use electrical equipment with a damaged power cord. Use fiberglass ladders when working with electrical equipment.

Source: The Occupational Safety and Health Administration (OSHA)

- Extension cords used with portable tools and equipment must be three-wire type and designated for hard or extra-hard use. Check the UL label for the cord's use designation.
- Avoid overheating an electrical cord. Make sure the cord is uncoiled and that it does not run under any covering materials such as tarps, insulation rolls, or lumber.
- Do not run a cord through doorways or holes in ceilings, walls, and floors that might pinch the cord. Also, check to see that there are no sharp corners along the cord's path. Any of these situations will lead to cord damage.
- Extension cords are a tripping hazard. They should never be left unattended and should always be put away when not in use.

Additional Resources

About the NEC®. National Fire Protection Association. 2018. Available at: **https://www.nfpa.org/NEC**

HAND and POWER TOOLS. Occupational Safety & Health Administration. 2018. Available at: **https://www.osha.gov/Publications/osha3080.html**

1.0.0 Section Review

1. To prevent tripping hazards, air hoses must be tied off at least _____ feet above floor level.
   a. 3
   b. 5
   c. 7
   d. 9

2. A GFCI protects you from all of the following, except:
   a. Line-to-line contact hazards
   b. Ground faults
   c. Fires and overheating
   d. Wiring insulation deterioration

SECTION TWO

2.0.0 CUTTING PIPE WITH SAWS

Objective

Identify and describe the use of power tools for cutting pipe.

a. Identify and describe the use of portable band saws.
b. Identify and describe the use of abrasive saws.

Performance Task

1. Cut pipe using a portable band saw (do not use a threading machine).

Trade Terms

Revolutions per minute (rpm): The number of complete revolutions an object will make in one minute.

Pipefitters select saws based on the material being cut and purposes for which each saw is designed. Selecting the proper blade for each saw is equally important, as is caring for and storing the saw correctly. Always wear personal protective equipment (PPE) when using saws and follow company policy for the safe use of equipment.

2.1.0 Portable Band Saws

Pipefitters use portable band saws to cut carbon steel pipe, galvanized pipe, and pipe made from other materials. The larger diameter pipes, usually 4 inches and larger, require a special procedure to ensure that the pipe is cut square. Saw blades are chosen based on hardness of the material being cut.

2.1.1 Selecting Band Saw Blades

Band saw blades are selected depending on the type and wall thickness of the material to be cut. Blades are classified by the material of the blade and the number of teeth per inch. A coarse blade has fewer teeth per inch and should be used on softer materials. A fine blade has more teeth per inch and should be used on harder materials. *Table 1* shows a blade selection chart and recommended usages.

2.1.2 Replacing Portable Band Saw Blades

The portable band saw blade must be replaced if it becomes worn and dull.

To replace a band saw blade:

Step 1 Make sure the band saw is unplugged. Rotate the band adjust knob 180 degrees to release the tension on the blade. *Figure 3* shows portable band saw components.

Step 2 Turn the saw upside down on the work table.

Step 3 Remove the blade from the blade pulleys underneath the saw.

> **WARNING!**
> Be careful when removing the blade from the blade pulleys, as it tends to be forcefully ejected.

Step 4 Slip a new blade around the blade pulleys.

Step 5 Slip the new blade into the blade guide and the back stop.

Step 6 Turn the saw over and rotate the band adjust knob to put tension on the blade.

2.1.3 Cutting Pipe with a Portable Band Saw

When cutting small pipes (4 inches in diameter or less), it is important not to cut straight through the pipe because the blade may bend and it might not produce a straight cut. The procedure for cutting pipe using a portable band saw consists of cutting around the pipe instead of straight through it.

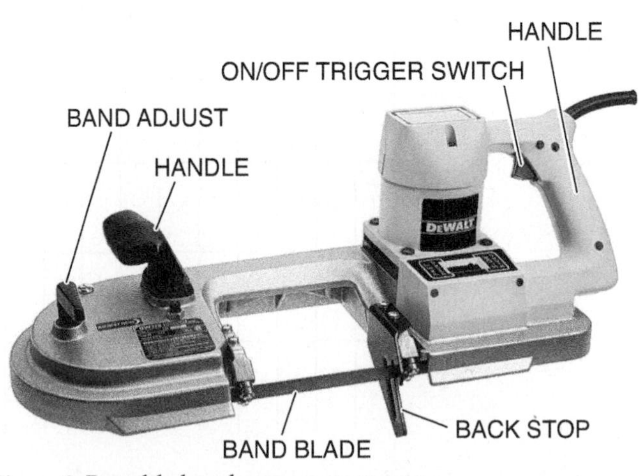

Figure 3 Portable band saw components.

Table 1 Blade Selection Chart and Recommended Usages

Band Saw Speed		Use Single-Speed Band Saw or Higher Speeds on Two-Speed and Variable-Speed Band Saws									Use Lower Speeds on Two-Speed or Variable-Speed Band Saws							
Material to be Cut		Aluminum, Brass, Copper, Bronze, Mild Steel					Angle Iron, Cast Iron, Galvanized Pipe				Stainless Steel				Fiberglass, Asbestos, Plastics			
Wall or Material Thickness		3/32 to 1/8	1/8 to 1/4	5/32 to 1/2	3/16 to 3/4	11/32 & over	3/32 to 1/8	1/8 to 1/4	5/32 to 1/2	3/16 & over	3/32 to 1/8	1/8 to 1/4	5/32 to 1/2	3/16 & over	3/32 to 1/8	1/8 to 1/4	5/32 to 1/2	3/16 & over
	Teeth per inch																	
Carbon Steel Blades	6					X												
	8					X												
	10				X					X								
	14			X					X									
	18		X					X										
	24	X					X											
Alloy Steel Blades	10				X					X								X
	14			X					X								X	
	18		X					X								X		
	24	X					X								X			
High-Speed Steel Blades	10				X					X				X				X
	14			X					X				X				X	
	18		X					X				X				X		
	24										X							

To cut pipe using a portable band saw:

Step 1 Select the pipe to be cut and determine what kind of blade will be needed.

Step 2 Replace the blade in the portable band saw if necessary.

Step 3 Measure and mark the desired length on the pipe.

Step 4 Use a wraparound to mark a straight line around the pipe at the cut mark.

Step 5 Start the band saw and cut the top of the pipe, keeping the blade on the cut line and the pipe snug against the back stop of the saw.

Step 6 Continue cutting on the cut line by rotating the saw over the top of the pipe.

Step 7 Stop cutting and remove the blade from the pipe when approximately ¾ of the blade between the front blade guide and back stop has cut into the pipe.

Step 8 Reposition the pipe in the vise so that the next section of uncut pipe is facing you.

Step 9 Repeat steps 5 through 8 until the pipe is cut completely through.

WARNING!

Avoid letting the pipe fall to the ground. It can fall on your foot or bounce back into your leg, causing injury. The freshly cut pipe end is extremely sharp and may be hot. Use care to avoid cutting or burning yourself on the pipe end.

Step 10 Remove the pipe from the vise.

2.2.0 Abrasive Saws

Abrasive saws use a special wheel that can slice through metal and masonry. The most common types of abrasive saws are the demolition saw and the chop saw (*Figure 4*). The main difference between the two is that the demolition saw is not mounted on a base.

2.2.1 Demolition Saw

A demolition saw (*Figure 4*) may run on electricity or gasoline. These saws are effective for cutting through most materials found at construction sites. They are often used to cut ductile iron underground pipe. Although all saws are potentially hazardous, the demolition saw is a particularly dangerous tool that requires the full attention and concentration of the operator.

WARNING!

When sawing, be sure to use the appropriate PPE for your eyes, ears, and hands. If you have long hair, be sure to tie it back or cover it properly.

WARNING!

Gas-powered demolition saws must be handled with caution. Using a gas-powered saw where there is little ventilation can cause carbon monoxide poisoning. Carbon monoxide gas is deadly and hard to detect. Use fans to circulate air and have a trained person monitor the air quality. *Source: Centers for Disease Control and Prevention website.* **www.cdc.gov**, *reviewed March 5, 2004.*

WARNING!

Because a demolition saw is handheld, there is a tendency to use it in positions that can easily compromise your balance and footing. Be very careful not to use the saw in any awkward position.

Choosing the Proper Saw Blade

Various saw blades are available, including steel and carbide-tipped. Steel blades are generally the least expensive, but they dull faster than others. They present a reasonable option for power saws that are used infrequently. Carbide-tipped blades are a better choice for power saws that are used on a regular basis. They are more expensive than steel blades, but they also last longer.

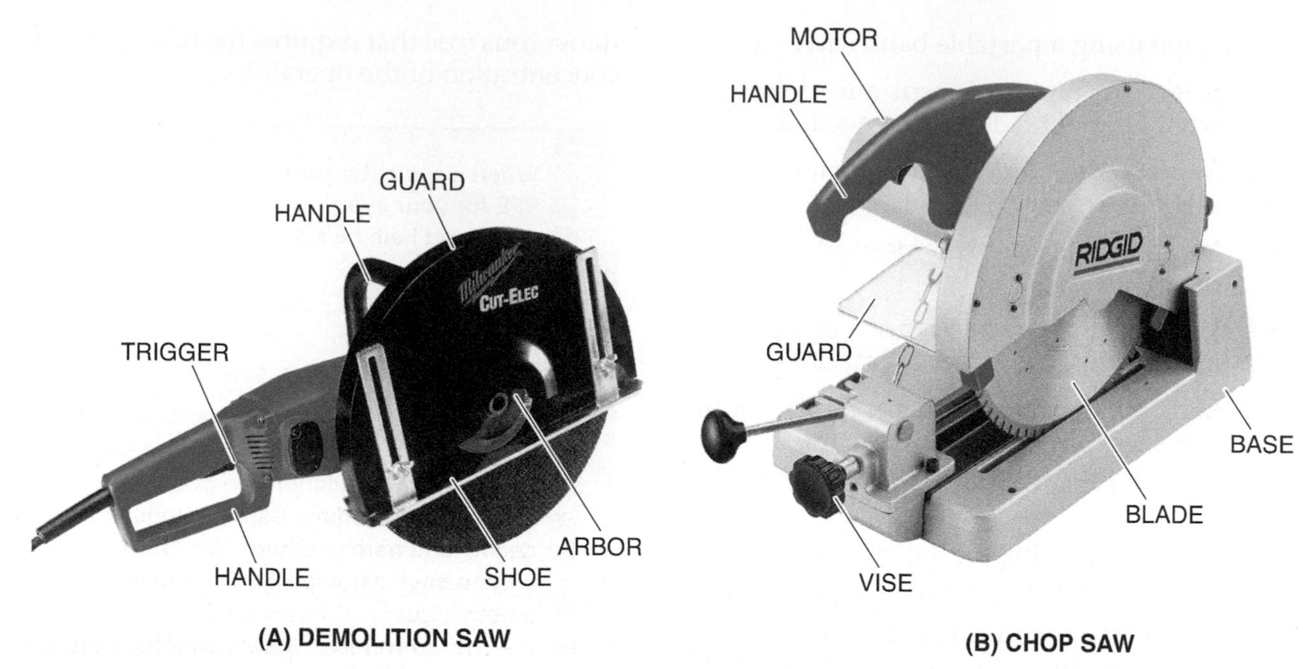

Figure 4 Abrasive saws.

Pneumatic Saw

This pneumatic reciprocating saw can be used to cut pipe and other materials. It uses hacksaw blades or modified reciprocating saw blades. It is ideal for use in situations where there is no electricity available to operate electrical power tools.

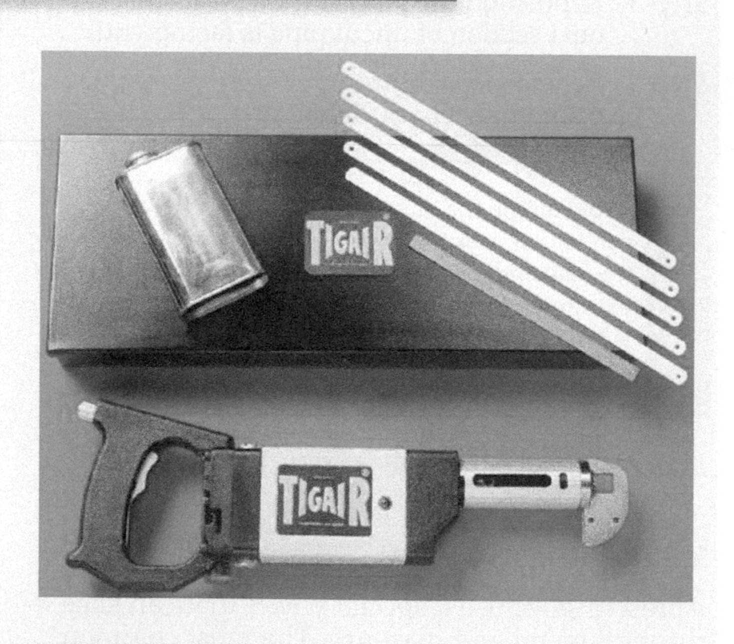

Storing Gasoline

OSHA 29 CFR 1910.106 regulates the amount of flammable or combustible liquids that may be located outside or inside a gas storage cabinet or in any fire area of a building.

Gas storage cabinets must be designed to limit the internal temperature to 325°F. Metal cabinets should be double-walled and constructed of 18-gauge sheet metal. For added safety, they should be self-closing. Gas storage cabinets must be labeled in conspicuous lettering with the statement "Flammable—Keep Fire Away."

Source: Occupational Safety and Health Administration website. "Regulations (Standards–29 CFR), *Flammable and Combustible Liquids–1910.106.*" **www.osha.gov/pls/oshaweb/owadisp.show_document?p_id=9752&p_table=STANDARDS**, reviewed March 30, 2004.

2.2.2 Chop Saw

The chop saw is a versatile and accurate tool. It is a lightweight, circular saw mounted on a spring-loaded arm that pivots and is supported by a metal base. This is a good saw to use to get exact square or angled cuts. It uses a blade (generally abrasive) to cut pipe, channel, tubing, conduit, and other light-gauge materials. The most common blade is an abrasive metal blade. Blades are available for other materials such as PVC or wood. A vise on the base of the chop saw holds the material securely and pivots to allow miter cuts. Some can be pivoted past 45-degrees in either direction.

Chop saws are sized according to the diameter of the largest abrasive blade they accept. Two common sizes are 12 inches and 14 inches. Each blade has a maximum safe speed. Never exceed that speed. Speed is measured as the number of complete revolutions an object will make in one minute, also known as **revolutions per minute** (rpm). Typically, the maximum safe speeds are 5,000 rpm for 12-inch blades and 4,350 rpm for 14-inch blades.

2.2.3 Abrasive Saw Safety

Abrasive saws, such as demolition and chop saws, require extreme care during use. To avoid injuring yourself, follow these guidelines:

- Wear PPE to protect your eyes, ears, and hands.
- Use abrasive wheels that are rated for higher rpms than the tool can produce. This way, you will never exceed the maximum safe speed for the wheel you are using. The blades are matched to a safety rating based on their rpms.
- Be sure the wheel is secured on the arbor before using a demolition saw. Point the wheel of a demolition saw away from yourself and others before starting it.
- Use two hands when operating a demolition saw.
- Use a vise to secure the materials being cut. The jaws of the vise hold the pipe, tube, or other material and prevent it from turning while being cut.
- Be sure the guard is in place and the adjustable shoe is secured before using a demolition saw.
- Keep the saw and blades clean.
- Inspect the saw before each use. Never operate a damaged saw. Ask your supervisor if you have a question about the condition of a saw.
- Inspect the blade before using a saw. If it's damaged, throw it away.

Additional Resources

Hazard Alert: Safe Work with Power Saws. 2018. Electronic Library of Construction Occupational Safety & Health. Available at: **http://www.elcosh.org/document/995/d000021/Hazard+Alert%3A+Safe+Work+with+Power+Saws.html?show_text=1**

Safe Operation of Portable Circular Power Saws. 2018. National Ag Safety Database. Available at: **http://nasdonline.org/1756/d001731/safe-operation-of-portable-circular-power-saws.html**

2.0.0 Section Review

1. A fine band saw blade has more teeth per inch than a coarse blade, so it should be used on _____ materials.
   a. porous
   b. alloy
   c. harder
   d. softer

2. Which of the following is recommended when exact square or angled cuts are necessary?
   a. Demolition
   b. Hack saw
   c. Pneumatic
   d. Chop saw

SECTION THREE

3.0.0 PORTABLE GRINDERS AND BEVELERS

Objective

Identify and describe the use of power tools for grinding and beveling pipe.

a. Identify and describe the use of pipe grinding tools.
b. Identify and describe the use of pipe beveling tools.

Performance Tasks

2. Use an end grinder/die grinder.
3. Operating a portable grinder, properly prep and bevel the end of a pipe.
4. Identify several types of pipe bevelers.

Trade Terms

Bevel: An angle cut or ground on the end of a piece of solid material.

Chamfer: An angle cut or ground only on the edge of a piece of material.

Die: A tool used to make male threads on a pipe or bolt.

Fabrication: The act of putting component parts together to form an assembly.

Horsepower (hp): A unit of power equal to 745.7 watts or 33,000 foot-pounds per minute.

Grinders and bevelers of varying types and sizes are available for the work of pipefitters. The grinding and beveling of pipe introduces fire hazards and other safety concerns. Hot work permits may be required for work involving grinding and beveling of pipe. Equipment must be inspected before each use, and PPE must be worn.

3.1.0 Portable Grinders

Pipefitters use portable grinders for a variety of tasks, such as cleaning scale from pipe before welding, beveling pipe ends, and grinding away welding spatter. If not used properly, the portable grinder can cause severe injury.

> **WARNING!** Compare the grinder wheel speed rating to the grinder operating speed. The wheel speed rating must meet or exceed the grinder operating speed.

Follow these safety guidelines when using a portable grinder:

- Visually inspect the grinding wheel before mounting it to the grinder. Do not mount a damaged wheel.
- Examine the grinder wheel guard for damage. It must be in place before starting the grinder.
- Inspect the power cord or air line for damage.
- Wear gloves and a safety shield to protect yourself from flying sparks and metal fragments when using a grinder.
- Run the grinder with the safety guard in place for a moment before grinding.

> **NOTE** A circular saw with an abrasive blade can be used to cut stainless steel pipe.

3.1.1 Types of Portable Grinders

Portable grinders can be electrically or pneumatically powered. They are rated in terms of **horsepower (hp)**, which refers to a unit of power equal to 745.7 watts or 33,000 foot-pounds per minute. Portable grinders are generally rated up to 5 hp and rotate up to 20,000 rpm at no load. Portable grinders are usually identified by the angle of the grinding wheel in relation to the motor housing. Several types of portable grinders are available for various uses. *Figure 5* shows a right angle grinder and die grinder, along with a selection of attachments that are used with them. A **die** is used to make male threads on a pipe or bolt.

Right angle grinders feature a motor and spindle which are set at a 90-degree angle from the handle. End grinders, which are similar to die grinders but much larger, are also available.

Die grinders and small angle grinders work very well for small areas such as weld grooves and beveled edges (*Figure 6*). A **bevel** is an angle cut or ground on the end of a piece of solid material.

Grinding disks and wheels are made for specific types of metals; be sure to choose the correct disk or wheel for the type of metal being ground. Always use aluminum oxide disks for grinding aluminum or stainless steel. Do not use wheels that have been used on other metals when

Figure 5 Grinders and grinder attachments.

Figure 6 Handheld grinder in use.

grinding stainless steel or aluminum, as this will contaminate the surface of the aluminum or the stainless steel.

When using wire brush attachments on stainless steel or aluminum, brushes with stainless steel bristles must be used. Do not use the same stainless steel brush for both aluminum and stainless steel because mild steel bristles will contaminate stainless steel, and cause weld defects. Use one brush for stainless steel, and another for aluminum.

When grinding, avoid grinding the base metal below the minimum allowable base metal thickness. If the base metal is ground below the minimum allowable thickness, the base metal will have to be replaced or, if allowable, built up with weld. These are expensive alternatives.

> **WARNING!**
>
> Never grind without full face protection. Avoid standing in the orbital path of the wheel or disk in case it disintegrates. If a grinding wheel or disk shatters while in use, the pieces of the wheel or disk become deadly projectiles as the weights of the pieces are multiplied by the centrifugal force and the revolutions per minute (rpm) of the grinding motor. The force with which a piece of wheel or disk is released is equivalent to that of an armor-piercing bullet. Follow company policies and procedures regarding the use of cutting equipment.

3.1.2 Inspecting Grinders Before Use

Whether the portable grinder is electric or pneumatic, a thorough inspection of the equipment is required before it is used.

To inspect a grinder:

Step 1 Check the air inlet and the air line of pneumatic grinders to make sure there is no sign of damage that could cause a bad connection or loss of air.

Step 2 Inspect the power cord and plug on electric models to ensure there are no signs of damage.

Step 3 Inspect the handle to make sure it is not loose, which could cause a loss of control.

Step 4 Inspect the grinder housing and body for defects.

Step 5 Ensure that the trigger switch works properly and does not stick in the ON position.

Step 6 Ensure that the safety guard is in good condition and securely attached to the grinder.

Step 7 Check the oil level in pneumatic grinders.

Step 8 Ensure that the maximum rotating speed of the grinding wheel is higher than the maximum rotating speed of the grinder.

Step 9 Start the grinder and allow it to run for a moment while checking for visual abnormalities, excessive vibration, extreme temperature changes, or noisy operation.

Step 10 Inspect the work area to make sure the heat and sparks generated by the grinder cannot start any fires.

3.1.3 Operating Grinders

When operating grinders, you must pay full attention to the grinder, the work being performed, and the flow of sparks and metal bits coming off the wheel. Obtain a hot work permit, if required, before starting this or any work that introduces fire hazards.

Each grinding accessory is designed to be used in only one way. When a flat grinding disc is being used, only put the flat surface of the disc against the work. If a cutoff wheel is being used, only use the edge of the wheel to do the cutting.

To safely operate a grinder:

Step 1 Inspect it before each use.

Step 2 Put on PPE, including gloves, a face shield, ear plugs and safety shoes. Wear long sleeves to cover exposed skin.

Step 3 Tuck in all loose clothing.

Step 4 Use a vise to secure the object so that it does not move.

Step 5 Attach the grinder to the power source.

Step 6 Position yourself with good footing and balance and establish a firm hold on the grinder to avoid kickback.

Step 7 Verbally warn bystanders before pulling the trigger to start the grinder.

Step 8 Start the grinder, making sure to direct the sparks to the ground whenever possible and away from any hazards in the area such as combustible debris or acetylene tanks

.WARNING!

Special precautions must be taken to avoid kickback while prepping fittings and pipe. Remember that the wheel rotates clockwise, and you must position yourself and the wheel on the pipe end correctly to avoid the wheel grabbing the pipe end and kicking back toward you. Always grind with the direction of the bevel and position the wheel to the bevel, so that if it kicks off, it will kick off away from you. The wheel is properly positioned on the bevel when the torque pulls the wheel in the direction that the wheel would leave the pipe end if it grabbed. Reposition the wheel on the pipe end when moving from the top half to the bottom half of the pipe.

CAUTION

Protect nearby alloy metals, such as stainless steel tanks, from cross contamination of other metals being ground in the area.

Step 9 Apply proper force to the grinder on the grinding surface. If too much force is applied, the motor strain loudly enough to hear over sound of the grinder.

Step 10 From time to time, stop grinding and inspect the work and the grinding wheel. Replace the wheel if necessary.

Step 11 Turn off and disconnect the grinder from the power source when grinding is complete.

Step 12 Inspect the grinder for any signs of damage.

Step 13 Return the grinder and any accessories to the storage area.

WARNING!

Always follow all manufacturers' safety procedures when using grinders. Failure to follow the manufacturer's safety recommendations could result in serious personal injury.

Grinding Wheels

A grinding wheel must be replaced if it is worn or cracked. Cracks are not always visible, however. One way to check for cracks is to hold the wheel by the center hole and tap the edge with a screwdriver handle. If the wheel is cracked, you will hear a dull sound. If the wheel is good, you will hear a clear ring

3.2.0 Power Bevelers

Power bevelers are used to prepare pipe for butt weld **fabrication**, or putting together component parts to form an assembly. Bevelers place the correct angle, or bevel, on the end of a pipe before it can be welded.

There are many different ways to bevel the end of a piece of pipe to be butt welded. The joint can be beveled mechanically using grinders, nibblers, or cutters, or it can be beveled thermally using an oxyfuel cutting torch. Mechanical joint beveling is used most often on alloy steel, stainless steel, and nonferrous metal piping. It is often required for materials that could be affected by the heat of the thermal process. Mechanical joint beveling is slower than thermal methods, such as oxyfuel cutting, but it offers high precision with low heat input. It also has the advantage of not leaving oxides, which commonly occurs when using thermal methods. The method used to bevel a pipe generally depends on the type of base material involved, as well as its ease of use and code or procedure specifications. *Figure 7* shows mechanically beveled pipe ends.

3.2.1 Pipe Beveling Machines

Welded piping is found extensively in the utilities and petrochemical industries. Before nearly every piece of pipe is welded to a fitting, to another piece of pipe, or to any other connection, its edge must

Figure 7 Beveled pipe ends.

be cut square and beveled according to specifications. If a mechanical cutting and beveling process is used, the procedure may be done using an electrically or pneumatically powered beveling machine (*Figure 8*). Many of these are portable machines that operate much like a lathe. They have mandrels that hold various cutting tools. Numerous adjusting mechanisms make it easy to set the bevel angle and depth. Various models are available to cut and bevel 2" to 60" pipe. When specifications call for tubing to be welded, machines similar to pipe beveling machines are available to face, square, and **chamfer** the ends of tubes in preparation for welding. Boiler tube ends are typically prepared using these smaller machines before welding the tubes in place.

Special cutoff machines are also made to be mounted on the outside of the pipe. They can be mounted with a ring or a special chain with rollers. An electrically or pneumatically operated grinder with a cutoff blade can be mounted on the pipe. It is then manually or electrically powered around the pipe to cut it off.

WARNING!

Always follow all manufacturers' safety procedures when using grinders. Failure to follow the manufacturer's safety recommendations could result in serious personal injury.

Automatic Pipe Bevelers

Automatic versions of pipe bevelers and cutters are an improvement over pattern cutters and other similar equipment. This is because there is no need to reset the preheat flame before each cut. Once an automatic pipe beveler is initially set up, all subsequent cuts can be made with the same settings. These systems also save time and keep the worker from cutting gases because all gases are immediately extinguished when the system is switched off.

Handheld Power Pipe Threaders

A handheld power pipe threader can be used in conjunction with a portable stand when a number of pipes must be threaded in the field. The rotating power head of this threader turns at about 30 rpm and uses the same dies as an equivalent manual threader. When a pipe that is 2½ inches or larger is being threaded, a support arm (not shown) is clamped to the pipe and the threader rests against it to counteract the torque of the threader.

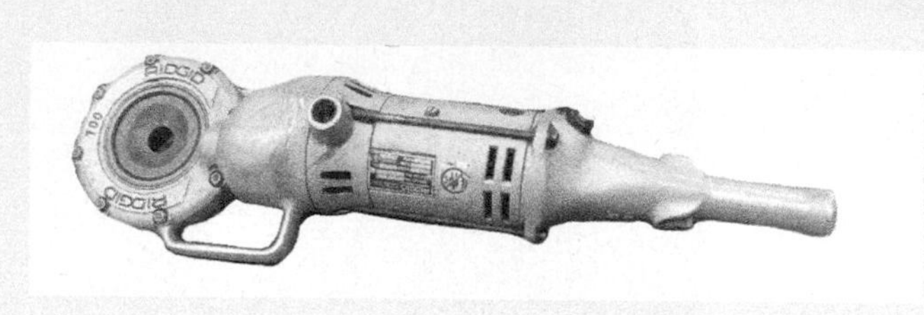

Figure 8 Pipe beveling machine.

Portable Pipe-Beveling Machine Floor Stands

Floor stands are available to convert portable pipe-beveling machines to more permanent stationary devices.

3.2.2 Cutters and Beveling Tools

Handheld electric or air-operated grinders are used in welding shops, and even more often in the field, to cut and bevel pipe and plate to prepare them for welding. *Figure 9* shows a handheld electric grinder being used to prepare a root bead on a test pipe.

WARNING!

Cutoff and grinding wheels must be selected based on the material to be cut or beveled. Using an improper wheel may damage the wheel or workpiece. Improper use can create hazardous conditions for the operator should the wheel shatter during operation. Refer to the manufacturer's recommendations and warnings when selecting grinding wheels.

3.2.3 Thermal Joint Preparation

Thermal joint preparation is done using the oxyfuel, plasma arc, or carbon arc cutting process. For beveling plate and pipe, it is best to use the oxyfuel or plasma arc cutting processes. The carbon arc cutting process is best for gouging seams, repairing cracks, and making weld repairs.

The torch used for oxyfuel or plasma arc cutting can be handheld or mounted on a motorized carriage. The motorized carriage for plate cutting runs on flat tracks positioned on the surface of the plate to be cut. The carriage has an ON/OFF switch, FORWARD/REVERSE switch, and a speed adjustment. A handwheel on top of the carriage adjusts the torch holder transversely, and a handwheel at the torch holder adjusts the torch vertically. The bevel angle is set by pivoting the torch holder. The carriage can carry a special oxyfuel or plasma arc cutting torch designed to fit

Figure 9 Handheld grinder.

into the torch holder. *Figure 10* shows a motorized carriage for cutting and beveling plate.

Special equipment is used for cutting pipe. A steel ring or special chain with rollers is attached to the outside. The torch is carried around the pipe on a torch holder that is powered by electricity, air, or a hand crank. The bevel angle is set by pivoting the torch holder, and the torch is adjusted vertically with a handwheel on the torch holder. An out-of-round attachment can be used to compensate for pipe that is out-of-round. For large-diameter pipe (54 inches or larger), special internal equipment is available. With this special equipment, the torch mechanism is mounted on the inside of the pipe. Special plasma or oxyfuel torches can be mounted in the torch holders of the external or internal pipe cutting equipment.

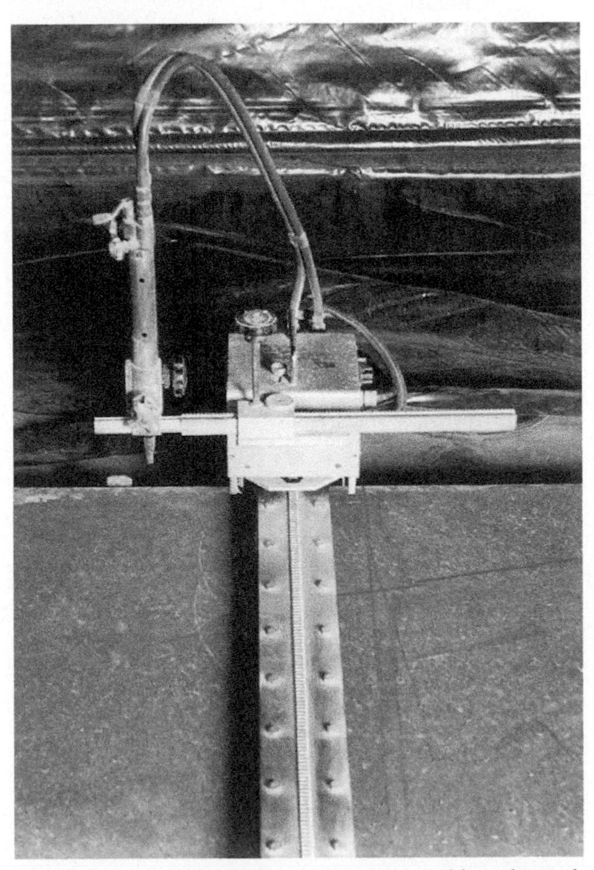

Figure 10 Motorized carriage for cutting and beveling plate.

3.2.4 Thermal Joint Preparation Precautions

When preparing a pipe joint with the oxyfuel, plasma arc, or carbon arc cutting process, all dross (metal expelled during cutting) must be removed prior to welding. Any dross remaining on the joint during welding will cause weld defects. Joints prepared with the carbon arc cutting torch must be carefully inspected for carbon deposits. It is common for small carbon deposits to be left behind, as the carbon electrode is consumed during the cutting or gouging operation. These carbon deposits will cause defects such as hard spots and cracking in the weld. Before welding, use a grinder to clean surfaces prepared with the carbon arc torch.

3.0.0 Section Review

1. _____ work permits may be required for work involving the grinding and beveling of pipe, especially when fire hazards are introduced.
   a. Hot
   b. Temporary
   c. Causal
   d. Differential

SECTION FOUR

4.0.0 PIPE THREADING MACHINES

Objective

Identify and describe the use of power tools for pipe threading.

a. Explain how to use a threading machine to load, cut, and ream pipe.
b. Explain how to perform pipe threading operations.
c. Identify and describe the use of portable power drives for pipe threading.

Performance Tasks

5. Replace dies in a threading machine.
6. Cut, ream, and thread pipe using a threading machine.
7. Cut and thread nipples using a nipple chuck.
8. Thread pipe using a portable power drive.

Trade Terms

Chucks: Parts of a machine that hold a piece of work tightly in the machine. A chuck is normally used only when the pipe or cutter will be rotated.

Pipe threading machines are multi-purpose power tools that secure and rotate pipe, conduit, or bolt stock while cutting, reaming, and threading operations are performed (*Figure 11*). All pipe threading machines are designed with a gear drive and **chucks**, which hold and turn the pipe, as well as a three-position power switch consisting of forward, off, and reverse drive positions. Depending on the size and complexity of the machine, pipe threading machines can be mounted on tripods, tables, or special stands (*Figure 12*).

A variety of vises and wrenches may be used to hold pipe while threading it. Vises have jaws that hold the pipe firmly and prevent it from turning. The teeth on vises may leave marks on the pipe, so only use a vise on pipe that will not be visible after installation. Use the standard yoke vise to hold pipe that is small in diameter. Use the chain vise to hold much larger pieces of pipe. Check the manufacturer's recommendations and use the correct oil on this vise to prevent the chain from becoming stiff, otherwise it will not work very well. Use pipe wrenches to grip and turn around stock.

4.1.0 Loading, Cutting, and Reaming Pipe

With the knowledge of proper power tool use and the application of safety practices, pipefitters may load, cut, and ream pipe with confidence. As with other equipment, threading machines must be used with caution due to the force that's exerted when in operation. Proper maintenance of all pipefitting tools and equipment is essential for both safety and economic considerations.

4.1.1 Loading Pipe into a Threading Machine

On most threading machines, pipe can be inserted from either end of the chuck. Some models, however, must be loaded from the rear. Either way, the operator must position the pipe so that the end to be threaded extends out and away from the chuck. On most threading machines, there is a stop to help position the pipe properly.

To load pipe into the threading machine:

Step 1 Measure and mark the length of pipe to be worked.

Step 2 Before inserting the pipe into the threading machine, determine if the pipe is long enough to be held by both chucks while inside the threader. If so, insert the pipe through the back. If the pipe is longer, use a jack stand to help support it and keep it as level as possible to the machine. If the pipe is shorter and will not be held by both chucks, insert it through the front.

Step 3 Extend the pipe into the working area of the threader enough to ream, cut thread, and measure. The pipe should not extend past the die head on the threader. Make sure the pipe is centered in the centering device.

Step 4 Once you have set the length of pipe needed to work with, tighten the back and front chucks in a counterclockwise motion. Knock the front chuck toward the operator two times after it has been tightened to ensure tightness. Dry run the machine using the foot pedal to make sure the pipe is rotating around evenly and not wobbling. If wobbling occurs,

Figure 11 Power threading machine.

loosen the chucks in a clockwise motion by knocking the front chuck twice in a clockwise motion to loosen the pipe and retighten it again. Repeat the dry run process.

Step 5 Spin the handwheel counterclockwise to tighten the chuck jaws. This tightens the jaws on the pipe. A clockwise spin releases the jaws.

WARNING!

When working with a power threading machine, do not wear loose clothing, jewelry, or gloves that could easily be caught in the rotating parts of the machine. Be sure to wear appropriate PPE. Be sure to tie back long hair to prevent it from being entangled in the machine during operation. Use a piece of cloth or cardboard to absorb any oil that spills during the machine's operation, so the oil does not get on the bottom of your work shoes and you track the slippery oil around the job site creating potential hazards.

4.1.2 Cutting and Reaming Pipe

To cut and ream pipe:

Step 1 Determine the length of pipe needed.

Step 2 Measure it to determine how much needs cut off.

Step 3 Mark the pipe at the point where it is to be cut.

Step 4 Swing the cutter, threader, and reamer up and out of the way.

Step 5 Load the pipe into the threading machine.

Step 6 Place pipe stands as needed.

Step 7 Turn the centering device to center the pipe in the machine.

Step 8 Spin the chuck handwheel counterclockwise to lock the pipe in place.

Step 9 Loosen the cutter until it fits over the pipe.

Step 10 Lower the cutter onto the pipe.

Step 11 Rotate the handwheel to move the cutter wheel to the cutting mark on the pipe.

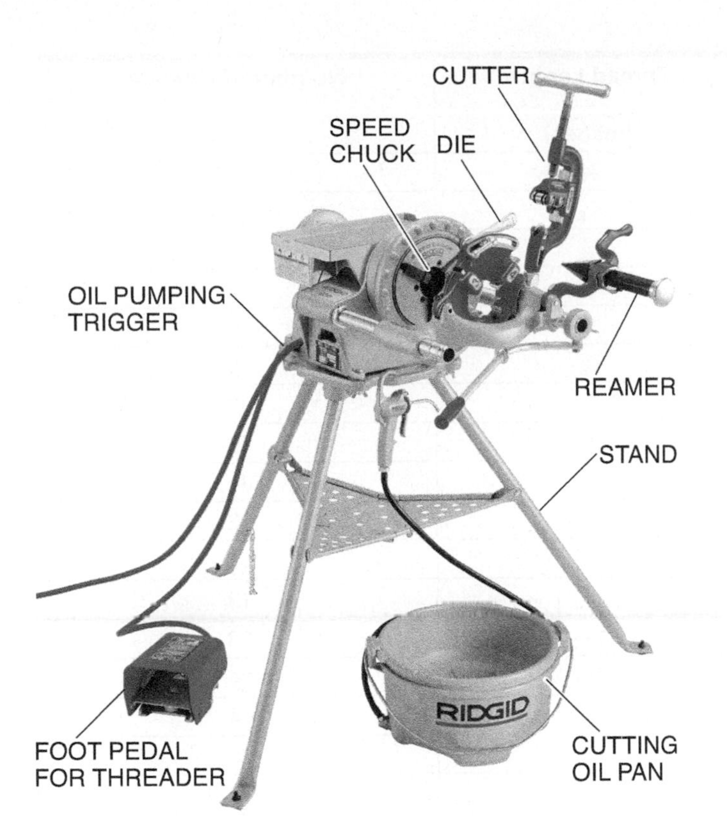

Figure 12 Cutting threads using a power threading machine.

Step 12 Rotate the cutter handle to tighten the cutter on the pipe.

Step 13 Direct the cutting oil to the cutting area.

Stpe 14 Turn the pipe machine switch to the FORWARD position.

Step 15 Step on the foot switch to start the machine and turn the pipe.

Step 16 Rotate the cutter handle to apply pressure to the cutter until the cut is complete.

Step 17 Release the foot switch.

Step 18 Raise the cutter to an out-of-the-way position.

CAUTION

Do not use a pipe cutter to cut pipe for a socket weld. Use a bandsaw. The pipe cutter will cause the pipe to flare. It may also interfere with the flow of material through the pipe.

Step 19 Lower the reamer into place.

Step 20 Press the latch on the reamer and slide the reamer bar toward the pipe until the latch catches.

Step 21 Step on the foot switch to start the machine.

Step 22 Rotate the handwheel to move the reamer into the end of the pipe.

Step 23 Ream the end of the pipe.

CAUTION

Do not over-ream the pipe. Only ream it enough to remove the burrs.

Step 24 Rotate the handwheel to move the reamer out of the pipe.

Step 25 Release the foot switch.

Step 26 Slide the power switch to the OFF position.

Step 27 Raise the reamer out of the way.

Step 28 Loosen the chuck handwheel and the centering device.

Step 29 Remove the pipe from the machine.

4.2.0 Performing Threading Operations

This section explains threading operations using a Ridgid® pipe threading machine. Before actually threading the pipe, the operator should check the thread cutting oil level and oil pump operation and prime the oil pump if necessary.

4.2.1 Replacing Dies in a Threading Machine

Four sets of dies are used with pipe threading machines that thread pipe up to 2 inches in diameter; another set is used for pipe 2½ inches and larger. Dies are classified according to the number of threads per inch. *Table 2* shows proper die selection.

Once selected, the pipe thread dies must be installed and adjusted for different sizes of pipe. Pipe threading machines have universal die heads that adjust the dies for different sizes of pipe. *Figure 13* shows the die head details.

Each set of dies, except those used on ⅛-inch pipe, can be used for several sizes of pipe. To adjust the die head for different sizes of pipe, loosen the clamp lever and move the size bar until the line underneath the desired pipe size lines up with the index line. Tighten the clamp lever after

Table 2 Proper Die Selection

Nominal Pipe Size (inches)	Drill Threads (per inch)	Thread Length (inches)	Number of Threads
⅛	27	⅜	10
¼	18	⅝	11
⅜	18	⅝	11
½	14	¾	10
¾	14	¾	10
1	11½	⅞	10
1¼	11½	1	11
1½	11½	1	11
2	11½	1	11
2½	8	1½	12
3	8	1½	12
3½	8	1⅝	13
4	8	1⅝	13
5	8	1¾	14
6	8	1¾	14
8	8	1⅞	15
10	8	2	16
12	8	2⅛	17

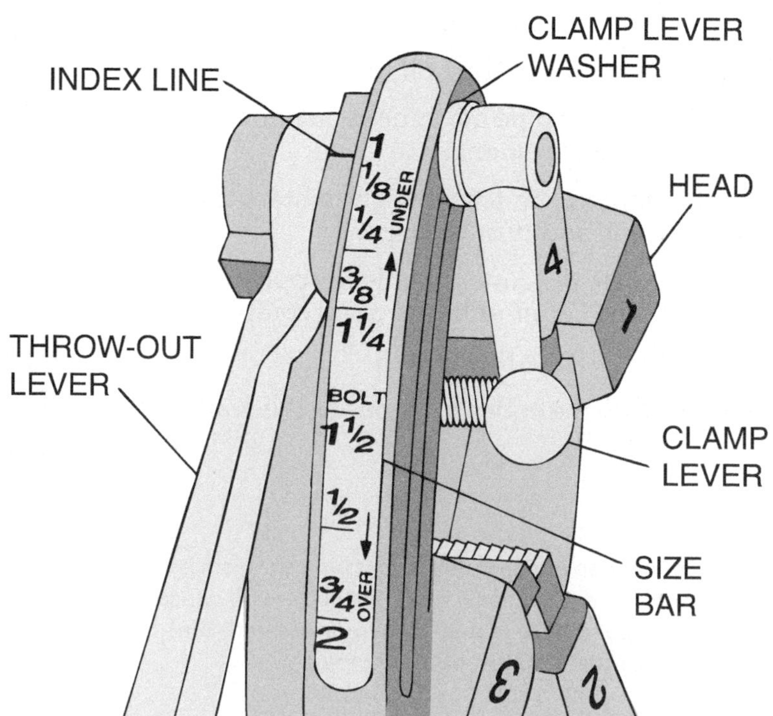

Figure 13 Die head details.

the adjustment has been made. If oversized or undersized threads are required, set the index line in the direction of the over or under mark on the size bar.

The throw-out lever, which is marked OPEN and CLOSE, is a quick-release handle used to close the dies into the threading position and open the dies to retract the dies away from the pipe. After threading a pipe, the operator can lift this handle to the open position and quickly remove the dies from the pipe. Before cutting threads with the pipe threading machine, the proper dies must be installed in the die head.

To replace the dies in a threading machine:

Step 1 Remove the die head from the machine and place it on a workbench with the numbers facing up.

Step 2 Open the throw-out lever.

Step 3 Loosen the clamp lever about three turns.

Step 4 Lift the tongue of the clamp lever washer up and out of the slot under the size bar.

Step 5 Slide the throw-out lever all the way to the end of the slot.

Step 6 Remove the old dies from the die head.

Step 7 Select a matched set of replacement dies.

Step 8 Clean and inspect the new dies.

Step 9 Insert the new dies in the die head.

NOTE

The numbers 1 through 4 on the dies must match the numbers on the slots of the die head.

Step 10 Slide the throw-out lever back to where the tongue of the clamp lever washer drops in the slot under the size bar.

Step 11 Adjust the die head size bar until the index line is lined up with the proper size mark on the size bar.

Step 12 Tighten the clamp lever.

Step 13 Install the die head into the machine.

4.2.2 Checking Thread Cutting Oil Level

Before any threading operations are performed, it is extremely important to check the level of the special thread cutting oil. Follow these steps to check the thread cutting oil level:

Step 1 Slide the chip pan out from the base of the threading machine.

Step 2 Check to see if the oil is up to the fill-level line in the reservoir. Fill the reservoir with cutting oil as necessary.

Step 3 Lower the lubrication arm over the open reservoir. Oil cannot flow from the lubrication arm in its upright position.

Step 4 Turn the power selector switch to the FORWARD position.

Step 5 Step on the foot switch. Oil should flow from the lubrication arm. If it does not, it may be necessary to prime the pump.

Step 6 Turn the power selector switch to the OFF position to stop the threading machine.

Step 7 Raise the lubrication arm and slide the chip pan back into the base of the threading machine.

4.2.3 Priming the Oil Pump

If the oil does not flow from the lubrication arm, it may be necessary to prime the pump. Follow these steps to prime the oil pump:

Step 1 Remove the button plug on the machine cover.

Step 2 Remove the primer screw through the opening.

Step 3 Fill the pump with cutting oil.

Step 4 Replace the primer screw and the button plug.

CAUTION

Failure to replace the primer screw and button plug causes the pump to drain itself immediately when the power is turned on.

Step 5 Turn the power on in the reverse direction and check the flow of oil from the lubrication arm. Running the machine in reverse primes the pump more efficiently than running the machine forward.

4.2.4 Threading Pipe

After checking the cutting oil level and priming the pump, the operator is ready to thread the pipe. Follow these steps to thread pipe:

Step 1 Load the pipe into the threading machine.

Step 2 Place pipe stands under the pipe as needed.

> **WARNING!**
>
> Do not wear loose clothing or jewelry. Keep hair pulled back while operating a threading machine. Wear a pair of properly fitting gloves to prevent cuts from sharp threads.

Step 3 Cut and ream the pipe to the required length.

Step 4 Make sure that the proper dies are in the die head.

Step 5 Loosen the clamp lever.

Step 6 Move the size bar to select the proper die setting for the size of pipe being threaded.

Step 7 Lock the clamp lever.

Step 8 Swing the die head down to the working position.

Step 9 Close the throw-out lever.

Step 10 Lower the lubrication arm and direct the oil supply onto the die.

Step 11 Turn the machine switch to the FORWARD position.

Step 12 Step on the foot switch to start the machine.

Step 13 Turn the carriage to bring the die against the end of the pipe.

Step 14 Apply light pressure on the handwheel to start the die.

Step 15 Release the handwheel once the dies have started to thread the pipe. The dies will automatically feed onto the pipe as they follow the newly cut threads.

> Make sure that the die is flooded with oil at all times while the die is cutting to prevent overheating the die and the pipe. Overheating can cause damage to the die and to the pipe threads.

Step 16 Open the throw-out lever as soon as two full threads extend from the back of the dies.

Step 17 Release the foot switch to stop the machine.

Step 18 Turn the carriage handwheel to back the die off the pipe.

Step 19 Swing the die and the oil spout up and out of the way.

Step 20 Screw a fitting that is the same size as the pipe you are threading onto the end of the pipe to check the threads.

> **NOTE**
>
> If the threads are cut correctly, you should be able to screw a fitting three to four revolutions by hand onto the new threads. If the threads are not deep enough, adjust the die using the clamp lever and repeat the threading operation.

Step 21 Turn the machine switch to the OFF position.

Step 22 Open the chuck and remove the pipe.

4.2.5 Cutting and Threading Nipples

A nipple is a piece of pipe less than 12 inches long that is threaded on each end. Threaded pipe longer than 12 inches is considered cut pipe. The nipple chuck is a useful tool for holding nipples in the threading machine when threading. A nipple chuck with inserts and adapters will thread nipples from $1/8$ inch to 2 inches in diameter. *Figure 14* shows a nipple chuck kit.

Follow these steps to cut and thread nipples:

Step 1 Load the pipe into the threading machine.

Step 2 Ream the end of the pipe.

Step 3 Thread the pipe.

Step 4 Measure from the threaded end the desired length of the nipple and mark a line on the pipe at this point.

Step 5 Cut the pipe at the mark.

Figure 14 Nipple chuck kit.

Sstep 6 Remove the pipe from the power drive chuck.

Step 7 Place the nipple chuck into the power drive chuck and tighten the jaws on the nipple chuck.

Step 8 Place the insert inside the end of the nipple chuck.

> **NOTE**
> Place the small end toward the outside if threading 1/8- to 3/4-inch pipe. Place the large end toward the outside if threading 1-inch pipe. Do not use an insert with pipe that is 1¼ inch and larger.

Step 9 Select a nipple chuck adapter of appropriate size, relevant to the size pipe being threaded.

> **NOTE**
> If threading 2-inch pipe, a nipple chuck adapter is not necessary.

Step 10 Screw the nipple chuck adapter into the nipple chuck.

Step 11 Tighten the nipple chuck adapter into the nipple chuck using a chuck wrench.

> **WARNING!**
> Do not start the threading machine with the chuck wrench over the nipple chuck adapter. The wrench could be thrown by the force of the machine.

Step 12 Screw the nipple into the end of the nipple chuck.

Step 13 Tighten the nipple chuck release collar using the chuck wrench.

Step 14 Ream the end of the nipple.

Step 15 Thread the nipple.

Step 16 Insert the chuck wrench into the release collar and turn the release collar to loosen the nipple from the nipple chuck.

Step 17 Unscrew the nipple from the nipple chuck.

> **WARNING!**
> Use gloves when handling the threads on the freshly cut nipple. These threads are extremely sharp and can cause cuts.

Step 18 Place the chuck wrench over the nipple chuck adapter and unscrew the adapter from the nipple chuck.

Step 19 Wipe off all excess cutting oil from the nipple chuck adapter and the insert. Store them with the nipple chuck kit.

Step 20 Release the power drive chuck to remove the nipple chuck from the threading machine.

Step 21 Wipe the nipple chuck and store it with the rest of the kit.

4.2.6 Threading Pipe Using a Geared Threader

A geared threader, also known as a mule, is used to thread pipe from 2½ to 6 inches in diameter (*Figure 15*). Two sets of dies are used with geared threaders. One set is used for pipe that is 2½ to 4 inches in diameter, and the other set is used for pipe that is 4 to 6 inches in diameter. Each die set contains five dies that are numbered just like the dies for the universal die head. The geared threader is connected to the pipe threading machine by a universal drive shaft.

> **WARNING!**
> A geared threader produces an enormous amount of torque while threading. It can easily tip over the threading machine if the machine is not properly secured. If possible, bolt the threading machine to a steel table and secure the pipe in a vise that is bolted to a steel table. Follow the manufacturer's instructions when using this tool.

4.2.7 Threading Machine Maintenance

Proper care of a threading machine will keep it in good working condition for a long period of time.

To optimize the length of service from a threading machine:

- Never reverse the rotation of the machine while it is running. Always turn off the power and wait until the drive has stopped rotating before changing the direction of the drive mechanism.

Figure 15 Geared threader.

- At the start of each day, clean the chuck jaws with a stiff brush to remove rust, scale, chips, pipe coating, or other foreign matter. Apply lubricating oil to the machine bedways, cutter rollers, and feed screw.
- Always use sharp dies, which produce smoother threads and require less motor power than dull ones.
- Refer to and follow the manufacturer's instructions for lubrication schedules for bearings and gears. The threading machine should be lubricated at least once for every 40 hours of running time.
- Make sure there is always plenty of cutting oil in the machine. The cutting oil should be periodically cleaned to remove accumulated sludge, chips, and other foreign matter. Replace the oil when it becomes dirty or contaminated.
- Remove the oil filter periodically and clean it with solvent. Blow the filter clean using compressed air. Do not operate the machine without the oil filter.
- Keep the chuck jaws in good condition and replace the jaw inserts when they become worn.

4.3.0 Portable Power Drives

A portable power drive (*Figure 16*) is a versatile handheld power unit that can be used for cutting threads, driving hoists and winches, and operating large valves. An electric motor provides the power to rotate a ring in the end of the tool. Threading dies and other accessories can be inserted into the circular ring, which has grooved spines to lock these accessories into place.

Threading die heads for pipe 2 inches and smaller either fit directly into the power drive or fit into a die head adapter and then into the power drive.

Usually, one hand can hold the power unit and control the power switch while oiling is done with the other hand. Follow these steps to thread pipe using a portable power drive:

Step 1 Use a vise to secure the pipe.

Step 2 Identify the size of the pipe to be threaded and insert the appropriate die into the portable power drive.

Step 3 Position the die head onto the end of the pipe with the face of the die stock facing away from the pipe (*Figure 17*). The drive unit should be positioned on the end of the pipe with the handle of the drive unit in a nearly vertical position above the pipe.

Step 4 Hold the handle of the drive unit with your right hand and center the die head onto the end of the pipe with your left hand.

Step 5 Pull the handle of the drive unit down sharply with your right hand while pushing the die stock against the end of the pipe with your left hand to make the dies bite into the end of the pipe.

Figure 16 Portable power drive.

Step 6 Continue to move the handle down to a position where you can straighten your elbow to provide a firm, strong grip on the drive unit. You should lower your right shoulder and balance your weight above the drive unit to resist the threading torque. If you are threading pipe larger than 1 inch, attach a torque arm to support the drive unit at this point. *Figure 18* shows the power drive support arm.

Step 7 Start the power drive and cut the thread.

CAUTION

Be sure to keep an adequate supply of cutting oil on the thread as it is being cut.

Step 8 Stop the power drive as soon as the thread has been cut to the required length.

Stpe 9 Turn the power switch to the reverse position and back the dies off the thread.

Step 10 Inspect the threads to make sure they have been cut properly.

Figure 17 Installing a die head.

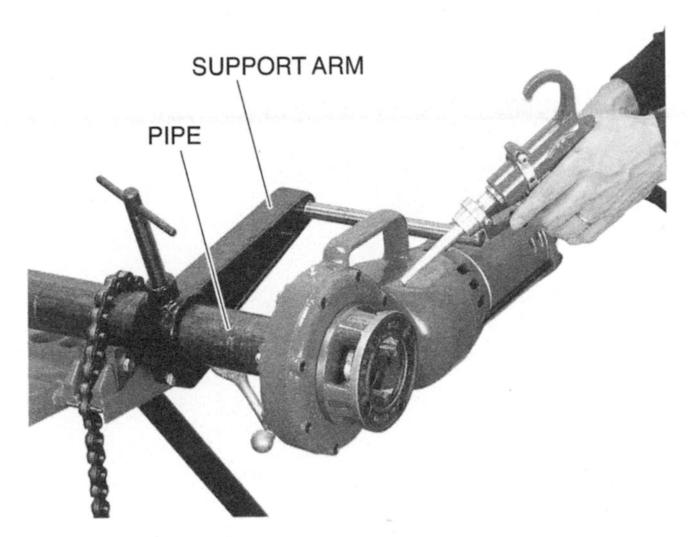

Figure 18 Power drive support arm.

4.0.0 Section Review

1. A chuck is normally used only when the pipe or cutter will be _____.
   a. powered off
   b. rotated
   c. disassembled
   d. positioned vertically

2. During a pipe threading operation, if oil does not flow from the lubrication arm, the pipefitter should _____.
   a. rethread the pipe
   b. install the die head
   c. prime the pump
   d. open the throw-out lever

SUMMARY

The safe use of power tools in pipefitting calls for understanding the task at hand and characteristics of available tools. Many times, the type of tool being used must match the type of material it is being used to cut, bend, or thread. Some pipe cutting processes involve the use of heat, which introduces additional safety measures. Pipe threading machines and other tools will last longer and operate more efficiently with proper care and maintenance, as well as proper use.

Review Questions

1. If a power cord has frayed insulation _____.
   a. it must be repaired with a special electrical tape
   b. fresh insulation must be melted into the opening
   c. it must be replaced by a licensed electrician
   d. it must be plugged into a GFCI outlet

2. Which of these statements regarding a GFCI is correct?
   a. It provides protection from all electrical shock hazards
   b. It does not provide protection from overheating
   c. It provides protection from line to line shocks
   d. It provides protection from ground faults caused by defective insulation

3. Which of the following materials would not be cut with a carbon steel blade?
   a. Aluminum
   b. Stainless steel
   c. Galvanized steel
   d. Copper

4. When grinding stainless steel, you should use a(n) _____ grinding wheel.
   a. stainless steel
   b. aluminum oxide
   c. carbon steel
   d. mild steel

5. A _____ should be used when cleaning a bevel on a pipe end.
   a. flapper wheel
   b. buffing wheel
   c. wire brush
   d. grinding wheel

6. When small-diameter pipe is being threaded, it should be secured with _____.
   a. a standard yoke vise
   b. a chain vise
   c. duct tape
   d. locking pliers

7. When preparing a pipe for socket welding, you should cut the pipe with a bandsaw rather than a _____.
   a. bandsaw
   b. pipe cutter
   c. chop saw
   d. demolition saw

8. Pipe threading dies are classified according to _____.
   a. weight
   b. pipe size
   c. threads per inch
   d. pipe material

9. Bevelers place the correct _____ on the end of a pipe before it can be welded.
   a. angle
   b. joint
   c. threading
   d. protection

10. To prime the cutting oil pump on a pipe threading machine, you should _____.
    a. run the machine forward
    b. run the machine in reverse
    c. simply fill the reservoir
    d. squirt oil on the die

11. If threads are cut correctly, you should be able to screw a fitting _____ turns onto a pipe by hand.
    a. ½
    b. one to two
    c. three to four
    d. five to six

12. A threading machine should be lubricated at least once for every _____ hours of running time.
    a. 8
    b. 40
    c. 80
    d. 100

13. A nipple is a piece of pipe that is less than _____ inches long and threaded on both ends.
    a. 2
    b. 6
    c. 8
    d. 12

14. A nipple chuck is used to _____.
    a. cut threads on nipples
    b. hold the nipple in place during threading
    c. ream nipples
    d. measure nipples

15. A mule is another name for a _____.
    a. geared pipe threader
    b. nipple chuck
    c. pipe stand
    d. portable power drive

Figure RQ01

16. The device shown in *Figure RQ01* is a _____.
    a. geared pipe threader
    b. ratcheting pipe threader
    c. threading collar
    d. portable power drive

17. Which of the following tasks cannot be done with a portable power drive?
    a. Cutting threads
    b. Driving a winch
    c. Operating valves
    d. Cutting pipe

18. Pipe bevelers are used to _____.
    a. prepare pipe ends for threading
    b. remove burrs from pipe ends
    c. prepare pipe ends for welding
    d. cut stainless steel pipe

19. Which of the following thermal joint preparation processes should not be used for beveling plate and pipe?
    a. Oxyfuel
    b. Carbon arc
    c. Plasma arc
    d. Both oxyfuel and plasma arc

Figure RQ02

20. The device shown in *Figure RQ02* is used to _____.
    a. locate the center of a pipe
    b. measure the outside diameter of a pipe
    c. secure pipe ends for welding
    d. cut and bevel pipe

Trade Terms Quiz

Fill in the blank with the correct term that you learned from your study of this module.

1. An angle cut or ground only on the edge of a piece of material is called a(n) _______.
2. A pipe or cutter is held in the machine by _______.
3. To cut male threads on a pipe, you would use a(n) _______.
4. The speed of revolutions of an object is measured in _______.
5. Power can be measured in _______, a unit equal to 745.7 watts.
6. An angle cut on the end of a piece of material is a(n) _______.
7. A detailed plan for required equipment inspections is called a(n) _______.
8. _______ is the act of putting components together.
9. Power tools should always be used with a(n) _______ to protect against electric shock.

Trade Terms

Assured equipment grounding conductor program
Bevel
Chamfer
Chucks
Die
Fabrication
Ground fault circuit interrupter (GFCI)
Horsepower (hp)
Revolutions per minute (rpm)

Trade Terms Introduced in This Module

Assured equipment grounding conductor program: A detailed plan specifying an employer's required equipment inspections and tests and a schedule for conducting those inspections and tests.

Bevel: An angle cut or ground on the end of a piece of solid material.

Chamfer: An angle cut or ground only on the edge of a piece of material.

Chucks: Parts of a machine that hold a piece of work tightly in the machine. A chuck is normally used only when the pipe or cutter will be rotated.

Die: A tool used to make male threads on a pipe or a bolt.

Fabrication: The act of putting together component parts to form an assembly.

Ground fault circuit interrupter (GFCI): A fast-acting circuit breaker that senses small imbalances in the circuit caused by current leakage to ground and, in a fraction of a second, shuts off the electricity.

Horsepower (hp): A unit of power equal to 745.7 watts or 33,000 foot-pounds per minute.

Revolutions per minute (rpm): The number of complete revolutions an object will make in one minute.

Additional Resources

This module presents thorough resources for task training. The following reference material is recommended for further study.

About the NEC®. National Fire Protection Association. 2018. Available at: **https://www.nfpa.org/NEC**

Hazard Alert: Safe Work with Power Saws. 2018. Electronic Library of Construction Occupational Safety & Health. Available at: **https://www.nfpa.org/NEC**

HAND and POWER TOOLS. Occupational Safety & Health. Available at: **https://www.osha.gov/Publications/osha3080.html**

Safe Operation of Portable Circular Power Saws. 2018. Available at: **http://nasdonline.org/1756/d001731/safe-operation-of-portable-circular-power-saws.html**

Tools and Their Uses, Latest Edition. Naval Education and Training Program and Development Center. Washington, DC: US Government Printing Offices.

Figure Credits

Courtesy of Tower Manufacturing Corporation, Figure 2

Tigair®, Figure SA01

DeWalt Industrial Tool Company, Figure 3

Milwaukee Electric Tool Corporation, Figure 4(A)

Ridge Tool Company (RIDGID®), Figure 4(B), Figure 12, Figure 13, Figure 15-Figure 17, Figure RQ01

Reed Manufacturing Company, Figure 18, Figure 19

Mathey Dearman, Figure 8

Section Review Answer Key

Section 1.0.0

Answer	Section Reference	Objective
1. c	1.1.1	1a
2. a	1.2.1	1b

Section 2.0.0

Answer	Section Reference	Objective
1. c	2.1.1	2a
2. d	2.2.2	2b

Section 3.0.0

Answer	Section Reference	Objective
1. a	3.0.0	3a

Section 4.0.0

Answer	Section Reference	Objective
1. b	4.0.0	4a
2. c	4.2.3	4b

NCCER CURRICULA — USER UPDATE

NCCER makes every effort to keep its textbooks up-to-date and free of technical errors. We appreciate your help in this process. If you find an error, a typographical mistake, or an inaccuracy in NCCER's curricula, please fill out this form (or a photocopy), or complete the online form at **www.nccer.org/olf**. Be sure to include the exact module ID number, page number, a detailed description, and your recommended correction. Your input will be brought to the attention of the Authoring Team. Thank you for your assistance.

Instructors – If you have an idea for improving this textbook, or have found that additional materials were necessary to teach this module effectively, please let us know so that we may present your suggestions to the Authoring Team.

NCCER Product Development and Revision

13614 Progress Blvd., Alachua, FL 32615

Email: curriculum@nccer.org
Online: www.nccer.org/olf

❑ Trainee Guide ❑ Lesson Plans ❑ Exam ❑ PowerPoints Other ____________

Craft / Level: Copyright Date:

Module ID Number / Title:

Section Number(s):

Description:

Recommended Correction:

Your Name:

Address:

Email: Phone:

This page is intentionally left blank.

Oxyfuel Cutting

Overview

Oxyfuel cutting is a method for cutting metal that uses an intense flame produced by burning a mixture of a fuel gas and pure oxygen. It is a versatile metal cutting method that has many uses on job sites. Because of the flammable gases and open flame involved, there is a danger of fire and explosion when oxyfuel equipment is used. However, these risks can be minimized when the operator is well-trained and knowledgeable about the function and operation of each part of an oxyfuel cutting outfit.

Module 29102

Trainees with successful module completions may be eligible for credentialing through the NCCER Registry. To learn more, go to **www.nccer.org** or contact us at **1.888.622.3720**. Our website has information on the latest product releases and training, as well as online versions of our *Cornerstone* magazine and Pearson's product catalog.

Your feedback is welcome. You may email your comments to **curriculum@nccer.org**, send general comments and inquiries to **info@nccer.org**, or fill in the User Update form at the back of this module.

This information is general in nature and intended for training purposes only. Actual performance of activities described in this manual requires compliance with all applicable operating, service, maintenance, and safety procedures under the direction of qualified personnel. References in this manual to patented or proprietary devices do not constitute a recommendation of their use.

29102 V5

29102
OXYFUEL CUTTING

Objectives

When you have completed this module, you will be able to do the following:

1. Describe oxyfuel cutting and identify related safe work practices.
   a. Describe basic oxyfuel cutting.
   b. Identify safe work practices related to oxyfuel cutting.
2. Identify and describe oxyfuel cutting equipment and consumables.
   a. Identify and describe various gases and cylinders used for oxyfuel cutting.
   b. Identify and describe hoses and various types of regulators.
   c. Identify and describe cutting torches and tips.
   d. Identify and describe other miscellaneous oxyfuel cutting accessories.
   e. Identify and describe specialized cutting equipment.
3. Explain how to set up, light, and shut down oxyfuel equipment.
   a. Explain how to properly prepare a torch set for operation.
   b. Explain how to leak test oxyfuel equipment.
   c. Explain how to light the torch and adjust for the proper flame.
   d. Explain how to properly shut down oxyfuel cutting equipment.
4. Explain how to perform various oxyfuel cutting procedures.
   a. Identify the appearance of both good and inferior cuts and their causes.
   b. Explain how to cut both thick and thin steel.
   c. Explain how to bevel, wash, and gouge.
   d. Explain how to make straight and bevel cuts with portable oxyfuel cutting machines.

Performance Tasks

Under the supervision of your instructor, you should be able to do the following:

1. Set up oxyfuel cutting equipment.
2. Light and adjust an oxyfuel torch.
3. Shut down oxyfuel cutting equipment.
4. Disassemble oxyfuel cutting equipment.
5. Change empty gas cylinders.
6. Cut shapes from various thicknesses of steel, emphasizing:
   - Straight line cutting
   - Square shape cutting
   - Piercing
   - Beveling
   - Cutting slots
7. Perform washing.
8. Perform gouging.
9. Use a track burner to cut straight lines and bevels.

Trade Terms

Backfire
Carburizing flame
Drag lines
Dross
Ferrous metals
Flashback
Gouging
Kerf
Neutral flame
Oxidizing flame
Pierce
Soapstone
Washing

Industry Recognized Credentials

If you are training through an NCCER-accredited sponsor, you may be eligible for credentials from NCCER's Registry. The ID number for this module is 29102. Note that this module may have been used in other NCCER curricula and may apply to other level completions. Contact NCCER's Registry at 888.622.3720 or go to **www.nccer.org** for more information.

Contents

Topics to be presented in this module include:

Contents (continued)

Figures and Tables

Figures and Tables (continued)

SECTION ONE

1.0.0 OXYFUEL CUTTING BASICS

Objective

Describe oxyfuel cutting and identify related safe work practices.

a. Describe basic oxyfuel cutting.
b. Identify safe work practices related to oxyfuel cutting.

Trade Terms

Dross: The material (oxidized and molten metal) that is expelled from the kerf when cutting using a thermal process. It is sometimes called slag.

Ferrous metals: Metals containing iron.

Soapstone: Soft, white stone used to mark metal.

In order to perform oxyfuel cutting in a safe and effective manner, it is critical to understand the basic principles of oxyfuel cutting and be thoroughly familiar with safe work practices associated with the process.

1.1.0 The Oxyfuel Cutting Process

Oxyfuel cutting, also called flame cutting or burning, is a process that uses the flame and oxygen from a cutting torch to cut ferrous metals. The flame is produced by burning a fuel gas mixed with pure oxygen. The flame heats the metal to be cut to the kindling temperature (a cherry-red color); then a stream of high-pressure pure oxygen is directed from the torch at the metal's surface. This causes the metal to instantaneously oxidize or burn away. The cutting process results in oxides that mix with molten iron and produce dross, which is blown from the cut by the jet of cutting oxygen. This oxidation process, which takes place during the cutting operation, is similar to an accelerated rusting process. *Figure 1* shows an operator performing oxyfuel cutting.

The oxyfuel cutting process is usually performed only on ferrous metals such as straight carbon steels, which oxidize rapidly. This process can be used to quickly cut, trim, and shape ferrous metals, including the hardest steel.

Oxyfuel cutting can be used for certain metal alloys, such as stainless steel; however, the process requires higher preheat temperatures (white

29102-15_F01.EPS

Figure 1 Oxyfuel cutting.

heat) and about 20 percent more oxygen for cutting. In addition, sacrificial steel plate or rod may have to be placed on top of the cut to help maintain the burning process. Other methods, such as carbon arc cutting, powder cutting, inert gas cutting, and plasma arc cutting, are much more practical for cutting steel alloys and nonferrous metals. Some of these methods are covered in other modules.

1.2.0 Oxyfuel Safety Summary

Cutting activities present unique hazards depending upon the material being cut and the fuel used to power the equipment. The proper safety equipment and precautions must be used when working with oxyfuel equipment because of the potential danger from the high-pressure flammable gases and high temperatures used. The following is a summary of safety procedures and practices that must be observed while cutting or welding. Keep in mind that this is just a summary. Complete safety coverage is provided in the *Welding Safety* module. Trainees who have not completed that module should do so before continuing.

1.2.1 Protective Clothing and Equipment

Oxyfuel cutting produces intense light and heat. It can also produce flying sparks and toxic fumes. To avoid injury, operators must wear appropriate personal protective equipment (PPE) when performing oxyfuel cutting operations. The following is a list of safety practices related to protective clothing and equipment that operators should follow:

- Always use safety glasses with a full face shield or a helmet (*Figure 2*). The glasses, face shield, or helmet lens must have the proper light-reducing tint for the type of cutting to be performed.
- Wear proper protective leather and/or flame-retardant clothing along with welding gloves that protect from flying sparks and molten metal as well as heat.
- Wear high-top safety shoes or boots. Make sure that the tongue and lace area of the footwear will be covered by a pant leg. If the tongue and lace area is exposed or the footwear must be protected from burn marks, wear leather spats under the pants or chaps and over the top of the footwear.
- Wear a 100-percent cotton cap with no mesh material included in its construction. The bill of the cap points to the rear. If a hard hat is required for the environment, use one that allows the attachment of rear deflector material and a face shield. A hard hat with a rear deflector is generally preferred when working overhead, and may be required by some employers and job sites.

WARNING!

Do not wear a cap with a button in the middle. The conductive metal button beneath the fabric represents a safety hazard.

- Wear a face shield over safety glasses for cutting. Either the face shield or the lenses of the safety glasses must be an approved shade for the application. A welding hood equipped with a properly tinted lens is also acceptable. A shade 3 to 6 filter is recommended, depending on the thickness of the metal being cut, as required by *ANSI Z49.1*.
- Wear earplugs to protect ear canals from sparks. Wear hearing protection to protect against the consistent sound of the torch.

WARNING!

Ear protection is essential to protect ears from the noise of the torch. Other personal protective equipment (PPE) must be worn to protect the operator from hot metal and slag.

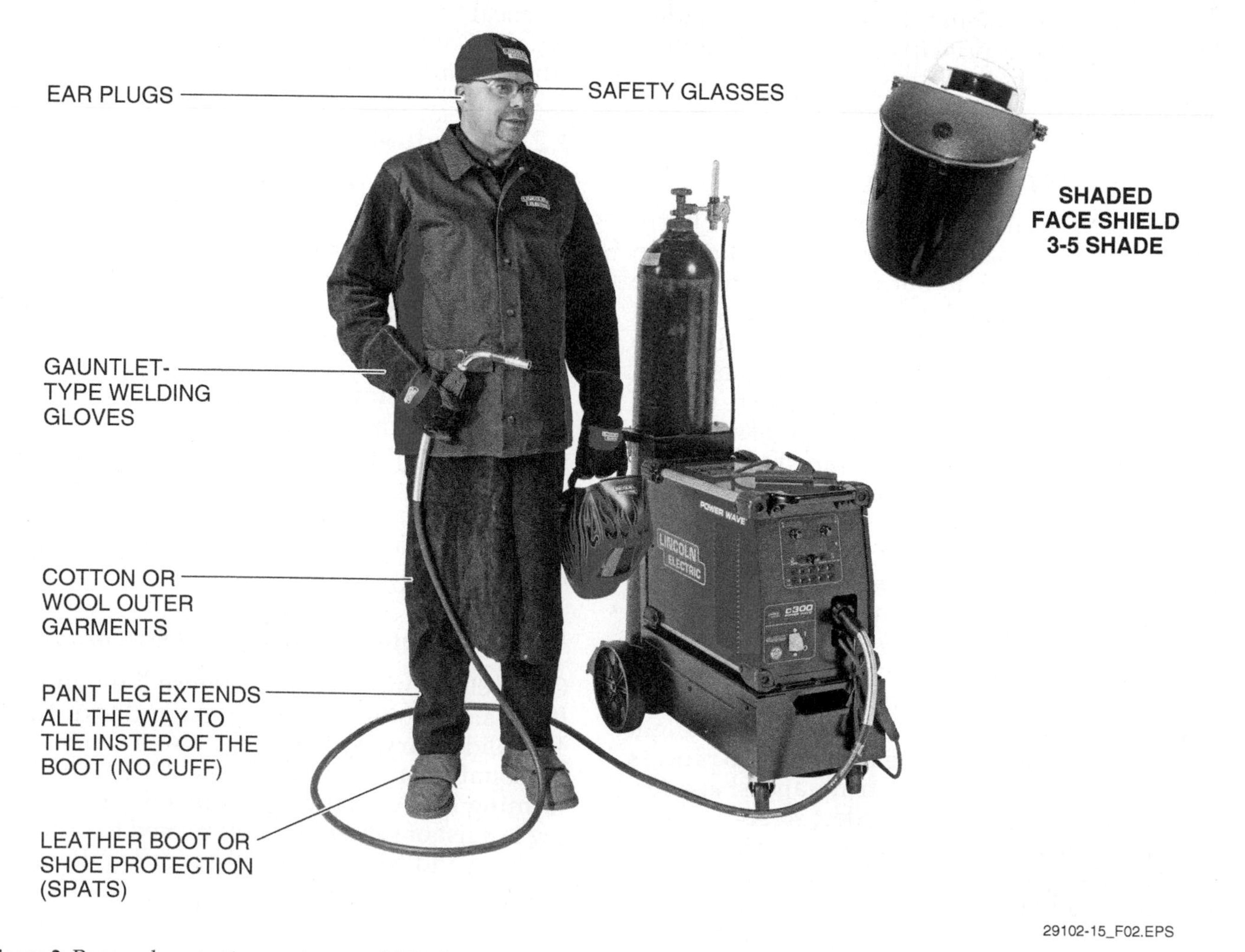

Figure 2 Personal protective equipment (PPE) for oxyfuel cutting.

- Cutting operations involving materials or coatings containing cadmium, mercury, lead, zinc, chromium, and beryllium result in toxic fumes. For long-term cutting of such materials, always wear an approved full face, supplied-air respirator (SAR) that uses breathing air supplied externally of the work area. For occasional, very short-term exposure to zinc or copper fumes, a high-efficiency particulate arresting (HEPA)-rated or metal-fume filter may be used on a standard respirator.

1.2.2 Fire/Explosion Prevention

Most welding environment fires occur during oxyfuel gas welding or cutting. To minimize fire and explosion hazards, all cutting should be done in designated areas of the shop if possible. These areas should be made safe for cutting operations with concrete floors, arc filter screens, protective drapes, and fire extinguishers. A welding blanket (*Figure 3*) can be used to protect items in the area that would otherwise be damaged. No combustibles should be stored nearby. The work area should be kept neat and clean, and any metal scrap or dross must be cold before disposal.

Operators should be well-trained in the function and operation of each part of an oxyfuel gas welding or cutting station. In addition, it is often required that at least one fire watch be posted with an extinguisher to watch for possible fires.

The following list contains other steps that operators should follow to help prevent fires and explosions.

- Never carry matches or gas-filled lighters. Sparks can cause the matches to ignite or the lighter to explode, resulting in serious injury.
- Always comply with any site requirement for a hot-work permit and/or a fire watch.
- Never use oxygen to blow off clothing. The oxygen can remain trapped in the fabric for a time. If a spark hits the clothing during this time, the clothing can burn rapidly and violently out of control.
- Never release a large amount of oxygen or use oxygen in place of compressed air. Its presence around flammable materials or sparks can cause rapid and uncontrolled combustion. Keep pure oxygen away from oil, grease, and other petroleum products.
- Make sure that any flammable material in the work area is moved or shielded by a fire-resistant covering.

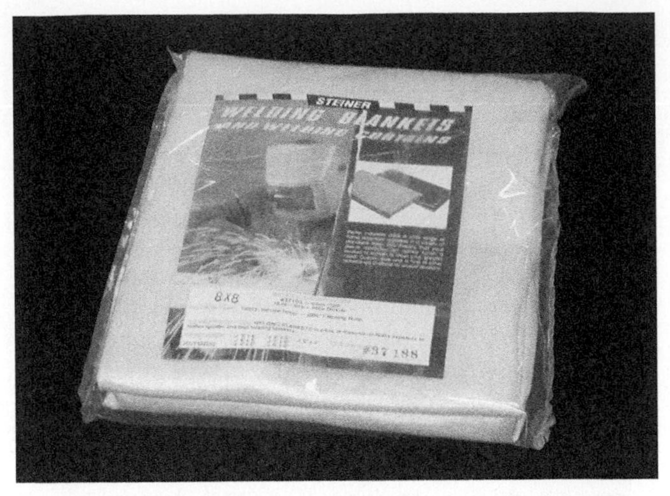

3,000°F (1,649°C) INTERMITTENT, 1,500°F (816°C) CONTINUOUS SILICON DIOXIDE CLOTH

29102-15_F03.EPS

Figure 3 Welding blanket.

- Approved fire extinguishers must be available before attempting any heating, welding, or cutting operations. Make sure the extinguisher is charged, the inspection tag is valid, and any individual that may be required to operate it knows how to do so.
- Never release a large amount of fuel gas, especially acetylene. Propylene and propane tend to concentrate in and along low areas and can ignite at a considerable distance from the release point. Acetylene is lighter than air but is even more dangerous. When mixed with air or oxygen, it will explode at much lower concentrations than any other fuel gas.
- To prevent fires, maintain a neat and clean work area, and make sure that any metal scrap or slag is cold before disposal.

Before cutting containers such as tanks or barrels, check to see if they have contained any explosive, hazardous, or flammable materials, including petroleum products, citrus products, or chemicals that decompose into toxic fumes when heated. Proper procedures for cutting or welding hazardous containers are described in the *American Welding Society (AWS) F4.1, Safe Practices for the Preparation of Containers and Piping for Welding and Cutting,* and *ANSI Z49.1.* As a standard practice, always clean and then fill any tanks or barrels with water, or purge them with a flow of inert gas such as nitrogen to displace any oxygen.

WARNING!

Welding or cutting must never be performed on drums, barrels, tanks, vessels, or other containers until they have been emptied and cleaned thoroughly, eliminating all flammable materials and all substances (such as detergents, solvents, greases, tars, or acids) that might produce flammable, toxic, or explosive vapors when heated. Do not assume that a container that has held combustibles is clean and safe until proven so by proper tests. Do not weld in places where dust or other combustible particles are suspended in air or where explosive vapors are present.

Containers must be cleaned by steam cleaning, flushing with water, or washing with detergent until all traces of the material have been removed.

WARNING!

Clean containers only in well-ventilated areas. Vapors can accumulate during cleaning, causing explosions or injury.

After cleaning the container, fill it with water (*Figure 4*) or a purge gas, such as carbon dioxide, argon, or nitrogen to displace the explosive fumes. Air, which contains oxygen, is displaced from inside the container by the water or inert gas. Without oxygen, combustion cannot take place.

A water-filled vessel is the best alternative. When using water, position the container to minimize the air space. When using an inert gas, provide a vent hole so the inert gas can push the air and other vapors out to the atmosphere. Keep in mind that these precautions do not guarantee the absence of flammable materials inside. For that reason, these types of activities should not be done without adequate supervision and the use of proper testing methods.

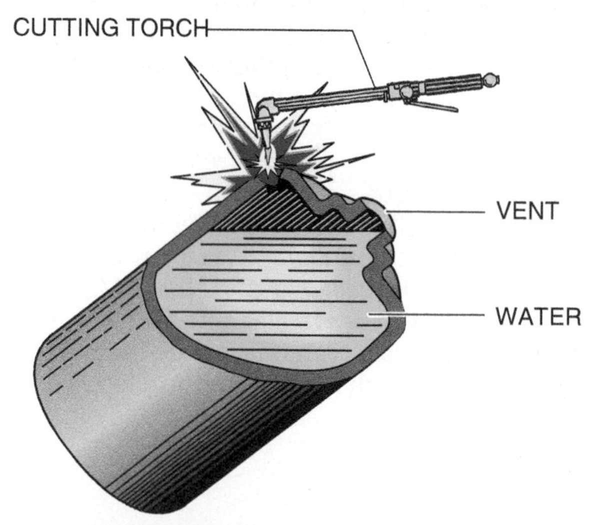

Figure 4 Eliminating/minimizing oxygen in a container.

1.2.3 Work Area Ventilation

Cutting operations should always be performed in a well-ventilated area. This greatly reduces the risks associated with toxic fumes. This rule is especially true for confined spaces. Confined-space procedures must be followed before cutting begins. Proper ventilation must be provided before any cutting procedures take place, but never use oxygen for ventilation purposes.

Work area safety rules apply to all workers in the area. In a typical work area, there might be numerous workers performing various tasks. Always remain aware of other personnel in the area and take the necessary precautions to ensure their safety as well as your own.

1.2.4 Cylinder Handling and Storage

Operators must be aware of the hazards involved in the use of fuel gas, oxygen, and shielding gas cylinders and know how to store these cylinders are stored safely. One basic rule is that only compressed gas cylinders containing the correct gas for the process should be used. Regulators must be correct for the gas and pressure and must function properly. All hoses, fittings, and other parts must be suitable and maintained in good condition.

Any cylinder that leaks, bad valves, or damaged threads must be identified and reported to the supplier. Use a piece of **soapstone** or a tag to write the problem on the cylinder. If closing the cylinder valve cannot stop the leak, move the cylinder outdoors to a safe location, away from any source of ignition, and notify the supplier. Post a warning sign and then slowly release the pressure.

In its gaseous form, acetylene is extremely unstable and explodes easily. For this reason it must remain at pressures below 15 pounds per square inch (psi) or 103 kilopascals (kPa). If an acetylene cylinder is tipped over, stand it upright and wait at least an hour before using it. If the cylinder has spent a significant period of time laying on its side,

Oxygen Consumption

Two-thirds of the oxygen in a neutral flame comes from the air you breathe inside the work space. This could be an issue in a heavily occupied, tight work space.

it is best to allow at least two hours to ensure all of the liquid has drained down to the base of the cylinder. Acetylene cylinders contain liquid acetone. If the liquid is withdrawn from a cylinder, it will foul the safety check valves and regulators and decrease the stability of the acetylene stored in the cylinder. For this reason, acetylene must never be withdrawn at a per-hour rate that exceeds one-seventh of the volume of the cylinder(s) in use. Acetylene cylinders in use should be opened no more than one and one-half turns and, preferably, no more than three-fourths of a turn.

Other precautions associated with gas cylinders include the following:

- Keep cylinders in the upright position and securely chained to an undercarriage or fixed support so that they cannot be knocked over accidentally. Even though they are more stable, cylinders attached to a gas distribution manifold should be chained or otherwise confined, as should cylinders stored in a special room used only for cylinder storage.
- Keep cylinders that contain combustible gases in one area of the building for safety. Cylinders must be at a safe distance from arc welding or cutting operations and any other source of heat, sparks, or flame.
- Cylinder storage areas must be located away from halls, stairwells, and exits so that in case of an emergency they will not block an escape route. Storage areas should also be located away from heat, radiators, furnaces, and welding sparks. The location of storage areas should be where unauthorized people cannot tamper with the cylinders. A warning sign that reads "Danger—No Smoking, Matches, or Open Lights", or similar wording, should be posted in the storage area.
- Oxygen and fuel gas cylinders or other flammable materials must be stored separately. The storage areas must be separated by 20 feet (6.1 m) or by a wall 5 feet (1.5 m) high with at least a 30-minute burn rating. The purpose of the distance or wall is to keep the heat of a small fire from causing the oxygen cylinder safety valve to release. If the safety valve releases the oxygen, a small fire could quickly become an inferno.
- Inert gas cylinders may be stored separately or with either fuel cylinders or oxygen cylinders. Empty cylinders must be stored separately from full cylinders, although they may be stored in the same room or area. All cylinders must be stored vertically and have the protective caps screwed on firmly.
- Never allow a welding electrode, electrode holder, or any other electrically energized parts to come in contact with the cylinder.
- When opening a cylinder valve, operators should stand with the valve stem between themselves and the regulator.
- If a cylinder is frozen to the ground, use warm water (not boiling) to loosen it.

Cylinders equipped with a valve protection cap must have the cap in place unless the cylinder is in use. The protection cap prevents the valve from being broken off if the cylinder is knocked over. If the valve of a full high-pressure cylinder (such as argon or oxygen) is broken off, the cylinder can take flight like a missile if it has not been secured properly. Never lift a cylinder by the safety cap or valve. The valve can easily break off or be damaged. When moving cylinders, the valve protection cap must be replaced, especially if the cylinders are mounted on a truck or trailer. Cylinders must never be dropped or handled roughly.

WARNING!

Using a wrench inserted through the cap can open the valve. If the cap is stuck, use a strap wrench to remove, or call the gas supplier.

Additional Resources

ANSI Z49.1, Safety in Welding, Cutting, and Allied Processes. Miami, FL: American Welding Society.

Uniweld Products, Inc. Numerous videos are available at **www.uniweld.com/en/uniweld-videos**. Last accessed: November 30, 2014.

The Harris Products Group, a division of Lincoln Electric. Numerous videos are available at **www.harrisproductsgroup.com/en/Expert-Advice/videos.aspx**. Last accessed: November 30, 2014.

1.0.0 Section Review

1. The oxyfuel cutting process creates oxides that mix with molten iron and produce dross, which is blown from the cut by a jet of ____ .
   a. pure nitrogen
   b. pure oxygen
   c. carbon dioxide
   d. cooling water

2. Carbon dioxide, argon, and nitrogen are all gases that are suitable for ____ .
   a. mixing with a fuel gas to enable oxyfuel cutting
   b. oxidizing dross to create reusable slag
   c. filling and pressurizing confined spaces
   d. purging explosive fumes from containers

Section Two

2.0.0 Oxyfuel Cutting Consumables and Equipment

Objective

Identify and describe oxyfuel cutting equipment and consumables.

a. Identify and describe various gases and cylinders used for oxyfuel cutting.
b. Identify and describe hoses and various types of regulators.
c. Identify and describe cutting torches and tips.
d. Identify and describe other miscellaneous oxyfuel cutting accessories.
e. Identify and describe specialized cutting equipment.

Trade Terms

Backfire: A loud snap or pop as a torch flame is extinguished.

Flashback: The flame burning back into the tip, torch, hose, or regulator, causing a high-pitched whistling or hissing sound.

Gouging: The process of cutting a groove into a surface.

Kerf: The gap produced by a cutting process.

Washing: A term used to describe the process of cutting out bolts, rivets, previously welded pieces, or other projections from the metal surface.

The equipment used to perform oxyfuel cutting includes oxygen and fuel gas cylinders, oxygen and fuel gas regulators, hoses, and a cutting torch. A typical movable oxyfuel (oxyacetylene) cutting outfit is shown in *Figure 5*.

2.1.0 Cutting Gases

Many oxyfuel cutting outfits use oxygen and acetylene. However, other fuel gases including natural gas and liquefied gases are also used with oxygen for cutting. This section examines some of the more common cutting gases.

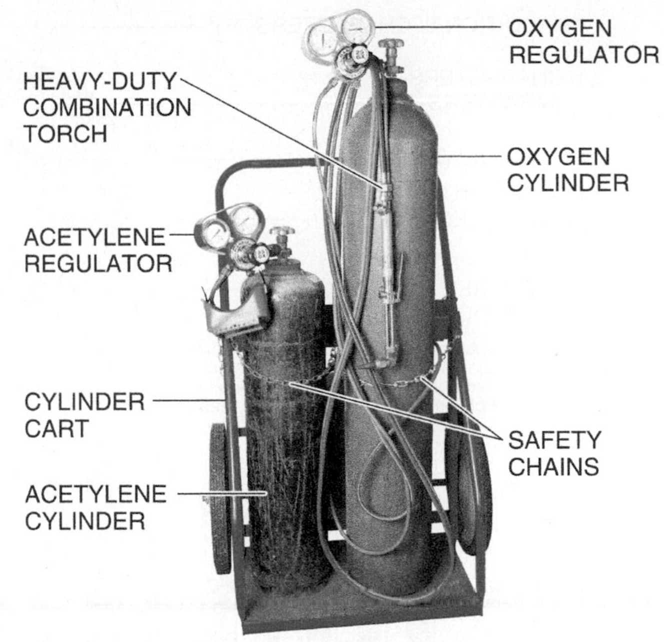

Figure 5 Typical oxyfuel welding/cutting outfit.

2.1.1 Oxygen

Oxygen (O_2) is a colorless, odorless, tasteless gas that supports combustion. It is not considered a fuel gas, but is necessary for combustion. Combined with burning material, pure oxygen causes a fire to flare and burn out of control. When mixed with fuel gases, oxygen produces the high-temperature flame required to flame-cut metals.

Oxygen is stored at more than 2,000 psi (13,790 kPa) in hollow steel cylinders. The cylinders come in a variety of sizes based on different international standards. Specific details about the cylinders are typically regulated by the government agency for the country in which the cylinders are transported, such as the US Department of Transportation (DOT), Transport Canada (TC), and the Department for Transport (DfT) in Europe. *Figure 6* shows high-pressure oxygen cylinder markings and capacities in cubic feet based on US DOT specifications. The smallest standard cylinder holds about 85 cubic feet (2.4 cubic meters) of oxygen, and the largest ultra-high-pressure cylinder holds about 485 cubic feet (13.7 cubic meters). The most common oxygen cylinder size used for cutting operations is the 227-cubic foot (6.4-cubic meter) cylinder. It is more than 4' (1.2 m) tall and 9" (10.2 cm) in diameter. Regardless of their size and capacity, oxygen cylinders must be tested every 10 years in the United States. Other locations may have a different testing standard.

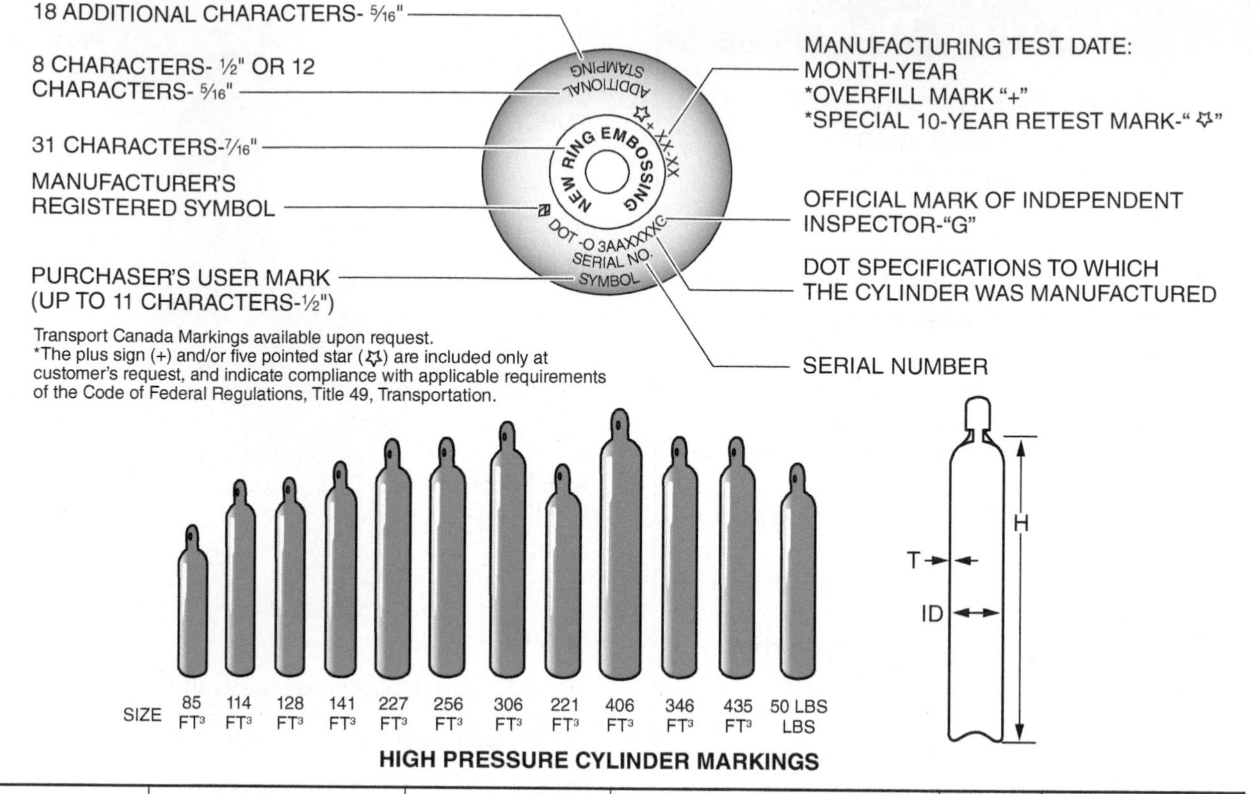

HIGH PRESSURE CYLINDER MARKINGS

DOT SPECIFICATIONS	O_2 CAPACITY (FT^3)		WATER CAPACITY (IN^3)		NOMINAL DIMENSIONS (IN)			NOMINAL WEIGHT (LB)	PRESSURE (PSI)	
	AT RATED SERVICE PRESSURE	AT 10% OVERCHARGE	MINIMUM	MAXIMUM	AVG. INSIDE DIAMETER "ID"	HEIGHT "H"	MINIMUM WALL "T"		SERVICE	TEST
STANDARD HIGH PRESSURE CYLINDERS[1]										
3AA2015	85	93	960	1040	6.625	32.50	0.144	48	2015	3360
3AA2015	114	125	1320	1355	6.625	43.00	0.144	61	2015	3360
3AA2265	128	140	1320	1355	6.625	43.00	0.162	62	2265	3775
3AA2015	141	155	1630	1690	7.000	46.00	0.150	70	2015	3360
3AA2015	227	250	2640	2710	8.625	51.00	0.184	116	2015	3360
3AA2265	256	281	2640	2710	8.625	51.00	0.208	117	2265	3775
3AA2400	306	336	2995	3060	8.813	55.00	0.226	140	2400	4000
3AA2400	405	444	3960	4040	10.060	56.00	0.258	181	2400	4000
ULTRALIGHT® HIGH PRESSURE CYLINDERS[1]										
E-9370-3280	365	NA	2640	2710	8.625	51.00	0.211	122	3280	4920
E-9370-3330	442	NA	3181	3220	8.813	57.50	0.219	147	3330	4995
ULTRA HIGH PRESSURE CYLINDERS[2]										
3AA3600	347[3]	374	2640	2690	8.500	51.00	0.336	170	3600	6000
3AA6000	434[3]	458	2285	2360	8.147	51.00	0.568	267	6000	10000
E-10869-4500	435[3]	NA	2750	2890	8.813	51.00	0.260	148	4500	6750
E-10869-4500	485[3]	NA	3058	3210	8.813	56.00	0.260	158	4500	6750

1. Regulators normally permit filling these cylinders with 10% overcharge, provided certain other requirements are met.
2. Under no circumstances are these cylinders to be filled to a pressure exceeding the marked service pressure at 70°F.
3. Nitrogen capacity at 70°F.

All cylinders normally furnished with 3/4" NGT internal threads, unless otherwise specified.
Nominal weights include neck ring but exclude valve and cap, add 2 lbs. (.91 kg) for cap and 1½ lb. (.8 kg) for valve.
Cap adds approximately 5 in. (127 mm) to height.
Cylinder capacities are approximately 5 in. (127 mm) to height.
Cylinder capacities are approximately at 70°F. (21°C).

29102-15_F06.EPS

Figure 6 High-pressure oxygen cylinder markings and sizes.

Oxygen cylinders have bronze cylinder valves on top (*Figure 7*). The cylinder valve controls the flow of oxygen out of the cylinder. A safety plug on the side of the cylinder valve allows oxygen in the cylinder to escape if the pressure in the cylinder rises too high. Although the escaping oxygen presents a hazard, the risk of explosion represents an even more significant hazard. Oxygen cylinders are usually equipped with Compressed Gas Association (CGA) Valve Type 540 valves for service up to 3,000 pounds per square inch gauge (psig), or 20,684 kPa. Some cylinders are equipped with CGA Valve Type 577 valves for up to 4,000 psig (27,579 kPa) oxygen service or CGA Valve Type 701 valves for up to 5,500 psig (34,473 kPa) service. Each CGA valve type is for a specific type of gas and pressure rating. Use care when handling oxygen cylinders because oxygen is stored at such high pressures. When it is not in use, always cover the cylinder valve with the protective steel safety cap and tighten it securely (*Figure 8*).

WARNING!

Do not remove the protective cap unless the cylinder is secured. If the cylinder falls over and the valve assembly breaks off, the cylinder will be propelled like a rocket, causing severe injury or death to anyone in its path.

2.1.2 Acetylene

Acetylene gas (C_2H_2), a compound of carbon and hydrogen, is lighter than air. It is formed by dissolving calcium carbide in water. It has a strong, distinctive, garlic-like odor, which is added to the gas intentionally so that it can be detected. In its gaseous form, acetylene is extremely unstable and explodes easily. Because of this instability, it cannot be compressed at pressures of more than 15 psi (103 kPa) when in its gaseous form. At higher pressures, acetylene gas breaks down chemically, producing heat and pressure that could result in a violent explosion. When combined with oxygen, acetylene creates a flame that burns hotter than 5,500°F (3,037°C), one of the hottest gas flames. Acetylene can be used for flame cutting, welding, heating, flame hardening, and stress relieving.

Because of the explosive nature of acetylene gas, it cannot be stored above 15 psi (103 kPa) in a hollow cylinder. To solve this problem, acetylene cylinders are specially constructed to store acetylene. The acetylene cylinder is filled with a porous material that creates a solid cylinder, instead of a hollow cylinder as used for all other common gases. The porous material is soaked with liquid acetone, which absorbs the acetylene gas, stabilizing it and allowing for storage at pressures above 15 psi (103 kPa).

Because of the liquid acetone inside the cylinder, acetylene cylinders must always be used in an upright position. If the cylinder is tipped over, stand the cylinder upright and wait at least one hour before using it. If liquid acetone is withdrawn from a cylinder, it will foul the safety check valves and regulators. It will also cause extremely unstable torch operation. Always take

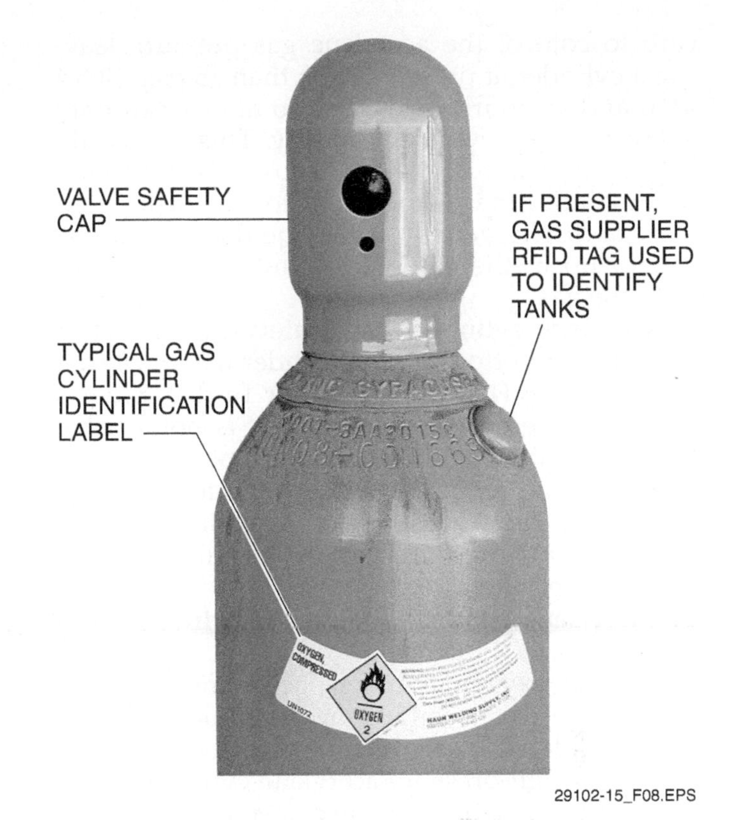

Figure 8 Oxygen cylinder with standard safety cap.

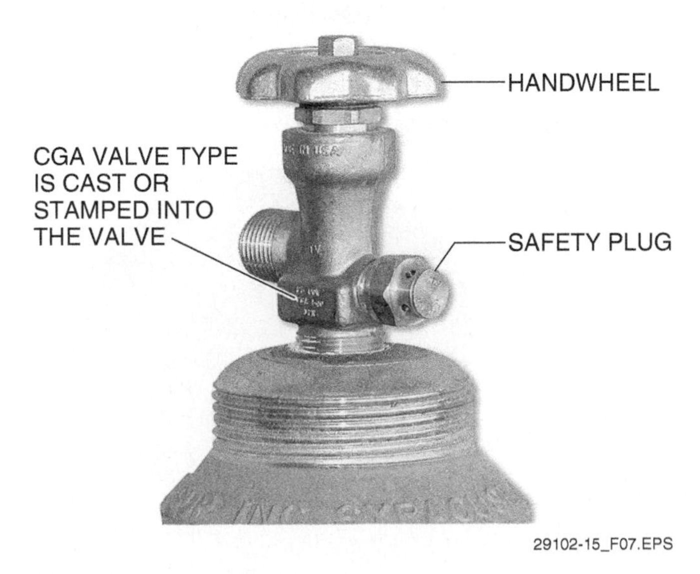

Figure 7 Oxygen cylinder valve.

care to control the acetylene gas pressure leaving a cylinder at pressures less than 15 psig (103.4 kPa) and at hourly rates that do not exceed one-seventh of the cylinder capacity. This can easily happen if a torch with a large nozzle is connected to a cylinder that is too small for the task. High rates of discharge may cause liquid acetone to be caught up in the gas stream and be carried out with the gas.

Acetylene cylinders have safety fuse plugs in the top and bottom of the cylinder (*Figure 9*) that melt at 212°F (100°C). A fuse, or fusible, plug is a type of pressure relief device. It is not a valve however; once it is activated, it cannot be reclosed and must be replaced. Fuse plugs, also known as rupture disks, are often used in place of relief valves for low-pressure applications like this one. In the event of a fire, the fuse plugs will release the acetylene gas, preventing the cylinder from exploding.

As with oxygen cylinders, acetylene cylinders are available in a variety of sizes and are typically regulated by government agencies. *Figure 10* shows high-pressure acetylene cylinder markings and capacities in cubic feet based on US DOT specifications. The smallest standard cylinder holds about 10 cubic feet (0.28 cubic meters) of gas. The largest standard cylinder holds about 420 cubic feet (11.9 cubic meters) of gas. A cylinder that holds about 850 cubic feet (24.1 cubic meters) is also available. Like oxygen cylinders, acetylene cylinders used in the United States must be tested every 10 years.

Acetylene cylinders are usually equipped with a CGA 510 brass cylinder valve. The valve controls the flow of acetylene from the cylinder into a regulator. Some acetylene cylinders are equipped with an alternate CGA 300 valve. Some obsolete valves still in use require a special long-handled wrench with a square socket end to operate the valve.

CYLINDER TOP FUSE PLUGS

VALVE HANDWHEEL

GAS SUPPLIER RFID TAG FOR CYLINDER IDENTIFICATION

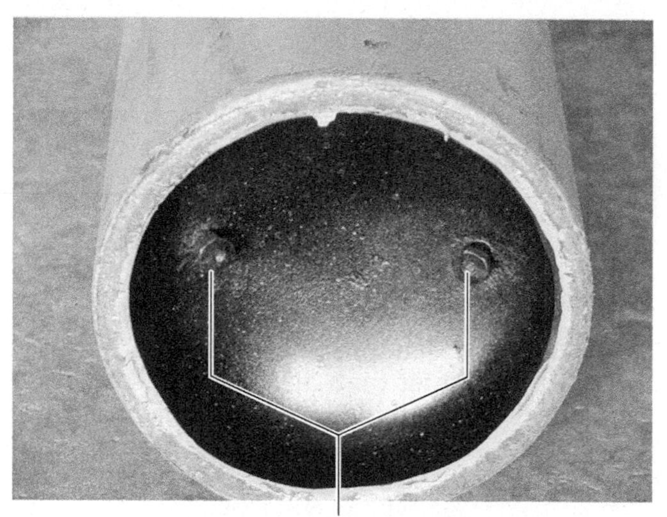

29102-15_F09.EPS

Figure 9 Standard acetylene cylinder valve and fuse plugs.

Lifetime Cylinder Management

Gas cylinders may be fitted with radio-frequency identification (RFID) tags so that they can be readily identified and tracked. The RFID tag is electronically scanned and a coded number is matched against the records for identification and tracking. This aids in quickly determining the identity of the purchaser or user of the cylinder, where it has been, and for determining the testing and maintenance records of the cylinder. The RFID tag may appear to be a button like the one shown here. However, it can also be concealed in the walls of the cylinder neck and covered for protection, or have a tight-fitting collar that snaps around the cylinder valve neck.

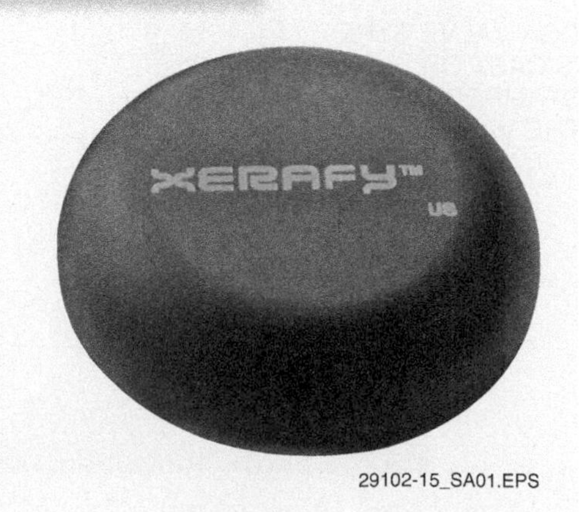

29102-15_SA01.EPS

Alternate High-Pressure Cylinder Valve Cap

High-pressure cylinders can be equipped with a clamshell cap that can be closed to protect the cylinder valve with or without a regulator installed on the valve. This enables safe movement of the cylinder after the cylinder valve is closed. This type of cap is usually secured to the cylinder body cap threads when it is installed so that it cannot be removed. When the clamshell is closed, it can also be padlocked to prevent unauthorized operation of the cylinder valve.

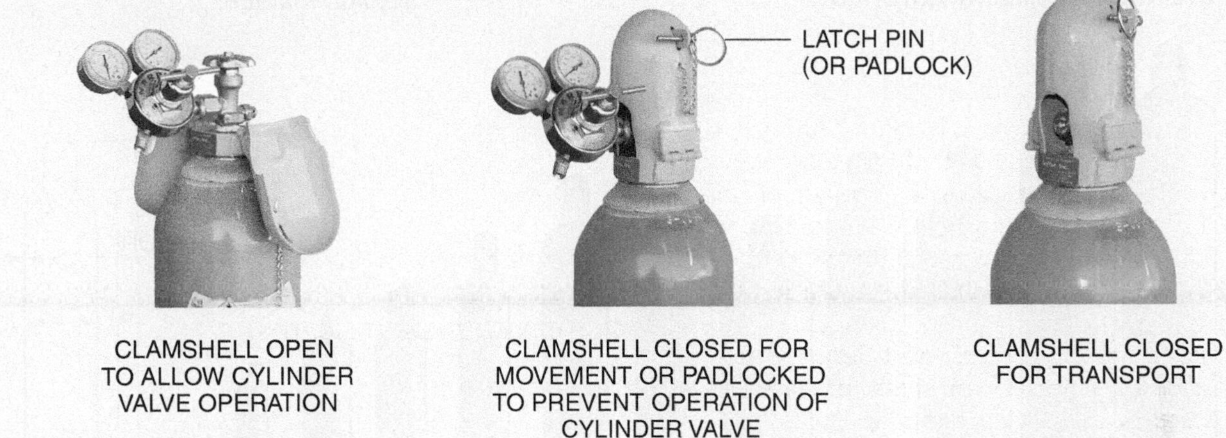

CLAMSHELL OPEN TO ALLOW CYLINDER VALVE OPERATION

CLAMSHELL CLOSED FOR MOVEMENT OR PADLOCKED TO PREVENT OPERATION OF CYLINDER VALVE

CLAMSHELL CLOSED FOR TRANSPORT

29102-15_SA02.EPS

Alternate Acetylene Cylinder Safety Cap

Acetylene cylinders can be equipped with a ring guard cap that protects the cylinder valve with or without a regulator installed on the valve. This enables safe movement of the cylinder after the cylinder valve is closed. This type of cap is usually secured to the cylinder body cap threads when it is installed so that it cannot be removed. Some other types of cylinders, such as propane cylinders, may also be fitted with this type of guard.

29102-15_SA03.EPS

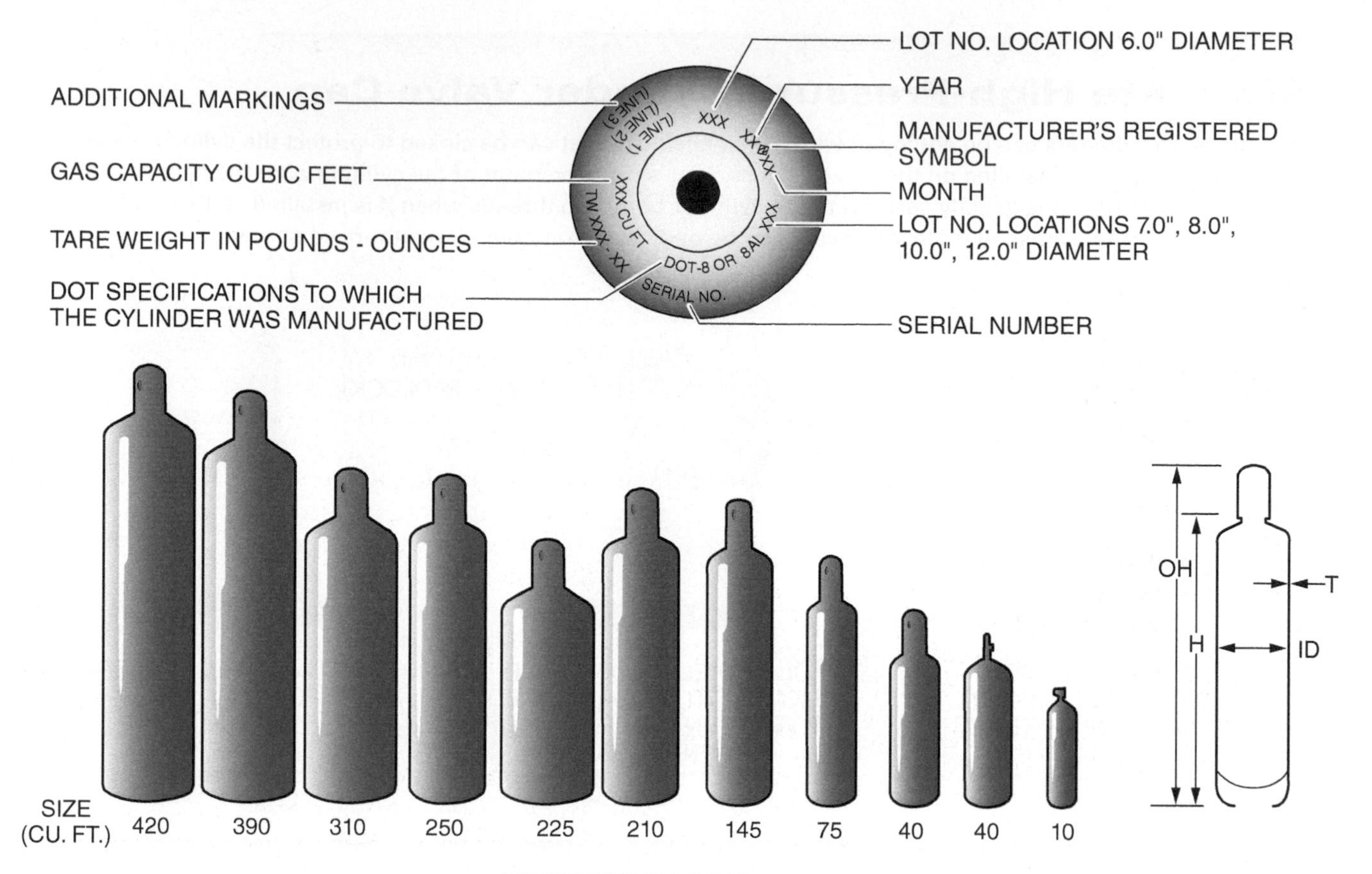

ACETYLENE CYLINDER MARKINGS

DOT SPECIFICATIONS	CAPACITY			NOMINAL DIMENSIONS (IN)				ACETONE (LB - OZ)	APPROXIATE TARE WEIGHT WITH VALVE WITHOUT CAP (LB)
	ACETYLENE	MIN. WATER							
	(FT^3)	(IN^3)	(LB.)	AVG. INSIDE DIAMETER "ID"	HEIGHT W/OUT VALVE OR CAP "H"	HEIGHT W/VALVE AND CAP "OH"	MINIMUM WALL "T"		
8 AL[1]	10	125	4.5	3.83	13.1375	14.75	0.0650	1-6	8
8[1]	40	466	16.8	6.00	19.8000	23.31	0.0870	5-7	25
8[2]	40	466	16.8	6.00	19.8000	28.30	0.0870	5-7	28
8[3]	75	855	30.8	7.00	25.5000	31.25	0.0890	9-8	45
8	100	1055	38.0	7.00	30.7500	36.50	0.0890	12-2	55
8	145	1527	55.0	8.00	34.2500	40.00	0.1020	18-10	76
8	210	2194	79.0	10.00	32.2500	38.00	0.0940	25-13	105
8AL	225	2630	94.7	12.00	27.5000	32.75	0.1280	29-6	110
8	250	2606	93.8	10.00	38.0000	43.75	0.0940	30-12	115
8AL	310	3240	116.7	12.00	32.7500	38.50	0.1120	39-5	140
8AL	390	4151	150.0	12.00	41.0000	46.75	0.1120	49-14	170
8AL	420	4375	157.5	12.00	43.2500	49.00	0.1120	51-14	187
8	60	666	24.0	7.00	25.79 OH		0.0890	7-11	40
8	130	1480	53.3	8.00	36.00 OH		0.1020	17-2	75
8AL	390	4215	151.8	12.00	46.00 OH		0.1120	49-14	180

1. Tapped for 3/8" valve but are not equipped with valve protection caps.
2. Includes valve protection cap.
3. Can be tared to hold 60 ft^3 (1.7 m^3) of acetylene gas.

Standard tapping (except cylinders tapped for 3/8") 3/4"-14 NGT.

Weight includes saturation gas, filler, paint, solvent, valve, fuse plugs. Does not include cap of 2 lb. (.91 kg.)
Cylinder capacities are based upon commercially pure acetylene gas at 250 psi (17.5 kg/cm^2), and 70°F (15°C).

29102-15_F10.EPS

Figure 10 Acetylene cylinder markings and sizes.

The smallest standard acetylene cylinder, which holds 10 cubic feet (0.28 cubic meters), is equipped with a CGA 200 small series valve, and 40 cubic foot (1.13 cubic meter) cylinders use a CGA 520 small series valve. As with oxygen cylinders, place a protective valve cap on the acetylene cylinders during transport (*Figure 11*).

WARNING!

Do not remove the protective cap unless the cylinder is secured. If the cylinder falls over and the nozzle breaks off, the cylinder will release highly explosive gas.

2.1.3 Liquefied Fuel Gases

Many fuel gases other than acetylene are used for cutting. They include natural gas and liquefied fuel gases such as propylene and propane. Their flames are not as hot as acetylene, but they have higher British thermal unit (Btu) ratings and are cheaper and safer to use. Job site policies typically determine which fuel gas to use. *Table 1* compares the flame temperatures of oxygen mixed with various fuel gases.

Propylene mixtures are hydrocarbon-based gases that are stable and shock-resistant, making them relatively safe to use. They are purchased under trade names such as High Purity Gas (HPG™), Apachi™, and Prestolene™. These gases and others have distinctive odors to make leak detection easier. They burn at temperatures around 5,193°F (2,867°C), hotter than natural gas and propane. Propylene gases are used for flame cutting, scarfing, heating, stress relieving, brazing, and soldering.

Propane is also known as liquefied petroleum (LP) gas. It is stable and shock-resistant, and it has a distinctive odor for easy leak detection. It burns at 4,580°F (2,526°C), which is the lowest temperature of any common fuel gas. Propane has a slight tendency toward **backfire** and **flashback** and is used quite extensively for cutting procedures.

Table 1 Flame Temperatures of Oxygen With Various Fuel Gases

Type of Gas	Flame Temperature
Acetylene	More than 5,500°F (3,038°C)
Propylene	5,130°F (2,832°C)
Natural gas	4,600°F (2,538°C)
Propane	4,580°F (2,527°C)

Natural gas is delivered by pipeline rather than by cylinders. Manifolds must be available on site for the connection of regulators and hoses. It burns at about 4,600°F (2,537°C). Natural gas is relatively stable and shock-resistant and has a slight tendency toward backfire and flashback. Because of its recognizable odor, leaks are easily detectable. Natural gas is used primarily for cutting on job sites with permanent cutting stations.

Liquefied fuel gases are shipped in hollow steel cylinders (*Figure 12*). When empty, they are much lighter than acetylene cylinders.

The hollow steel cylinders for liquefied fuel gases come in various sizes. They can hold from 30 to 225 pounds (13.6 to 102.1 kilograms) of fuel gas. As the cylinder valve is opened, the

29102-15_F11.EPS

Figure 11 Acetylene cylinder with standard valve safety cap.

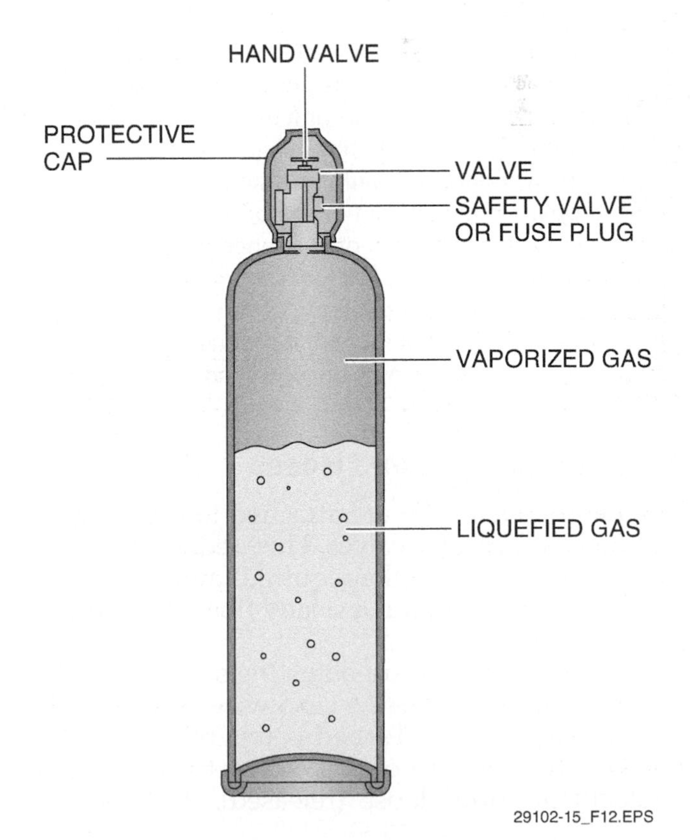

29102-15_F12.EPS

Figure 12 Liquefied fuel gas cylinder.

vaporized gas is withdrawn from the cylinder. The remaining liquefied gas absorbs heat and releases additional vaporized gas. The pressure of the vaporized gas varies with the outside temperature. The colder the outside temperature, the lower the vaporized gas pressure will be. If high volumes of gas are removed from a liquefied fuel gas cylinder, the pressure drops, and the temperature of the cylinder will also drop. A ring of frost can form around the base of the cylinder due to the cooling effect as the liquid vaporizes and absorbs heat. If high withdrawal rates continue, the regulator may also start to ice up. If high withdrawal rates are required, special regulators with electric heaters should be used.

WARNING!

Never apply heat directly to a cylinder or regulator. This can cause excessive pressure, resulting in an explosion.

The pressure inside a liquefied fuel gas cylinder is not an indicator of how full or empty the cylinder is. The weight of a cylinder determines how much liquefied gas is left. Liquefied fuel gas cylinders are equipped with CGA 510, 350, or 695 valves, depending on the fuel and storage pressures.

WARNING!

Do not remove the protective cap on liquefied fuel gas cylinders unless the cylinder is secured. If the cylinder falls over and the nozzle breaks off, the cylinder will release highly explosive gas. Cylinders containing a liquid such as propane must be kept in an upright position. If the valve is broken or is opened with the cylinder horizontal, the fuel can emerge as a liquid that will shoot a long distance before it vaporizes. If it is ignited, it produces an extremely dangerous, uncontrolled flame.

2.2.0 Regulators and Hoses

Regulators (*Figure 13*) are attached to the oxygen and fuel gas cylinder valves. They reduce the high cylinder pressures to the required lower working pressures and maintain a steady flow of gas from the cylinder.

A regulator's pressure-adjusting screw controls the gas pressure. Turned clockwise, it increases the pressure of gas. Turned counterclockwise, it reduces the pressure of gas. When turned counterclockwise until loose (released), it stops the flow of gas. When the adjusting screw feels very loose, it is an indication that the end of the screw is no longer in contact with the regulating spring. However, the regulator adjusting screw should never be considered a shut-off valve. When shut-off is desired, use the cylinder valve. The regulator can remain at its set position, allowing any gas remaining between the cylinder valve and the regulator to escape.

Most regulators contain two gauges. The high-pressure or cylinder-pressure gauge indicates the actual cylinder pressure (upstream); the low-pressure gauge indicates the pressure of the gas leaving the regulator (downstream).

Oxygen regulators differ from fuel gas regulators. Oxygen regulators may have green markings and always have right-hand threads on all connections. The oxygen regulator's high-pressure gauge generally reads up to 3,000 psi (20,684 kPa) and includes a second scale that shows the amount of oxygen in the cylinder in terms of cubic feet. The low-pressure or working-pressure gauge may read 100 psi (689 kPa) or higher.

Fuel gas regulators may have red markings and usually have left-hand threads on all the connections. As a reminder that the regulator has left-hand threads, a V-notch may be cut into the corners of the fitting nut. These notches are visible on the fitting nut of the fuel gas regulator shown in *Figure 13*. The fuel gas regulator's high-pressure gauge usually reads up to 400 psi (2,758 kPa). The low-pressure or working-pressure gauge may read up to 40 psi (276 kPa). Acetylene gauges, however, are always red-lined at 15 psi (103 kPa) as a reminder that acetylene pressure should not be increased beyond that point.

Single-stage and two-stage regulators will be discussed in the following sections.

Handling and Storing Liquefied Gas Cylinders

Liquefied fuel gas cylinders have a safety valve built into the valve at the top of the cylinder. The safety valve releases gas if the pressure begins to rise. Use care when handling fuel gas cylinders because the gas in cylinders is stored at significant pressures. Cylinders should never be dropped or hit with heavy objects, and they should always be stored in an upright position. When not in use, the cylinder valve must always be covered with the protective steel cap.

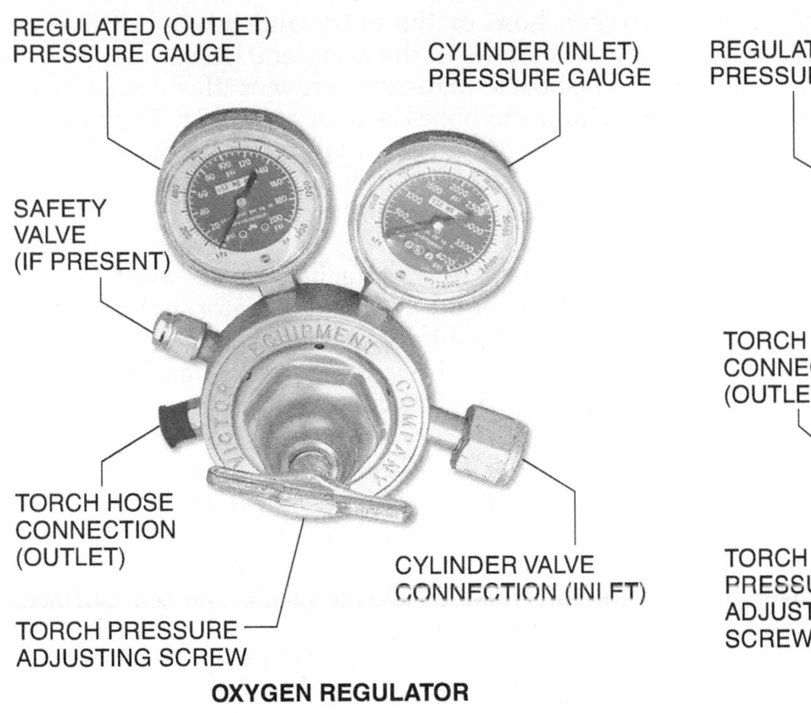

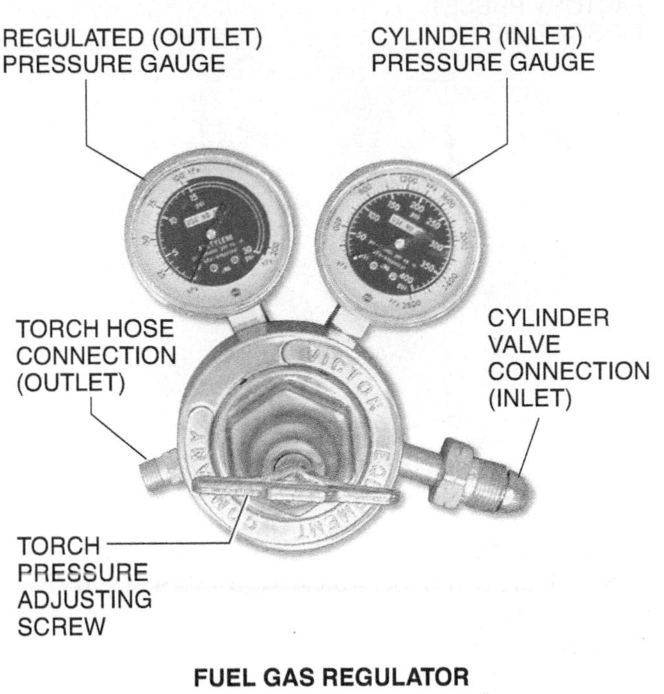

Figure 13 Oxygen and acetylene regulators.

WARNING!

To prevent injury and damage to regulators, always follow these guidelines:

- Never subject regulators to jarring or shaking, as this can damage the equipment beyond repair.
- Always check that the adjusting screw is fully released before the cylinder valve is turned on and when the welding has been completed.
- Always open cylinder valves slowly and stand with the valve stem between you and the regulator.
- Never use oil to lubricate a regulator. This can result in an explosion when the regulator is in use.
- Never use fuel gas regulators on oxygen cylinders, or oxygen regulators on fuel gas cylinders.
- Never work with a defective regulator. If it is not working properly, shut off the gas supply and have the regulator repaired by someone who is qualified to work on it.
- Never use large wrenches, pipe wrenches, pliers, or slip-joint pliers to install or remove regulators.

2.2.1 Single-Stage Regulators

Single-stage, spring-compensated regulators reduce pressure in one step. As gas is drawn from the cylinder, the internal pressure of the cylinder decreases. A single-stage, spring-compensated regulator is unable to automatically adjust for this decrease in internal cylinder pressure. Therefore, it becomes necessary to adjust the spring pressure periodically to modify the output gas pressure as the gas in the cylinder is consumed. These regulators are the most commonly used because of their low cost and high flow rates.

2.2.2 Two-Stage Regulators

The two-stage, pressure-compensated regulator reduces pressure in two steps. It first reduces the input pressure from the cylinder to a predetermined intermediate pressure. This intermediate pressure is then adjusted by the pressure-adjusting screw. With this type of regulator, the delivery pressure to the torch remains constant, and no readjustment is necessary as the gas in the cylinder is consumed. Standard two-stage regulators (*Figure 14*) are more expensive than single-stage regulators and have lower flow rates. There are also heavy-duty types with higher flow rates that are usually preferred for thick material and/or continuous-duty cutting operations.

2.2.3 Check Valves and Flashback Arrestors

Check valves and flashback arrestors (*Figure 15*) are safety devices for regulators, hoses, and torches. Check valves allow gas to flow in one direction only. Flashback arrestors stop fire from being able to travel backwards through the hose.

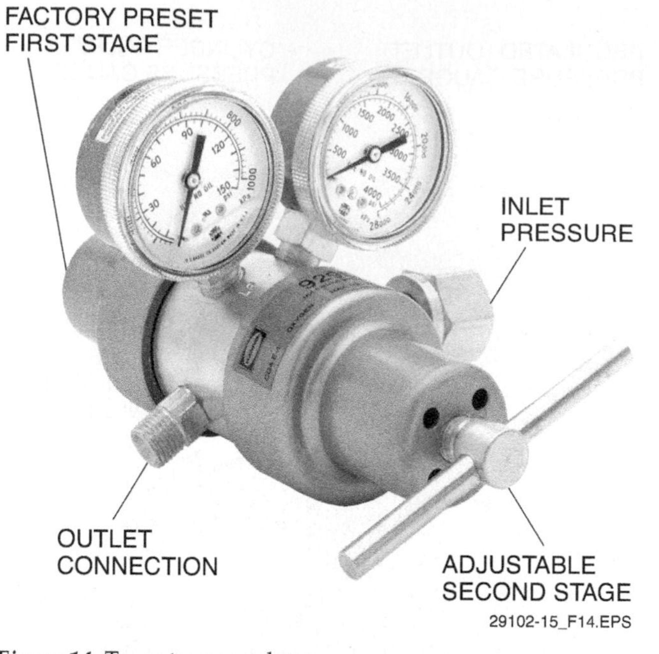

Figure 14 Two-stage regulator.

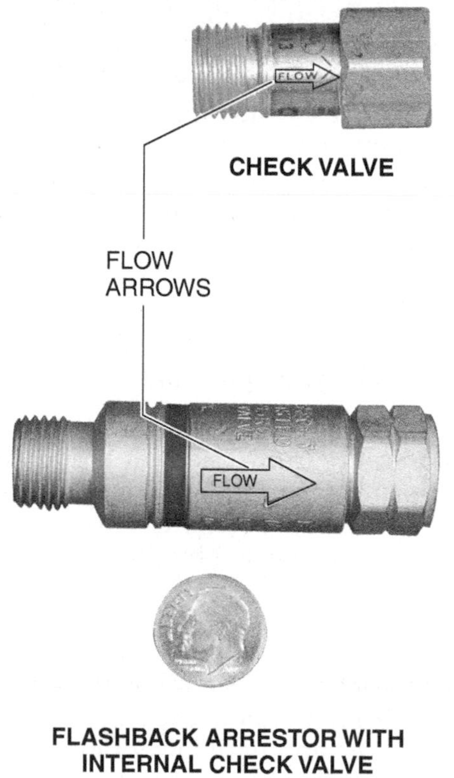

Figure 15 Add-on check valve and flashback arrestor.

Check valves consist of a ball and spring that open inside a cylinder. The valve allows gas to move in one direction but closes if the gas attempts to flow in the opposite direction. When a torch is first pressurized or when it is being shut off, back-pressure check valves prevent the entry and mixing of acetylene with oxygen in the oxygen hose or the entry and mixing of oxygen with acetylene in the acetylene hose.

Flashback arrestors prevent flashbacks from reaching the hoses and/or regulator. They have a flame-retarding filter that will allow heat, but not flames, to pass through. Most flashback arrestors also contain a check valve.

Add-on check valves and flashback arrestors are designed to be attached to the torch handle connections and to the regulator outlets. At a minimum, flashback arrestors with check valves should be attached to the torch handle connections. Both devices have arrows on them to indicate flow direction. When installing add-on check valves and flashback arrestors, be sure the arrow matches the desired gas flow direction.

The fittings for oxyfuel equipment are brass or bronze, and certain components are often fitted with soft, flexible, O-ring seals. The seal surfaces of the fittings or O-rings can be easily damaged by over-tightening with standard wrenches. For that reason, only a torch wrench (sometimes called a gang wrench) should be used to install regulators, hose connections, check valves, flashback arrestors, torches, and torch tips. Longer wrenches that provide more leverage should be avoided.

The universal torch wrench shown in *Figure 16* is equipped with various size wrench cutouts for use with a variety of equipment and standard CGA components. The length of a torch wrench is limited to reduce the chance of damage to fittings because of excessive torque. In some cases,

Serious Cutting

The cutting power of oxyfuel equipment can be very surprising. This 6" (15.2 cm) carbon steel block is no match for a worker with the right torch.

29102-15_SA04.EPS

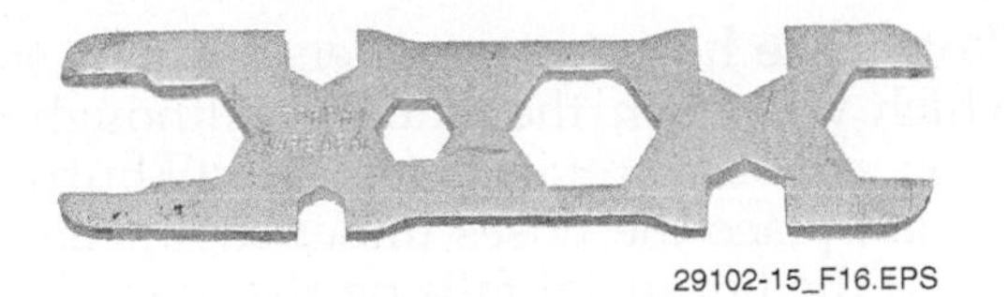
29102-15_F16.EPS

Figure 16 Universal torch wrench.

manufacturers specify only hand-tightening for certain fitting connections of a torch set (tips or cutting/welding attachments, for example). In any event, follow the manufacturer's specific instructions when connecting the components of a torch set.

2.2.4 Gas Distribution Manifolds

In some applications, cutting gases are used on a large scale. Instead of using a cylinder for each operator, a large bank of cylinders connected to a common manifold is used. *Figure 17* is a representation of such an arrangement. High-pressure hoses called pigtails are used to connect the gas supply tanks to the manifold. Regulators are provided at both the source and the workstation hookups.

Figure 18 shows a manifold setup that might be used in a pipe fabrication shop. Operators would tie-in with their hoses at the drops on the manifolds. Each of these manifolds has four drops. Each drop would be provided with a pressure gauge like the one shown. One function of this gauge is to provide an indication if there is a leak

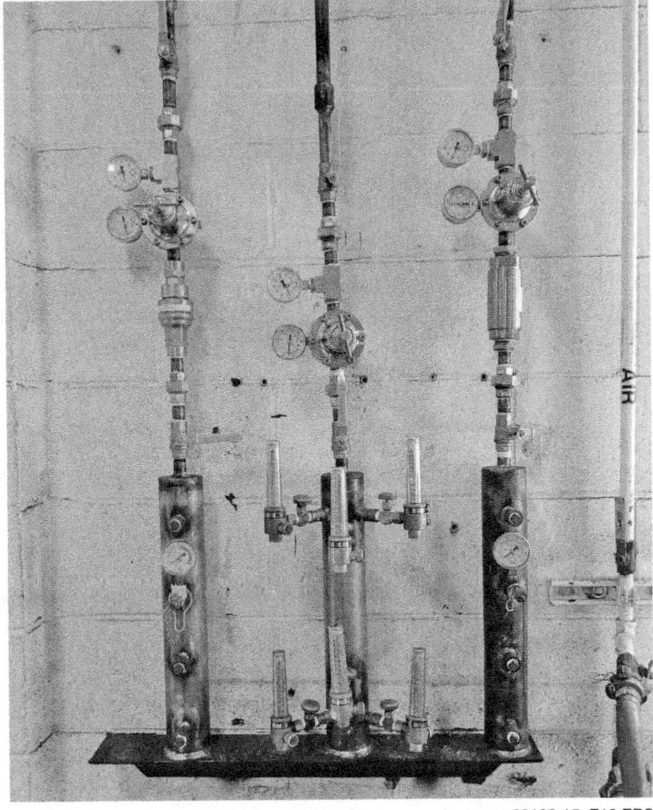

29102-15_F18.EPS

Figure 18 Operator hookups for cutting gases.

in a hose or in the torch. Hose length must be considered when using a manifold distribution system because the pressure drop increases with

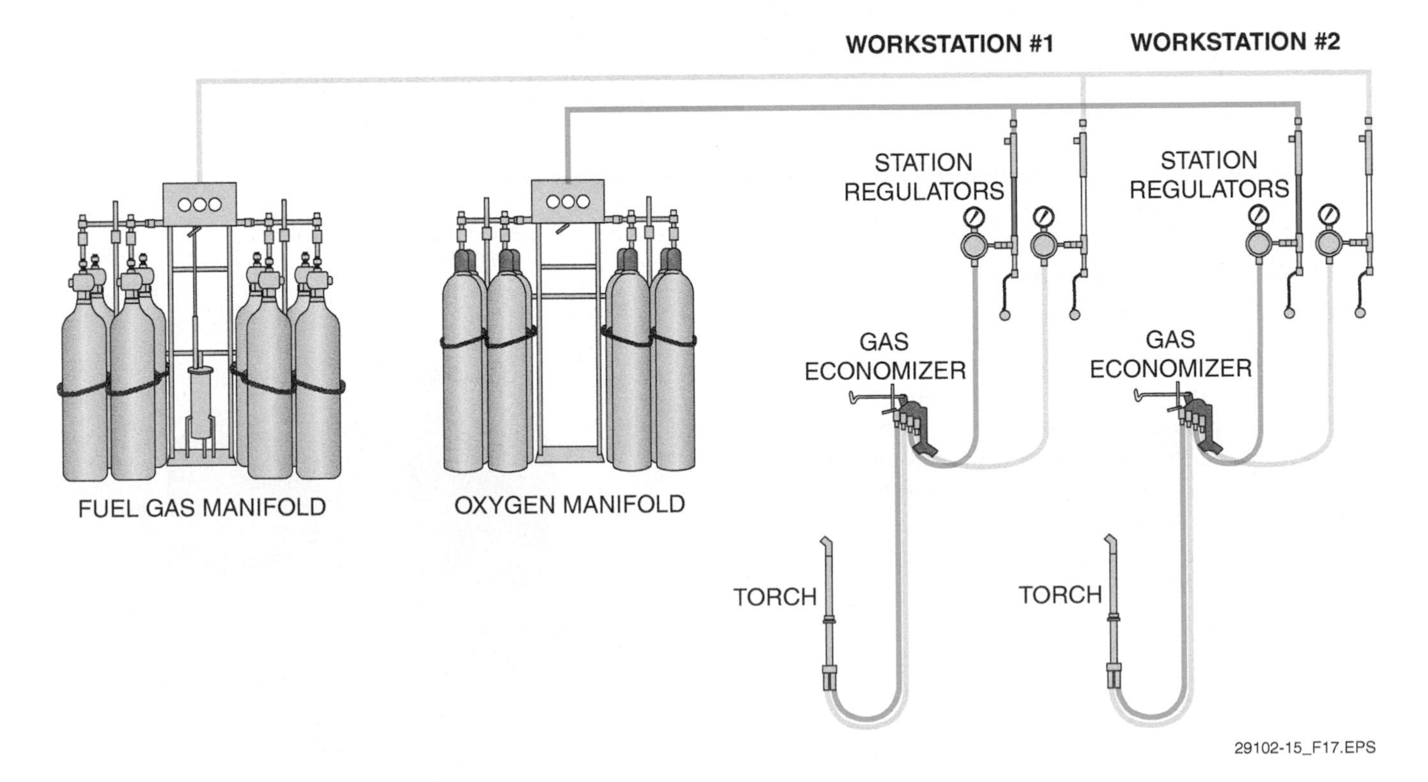

29102-15_F17.EPS

Figure 17 Gas distribution to stations through a manifold.

the length of the hoses. Hoses with a larger inside diameter are needed for long runs.

OSHA provides the following safety precautions specifically for manifold systems:

- Fuel gas and oxygen manifolds must bear the name of the substance they contain.
- Fuel gas and oxygen manifolds must not be placed in confined spaces; they must be placed in safe, well ventilated, and accessible locations.
- Hose connections must be designed so that they cannot be interchanged between fuel gas and oxygen manifolds and supply header connections. Adapters may not be used to interchange hoses.
- Hose connections must be kept free of grease and oil.
- Manifold and header hose connections must be capped when not in use.
- Nothing may be placed on a manifold that will damage the manifold or interfere with the quick closing of the valves.

2.2.5 Hoses

Hoses transport gases from the regulators to the torch. Oxygen hoses are usually green with right-hand threaded connections. Hoses for fuel gas are usually red and have left-hand threaded connections. The fuel gas connection fittings are grooved as a reminder that they have left-hand threads.

Proper care and maintenance of the hose is important for maintaining a safe, efficient work area. Remember the following guidelines for hoses:

- Protect the hose from molten dross or sparks, which will burn the exterior. Although some hoses are flame retardant, they will burn.
- Do not place the hoses under the metal being cut. If the hot metal falls on the hose, the hose will be damaged. Keep hoses as far away from the cutting activity as possible.
- Frequently inspect and replace hoses that show signs of cuts, burns, worn areas, cracks, or damaged fittings. The hoses are tough and durable, but not indestructible.
- Never use pipe-fitting compounds or lubricants around hose connections. These compounds often contain oil or grease, which ignite and burn or explode in the presence of oxygen.

Propane and propylene require hoses designed for the mixture of hydrocarbons present in these fuels. Ensure that any hoses used are appropriate for the fuel gas. Hose and fuel gas providers can help provide the correct hoses.

2.3.0 Torches and Tips

Cutting torches mix oxygen and fuel gas for the torch flame and control the stream of oxygen necessary for the cutting jet. Depending on the job site, either a one-piece or a combination cutting torch may be used.

2.3.1 One-Piece Hand Cutting Torch

The one-piece hand cutting torch, sometimes called a demolition torch or a straight torch, contains the fuel gas and oxygen valves that allow the gases to enter the chambers and then flow

Fuel and Oxygen Cylinder Separation for Fixed Installations

For fixed installations involving one or more cylinders coupled to a manifold, fuel and oxygen cylinders must be separated by at least 20' (6.10 m) or be divided by a wall 5' (1.52 m) or higher with a 30-minute burn rating, per American National Standards Institute *Z49.1.* This also applies to cylinders in storage. Special wheeled cradles designed to distribute gas from multiple cylinders are available.

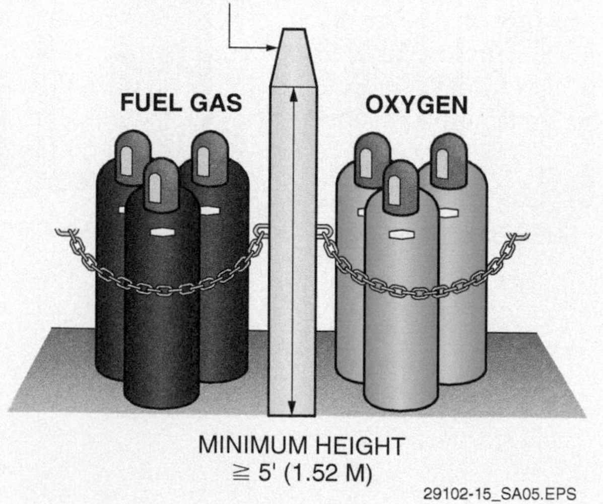

into the part of the torch where they are mixed. The main body of the torch is called the handle. The torch valves control the fuel gas and oxygen flow needed for preheating the metal to be cut. The cutting oxygen lever, which is spring-loaded, controls the jet of cutting oxygen. Hose connections are located at the end of the torch body behind the valves.

Figure 19 shows a three-tube, one-piece hand cutting torch in which the preheat fuel and oxygen are mixed in the tip. These torches are designed for heavy-duty cutting and little else. They have long supply tubes from the torch handle to the torch head to reduce radiated heat to the operator's hands. Cutting torches are generally available with sufficient capacity to cut steel up to 12" (≈30 cm) thick. Larger-capacity torches, with the ability to cut steel up to 36" (≈90 cm) thick, can also be obtained. Torches with this kind of capacity require a significant oxygen and fuel gas supply to perform.

Two different types of oxyfuel cutting torches are in general use. The positive-pressure torch (*Figure 19*) is designed for use with fuel supplied through a regulator from pressurized fuel storage cylinders. The injector torch is designed to use a vacuum created by the oxygen flow to draw the necessary amount of fuel from a very-low-pressure fuel source, such as a natural gas line or acetylene generator. The injector torch, when used, is most often found in continuous-duty, high-volume manufacturing applications. Both types may employ one of two different fuel-mixing methods:

- Torch-handle, or supply-tube, mixing
- Torch-head or tip mixing

The two methods can normally be distinguished by the number of supply tubes from the torch handle to the torch head. Torches that use three tubes from the handle to the head mix the preheat fuel and oxygen at the torch head or tip. This method tends to help eliminate any flashback damage to the torch head supply tubes and torch handle. One tube carries fuel gas to the head. The other two tubes carry oxygen; one carries the oxygen for the preheat flame while the other carries the oxygen for cutting.

The cutting torch with two tubes usually mixes the preheat fuel and oxygen in a mixing chamber in the torch body or in one of the supply tubes (*Figure 20*). Injector torches usually have the injector located in one of the supply tubes, and the mixing occurs in the tube between the injector and the torch head. Some older torches that have only two visible tubes are actually three-tube torches that mix the preheat fuel and oxygen in the torch head or tip. This is accomplished by using a separate preheat fuel tube inside a larger preheat oxygen tube.

2.3.2 Combination Torch

The combination torch consists of a cutting torch attachment that fits onto a welding torch handle. These torches are normally used in light-duty or medium-duty applications. Fuel gas and oxygen valves are on the torch handle. The cutting attachment has a cutting oxygen lever and another oxygen valve to control the preheat flame. When the cutting attachment is screwed onto the torch handle, the torch handle oxygen valve is opened all the way, and the preheat oxygen is controlled by an oxygen valve on the cutting attachment. When the cutting attachment is removed, brazing and heating tips can be screwed onto the torch handle. *Figure 21* shows a two-tube combination torch in which preheat mixing is accomplished in a supply tube. These torches are usually positive-pressure torches with mixing occurring in the attachment body, supply tube, head, or tip. These torches may be equipped with built-in flashback arrestors and check valves.

2.3.3 Cutting Torch Tips

Cutting torch tips, or nozzles, fit into the cutting torch and are either screwed in or secured with a tip nut. There are one- and two-piece cutting tips (*Figure 22*).

One-piece cutting tips are made from a solid piece of copper. Two-piece cutting tips have a separate external sleeve and internal section.

Torch manufacturers supply literature explaining the appropriate torch tips and gas pressures for various applications. *Table 2* shows a sample cutting tip chart that lists recommended tip sizes and gas pressures for use with acetylene fuel gas and a specific manufacturer's torch and tips.

CAUTION

Do not use the cutting tip chart from one manufacturer for the cutting tips of another manufacturer. The gas flow rate of the tips may be different, resulting in excessive flow rates. Different gas pressures may also be required. The cutting torch tip to be used depends on the base metal thickness and fuel gas being used. Special-purpose tips are also available for use in such operations as gouging and grooving.

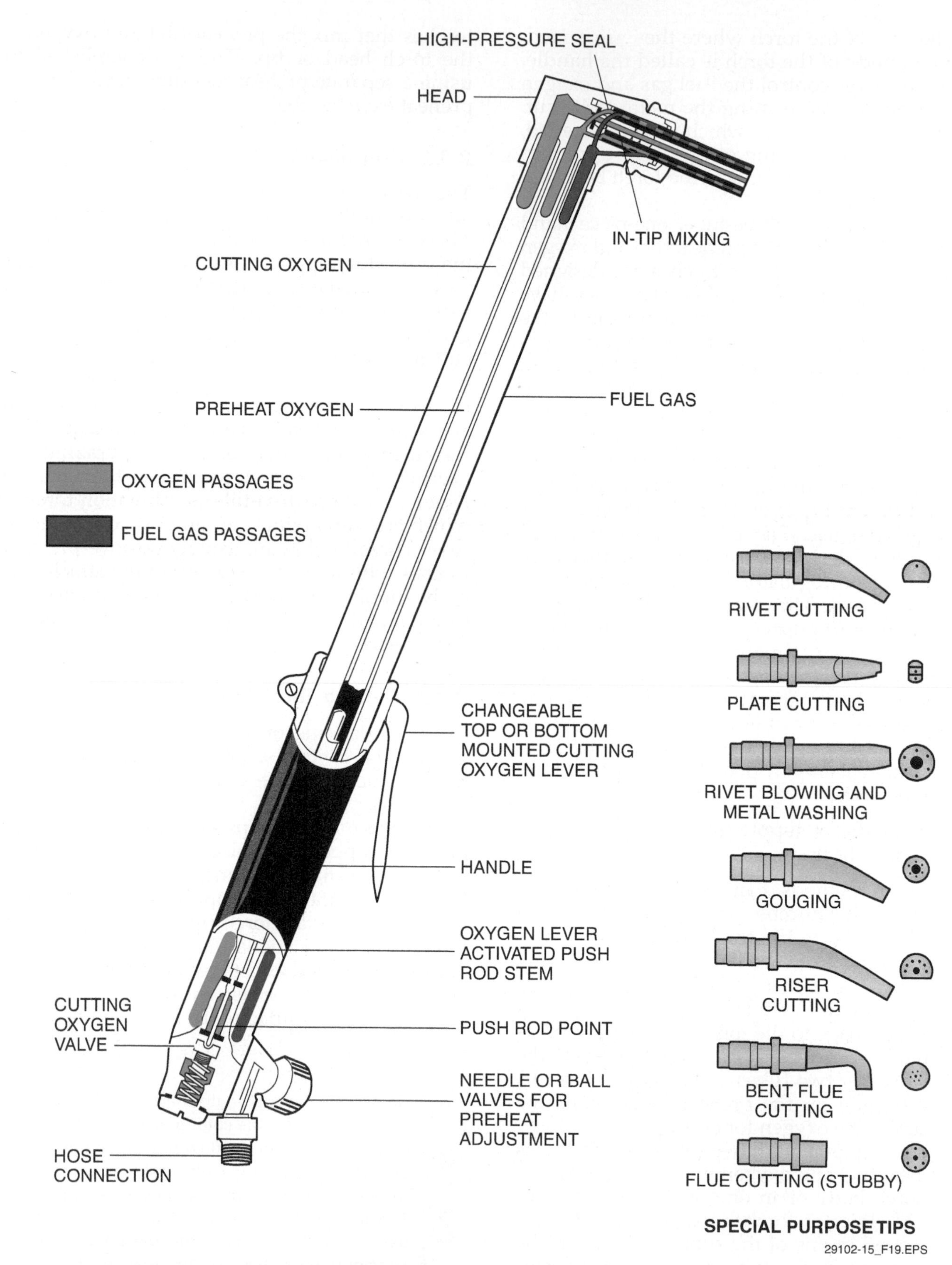

Figure 19 Heavy-duty three-tube one-piece positive-pressure hand cutting torch.

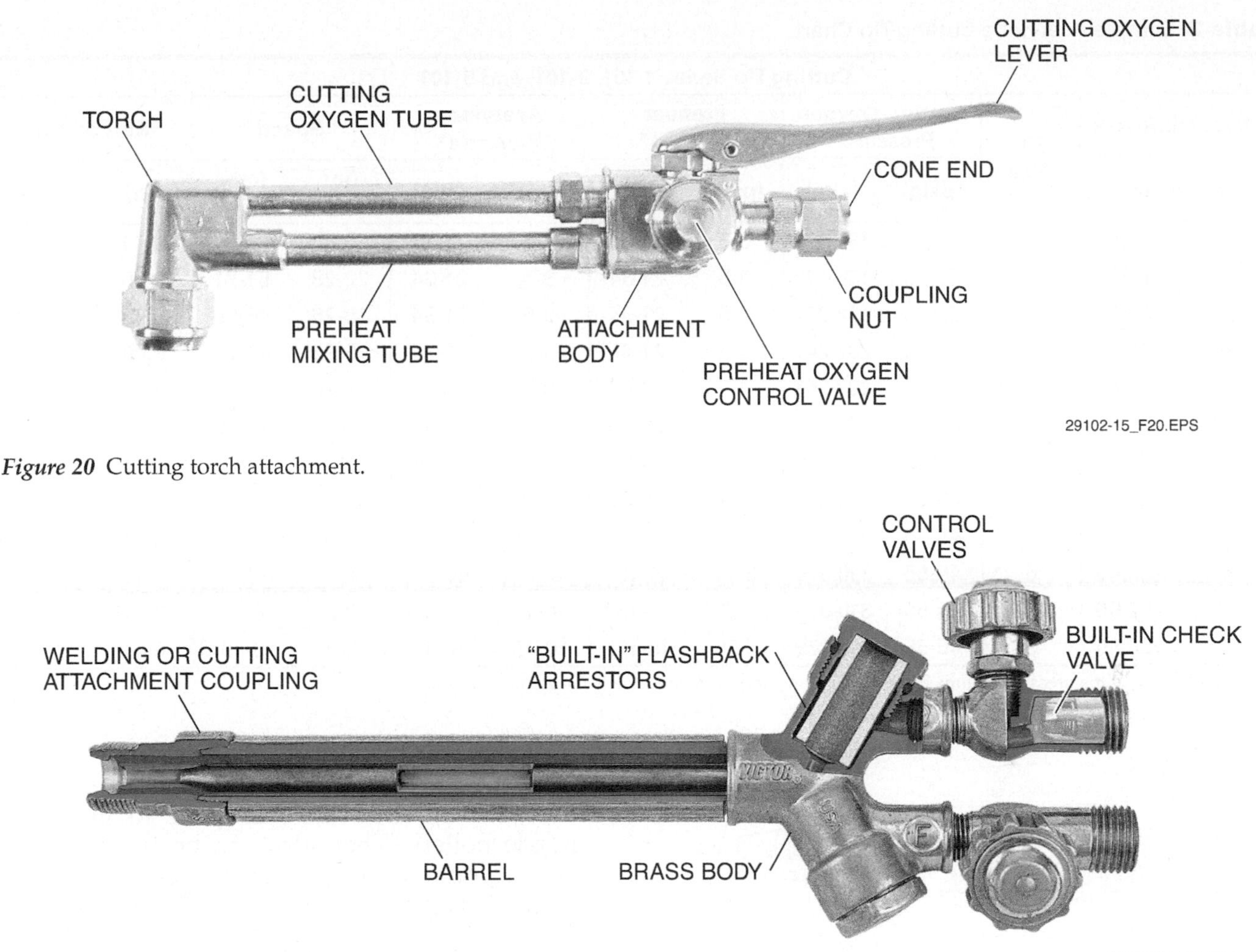

Figure 20 Cutting torch attachment.

Figure 21 Combination cutting torch

The cutting torch tip to be used depends on the base metal thickness and fuel gas being used. Special-purpose tips are also available for such operations as **gouging** and grooving.

One-piece torch tips are generally used with acetylene cutting because of the high temperatures involved. They can have four, six, or eight preheat holes in addition to the single cutting hole. *Figure 23* shows the arrangement of typical acetylene torch cutting tips.

Tips used with liquefied fuel gases must have at least six preheat holes (*Figure 24*). Because fuel gases burn at lower temperatures than acetylene, more holes are necessary for preheating. Tips used with liquefied fuel gases can be one- or two-piece cutting tips. *Figure 25* shows a typical two-piece cutting tip used with liquefied fuel gases.

Special-purpose tips are available for special cutting jobs such as cutting sheet metal, rivets, risers, and flues, as well as **washing** and gouging. *Figure 26* shows special-purpose torch cutting tips, which are described as follows.

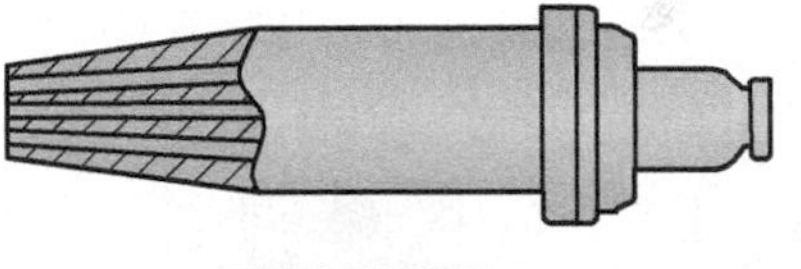

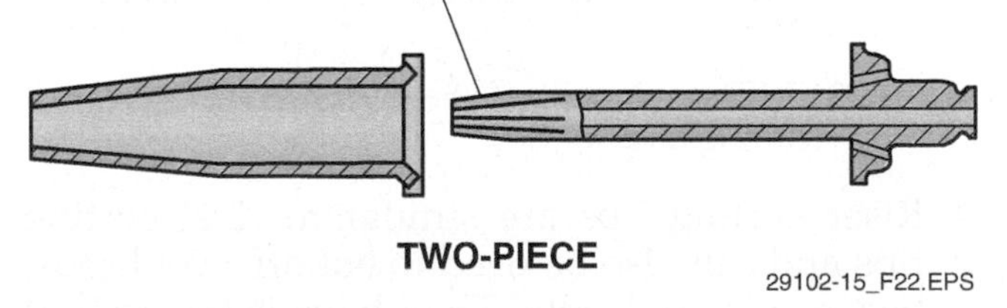

Figure 22 One- and two-piece cutting tips.

- The sheet metal cutting tip has only one preheat hole. This minimizes the heat and prevents distortion in the sheet metal. These tips are normally used with a motorized carriage, but can also be used for hand cutting.
- Rivet cutting tips are used to cut off rivet heads, bolt heads, and nuts.

Table 2 Sample Acetylene Cutting Tip Chart

Cutting Tip Series 1-101, 3-101, and 5-101												
Metal Thickness		Tip Size	Cutting Oxygen Pressure*		Preheat Oxygen*		Acetylene Pressure*		Speed		Kerf Width	
(in)	(mm)		(psig)	(kPa)	(psig)	(kPa)	(psig)	(kPa)	(in/min)	(cm/min)	(in)	(mm)
⅛	3.18	000	20-25	138-172	3-5	21-34	3-5	21-34	20-30	51-76	0.04	01.02
¼	6.35	00	20-25	138-172	3-5	21-34	3-5	21-34	20-28	51-71	0.05	01.27
⅜	9.52	0	25-30	172-207	3-5	21-34	3-5	21-34	18-26	46-66	0.06	01.52
½	12.70	0	30-35	207-241	3-6	21-41	3-5	21-34	16-22	41-56	0.06	01.52
¾	19.05	1	30-35	207-241	4-7	28-48	3-5	21-34	15-20	38-51	0.07	1.78
1	25.40	2	35-40	241-276	4-8	28-55	3-6	21-41	13-18	33-46	0.09	02.29
2	50.80	3	40-45	276-310	5-10	34-69	4-8	28-55	10-12	25-30	0.11	02.79
3	76.20	4	40-50	276-345	5-10	34-69	5-11	34-76	8-10	20-25	0.12	03.05
4	101.60	5	45-55	310-379	6-12	41-83	6-13	41-90	6-9	15-23	0.15	03.81
6	152.40	6	45-55	310-379	6-15	41-103	8-14	55-97	4-7	10-18	0.15	03.81
10	254.00	7	45-55	310-379	6-20	41-138	10-15	69-103	3-5	8-13	0.34	08.64
12	304.80	8	45-55	310-379	7-25	48-172	10-15	69-103	3-4	8-10	0.41	10.41

*The lower side of the pressure listings is for hand cutting and the higher side is for machine cutting.

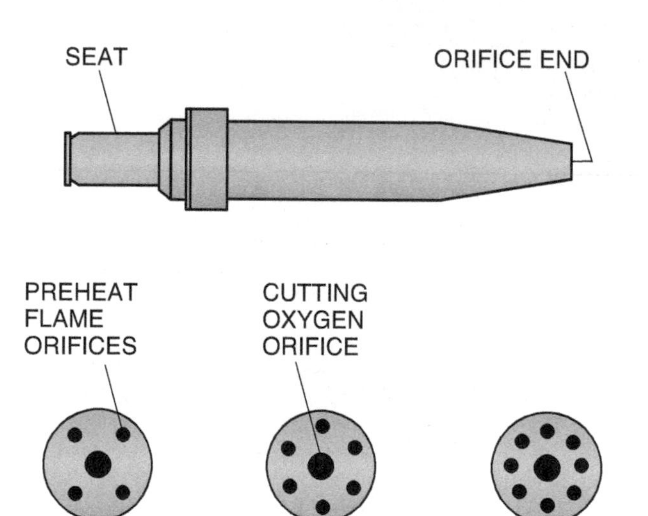

Figure 23 Orifice-end views of one-piece acetylene torch cutting tips.

- Riser cutting tips are similar to rivet cutting tips and can also be used to cut off rivet heads, bolt heads, and nuts. They have extra preheat holes to cut risers, flanges, or angle legs faster. They can be used for any operation that requires a cut close to and parallel to another surface, such as in removing a metal backing.
- Rivet blowing and metal washing tips are heavy-duty tips designed to withstand high heat. They are used for coarse cutting and for removing such items as clips, angles, and brackets.
- Gouging tips are used to groove metal in preparation for welding.
- Flue cutting tips are designed to cut flues inside boilers. They also can be used for any cutting operation in tight quarters where it is difficult to get a conventional tip into position.

Nearly all manufacturers use different tip-to-torch mounting designs, sealing surfaces, and diameters. In addition, tip sizes and flow rates are usually not the same between manufacturers even though the model number designations may be the same. This makes it impossible to safely interchange cutting tips between torches from different manufacturers. Even though some tips from different manufacturers may appear to be the same, do not interchange them. The sealing surfaces are very precise, and serious leaks may occur that

Acetylene Flow Rates

Manufacturers provide listings of the maximum fuel flow rate for each acetylene tip size in addition to recommended acetylene pressures. When selecting a tip, make sure that its maximum flow rate (in cubic feet or cubic meters per hour) does not exceed one-seventh of the total fuel capacity for the acetylene cylinder in use. Multiple cylinders must be manifolded together if the flow rate exceeds the cylinder(s) in use in order to prevent withdrawal of acetone along with acetylene.

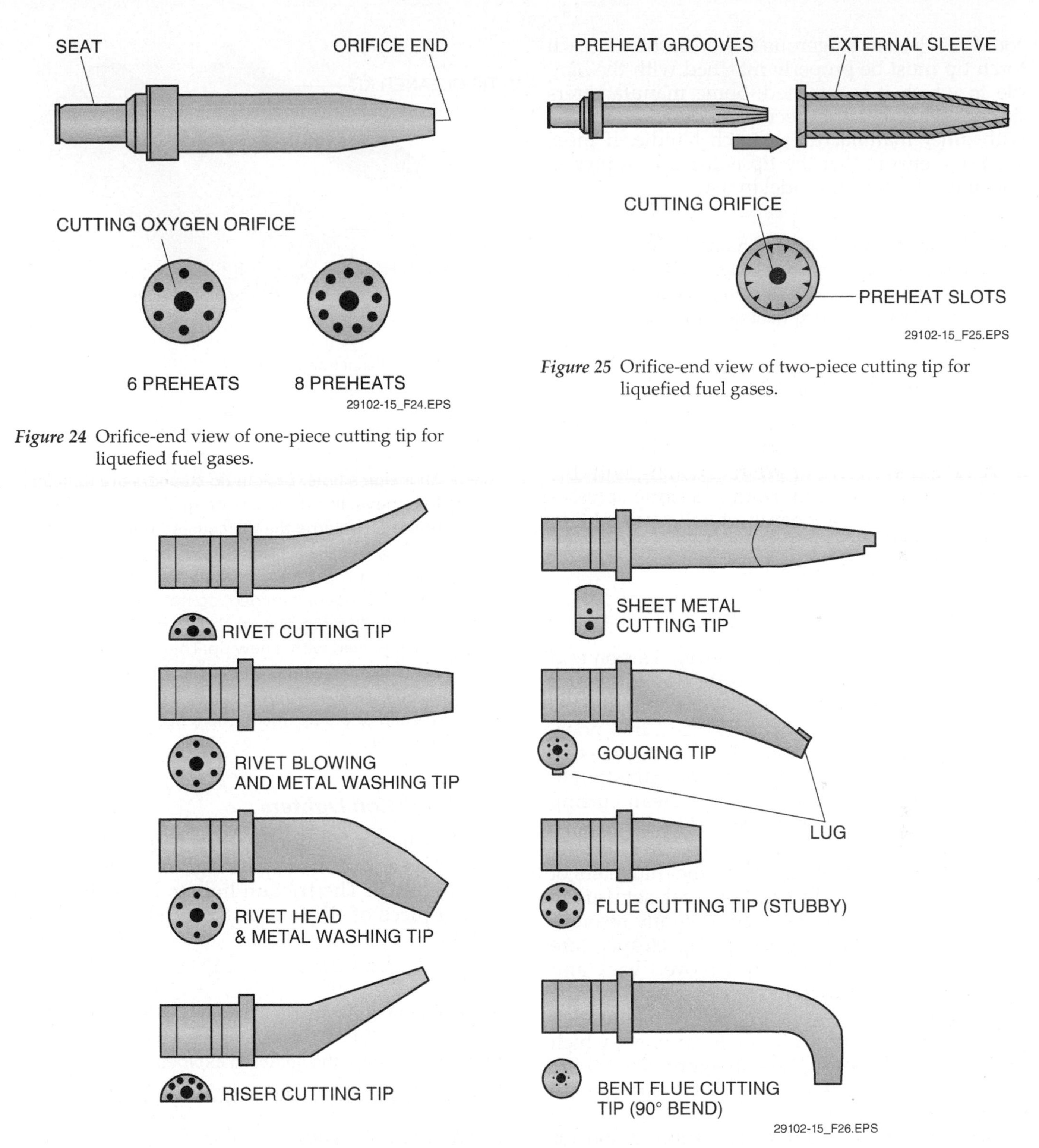

Figure 24 Orifice-end view of one-piece cutting tip for liquefied fuel gases.

Figure 25 Orifice-end view of two-piece cutting tip for liquefied fuel gases.

Figure 26 Special-purpose torch cutting tips.

could result in a dangerous fire or flashback. Each torch tip must be properly matched with the handle to which it is attached. Some manufacturers do make tips that are specifically designed for use with other manufacturer's torch handle. If these are used, ensure that the tip is listed as a precise match for the handle model in use.

CAUTION

Do not mix torch tips and handles from different manufacturers, unless they are specifically identified as being compatible in the manufacturer's documentation or catalog.

2.4.0 Other Accessories

In addition to the components that make up the actual oxyfuel cutting equipment, there are numerous accessories that are used along with the equipment. This section examines some accessories that are used for cleaning torch tips, lighting torches, transporting cylinders, and marking cylinders and metal workpieces.

2.4.1 Tip Cleaners and Tip Drills

With use, cutting tips become dirty. Carbon and other impurities build up inside the holes, and molten metal often sprays and sticks onto the surface of the tip. A dirty tip will result in a poor-quality cut with an uneven **kerf** and excessive dross buildup. To ensure good cuts with straight kerfs and minimal dross buildup, clean cutting tips with tip cleaners or tip drills (*Figure 27*).

Tip cleaners are tiny round files. They usually come in a set with files to match the diameters of the various tip holes. In addition, each set usually includes a file that can be used to lightly recondition the face of the cutting tip. Tip cleaners are inserted into the tip hole and moved back and forth a few times to remove deposits from the hole. The small files are not made from hard metals. Torch tips are typically made of brass, which is also relatively soft. Using an aggressive tip file made from hard metals could easily damage the precise opening.

Tip drills are used for major cleaning and for holes that are plugged. Tip drills are tiny drill bits that are sized to match the diameters of tip holes. The drill fits into a drill handle for use. The handle is held, and the drill bit is turned carefully inside the hole to remove debris. They are more brittle than tip cleaners, making them more difficult to use. If a torch tip is properly cared for, tip drills are rarely needed.

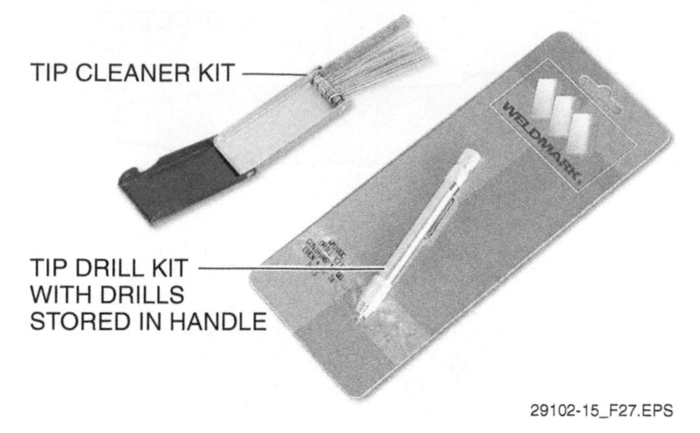

Figure 27 Tip cleaner and drill kits.

CAUTION

Tip cleaners and tip drills are brittle. Care must be taken to prevent these devices from breaking off inside a hole. Broken tip cleaners are difficult to remove. Improper use of tip cleaners or tip drills can enlarge the tip, causing improper burning of gases. If this occurs, the tip must be discarded. If the end of the tip has been partially melted or deeply gouged, do not attempt to cut it off or file it flat. The tip should be discarded and replaced with a new tip. This is because some tips have tapered preheat holes, and if a significant amount of metal is removed from the end of the tip, the preheat holes will become too large.

2.4.2 Friction Lighters

Always use a friction lighter (*Figure 28*), also known as a striker or spark-lighter, to ignite the cutting torch. The friction lighter works by rubbing a piece of flint on a steel surface to create sparks.

WARNING!

Do not use a match or a gas-filled lighter to light a torch. This could result in severe burns and/or could cause the lighter to explode.

2.4.3 Cylinder Cart

The cylinder cart, or bottle cart, is a modified hand truck that has been equipped with chains or straps to hold cylinders firmly in place. Bottle carts help ensure the safe transportation of gas cylinders. *Figure 29* shows two cylinder carts used for oxyfuel cylinders. Some carts are equipped with tool trays or boxes as well as rod holders.

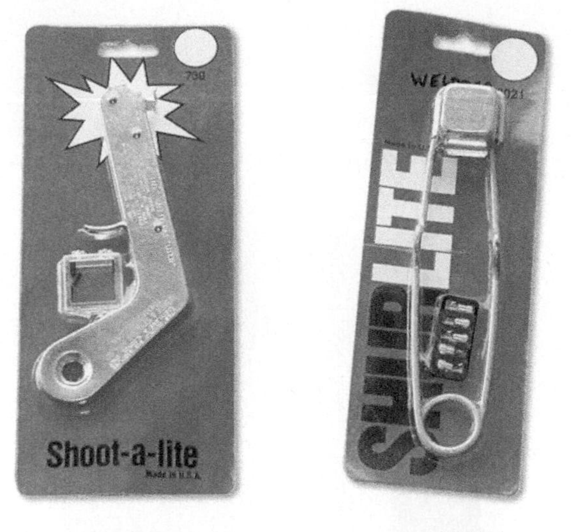

Figure 28 Friction lighters.

2.4.4 Soapstone Markers

Because of the heat involved in cutting operations, along with the tinted lenses that are required, ordinary pen or pencil marking for cutting lines or welding locations is not effective. The oldest and most common material used for marking is soapstone in the form of sticks or cylinders (*Figure 30*). Soapstone is soft and feels greasy and slippery. It is actually steatite, a dense, impure form of talc that is heat resistant. It also shows up well through a tinted lens under the illumination of an electric arc or gas cutting flame. Some welders prefer to use silver-graphite pencils (*Figure 30*) for marking dark materials and red-graphite pencils for aluminum or other bright metals. Graphite is also highly heat resistant. A few manufacturers also market heat-resistant paint/dye markers for cutting and welding.

2.5.0 Specialized Cutting Equipment

In addition to the common hand cutting torches, other types of equipment are used in oxyfuel cutting applications. This equipment includes mechanical guides used with a hand cutting torch, various types of motorized cutting machines, and oxygen lances. All of the motorized units use special straight body machine cutting or welding torches with a gear rack attached to the torch body to set the tip distance from the work.

2.5.1 Mechanical Guides

On long, circular, or irregular cuts, it is very difficult to control and maintain an even kerf with a hand cutting torch. Mechanical guides can help maintain an accurate and smooth kerf along the cutting line. For straight line or curved cuts, use a one- or two-wheeled accessory that clamps on the torch tip in a fixed position. The wheeled accessory maintains the proper tip distance while the tip is guided by hand along the cutting line.

The torch tip fits through and is secured to a rotating mount between two small metal wheels. The wheel heights are adjustable so that the tip distance from the work can be set. The radius of

Cup-Type Striker

When using a cup-type striker to ignite a welding torch, hold the cup of the striker slightly below and to the side of the tip, parallel with the fuel gas stream from the tip. This prevents the ignited gas from deflecting back from the cup and reduces the amount of carbon soot in the cup. Note that the flint in a striker can be replaced.

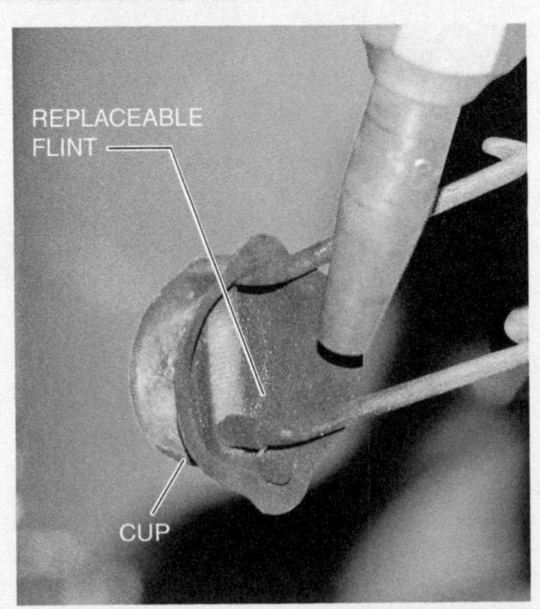

29102-15_SA06.EPS

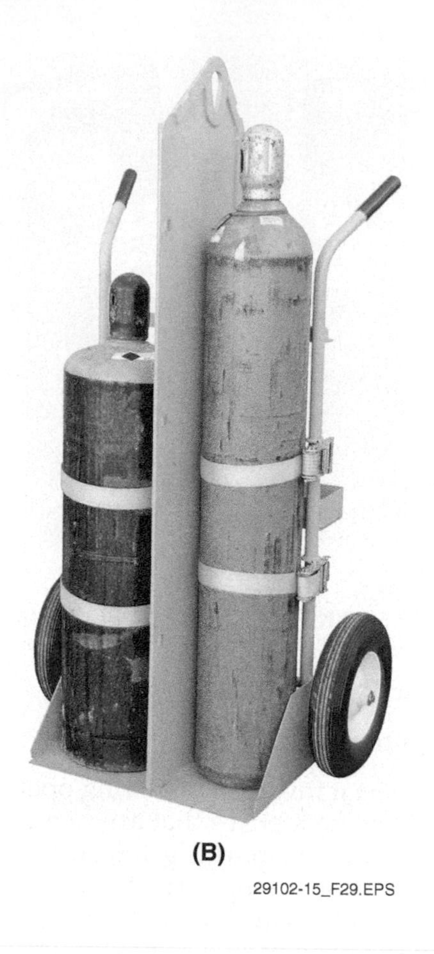

Figure 29 Cylinder carts.

SOAPSTONE STICK AND HOLDER

SOAPSTONE CYLINDER AND HOLDER

SILVER GRAPHITE PENCILS

29102-15_F30.EPS

Figure 30 Soapstone and graphite markers.

the circle is set by moving the pivot point on a radius bar. After a starting hole is cut (if needed), the torch tip is placed through and secured to the circle cutter rotating mount. Then the pivot point is placed in a drilled hole or a magnetic holder at the center of the circle. When the cut is restarted, the torch can be moved in a circle around the cut, guided by the circle cutter. The magnetic guide is used for straight line cuts. The magnets hold it securely in place during cutting operations.

When large work with an irregular pattern must be cut, a template is often used. The torch is drawn around the edges of the template to trace the pattern as the cut is made. If multiple copies must be cut, a metal pattern held in place and designed to allow space for tip distance from the pattern is usually used. For a one- or two-time

Sharpening Soapstone Sticks

The most effective way to sharpen a soapstone stick marker is to shave it on one side with a file. By leaving one side flat, accurate lines can be drawn very close to a straightedge or a pattern.

copy, a heavily weighted Masonite or aluminum template that is spaced off the workpiece could be carefully used and discarded.

2.5.2 Motor-Driven Equipment

A variety of fixed and portable motorized cutting equipment is available for straight and curved cutting/welding. The computer-controlled plate cutting machine (*Figure 31*) is a fixed-location machine used in industrial manufacturing applications. The computer-controlled machine can be programmed to pierce and then cut any pattern from flat metal stock. There is also an optical pattern-tracing machine that follows lines on a drawing using a light beam and an optical detector. They both can be rigged to cut multiple items using multiple torches operated in parallel. Both units have a motor-driven gantry that travels the length of a table and a transverse motor-driven torch head that moves back and forth across the table. Both units are also equipped to use oxyfuel or plasma cutting torches.

Other types of pattern-tracing machines use metal templates that are clamped in the machine. A follower wheel traces the pattern from the template. The pattern size can be increased or decreased by electrical or mechanical linkage to a moveable arm holding one or more cutting torches that cut the pattern from flat metal stock.

Portable track cutting machines, or track burners, can be used in the field for straight or curved cutting and beveling. *Figures 32* and *33* show units driven by a variable-speed motor. The unit shown in *Figure 32* is available with track extensions for any length of straight cutting or beveling, along with a circle cutting attachment. Some models use a single, centered track.

Machine oxyfuel gas cutters or track burners are basic guidance systems driven by a variable speed electric motor to enable the operator to cut or bevel straight lines at any desired speed. The device (*Figure 33*) is usually mounted on a track or used with a circle-cutting attachment to enable the operator to cut various circle diameters up to 96" (≈244 cm). It consists of a heavy-duty tractor unit fitted with an adjustable torch mount and gas hose attachments. It is also equipped with an On/Off switch, a Low/High speed range switch, a Forward/Reverse directional switch, and a speed-adjusting dial calibrated in inches (or metric equivalent) per minute.

The device shown in *Figure 33* offers the following operational features:

- Makes straight-line cuts of any length
- Makes circle cuts up to 96" (≈244 cm) in diameter
- Makes bevel or chamfer cuts
- Has an infinitely variable cutting speed from 1" to 110" (2.5 cm to 279 cm) per minute
- Has dual speed and directional controls to enable operation of the machine from either end

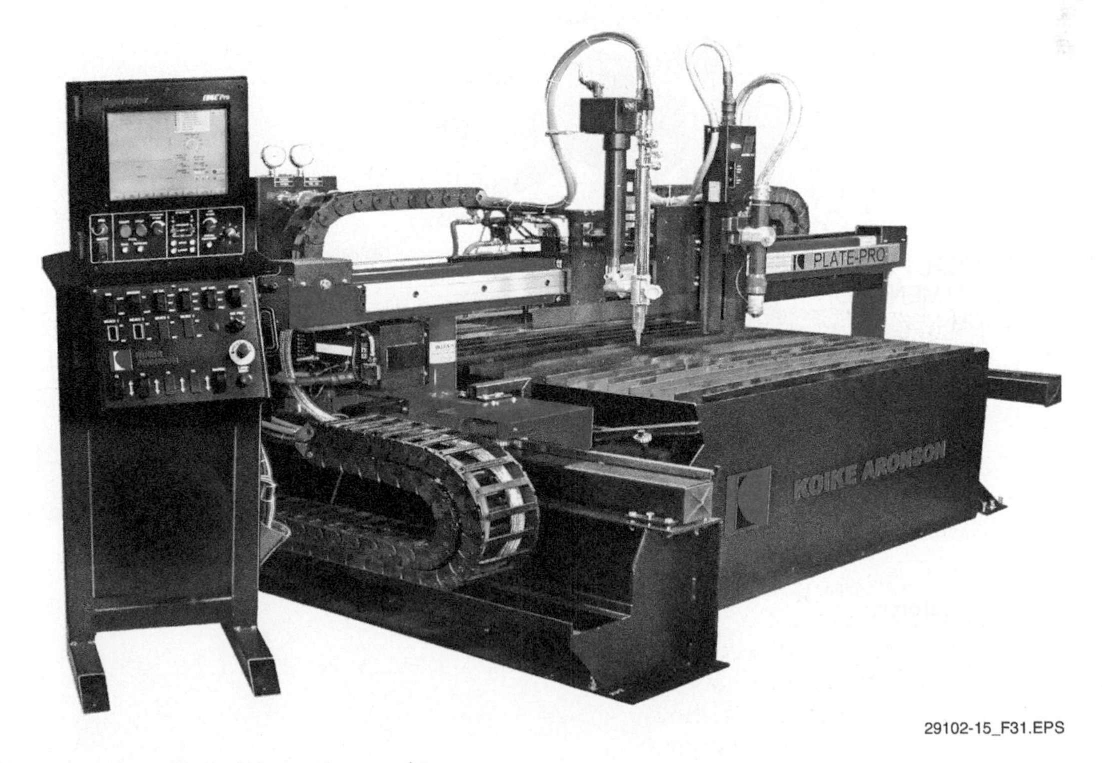

29102-15_F31.EPS

Figure 31 Computer-controlled plate cutting machine.

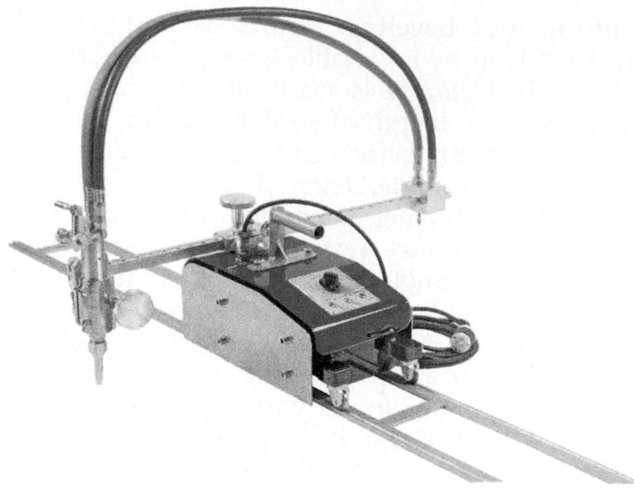

Figure 32 Track burner with oxyfuel machine torch.

Figure 34 shows the location of the following controls:

- *Power on/power off* – Turn the machine on and off by toggling the On/Off toggle switch.
- *Speed range control* – Set the machine's speed range by toggling the switch up or down. The use of two speed ranges allows for more precise speed control.
- *Directional control* – Set the machine's direction.
- *Speed control*– Turn the large knob to adjust the cutting speed based on the percentage of the selected speed range.

A portable, motor-driven band track or hand-cranked ring gear cutter/beveler can be set up in the field for cutting and beveling pipe with oxyfuel or plasma machine torches (*Figures 35*

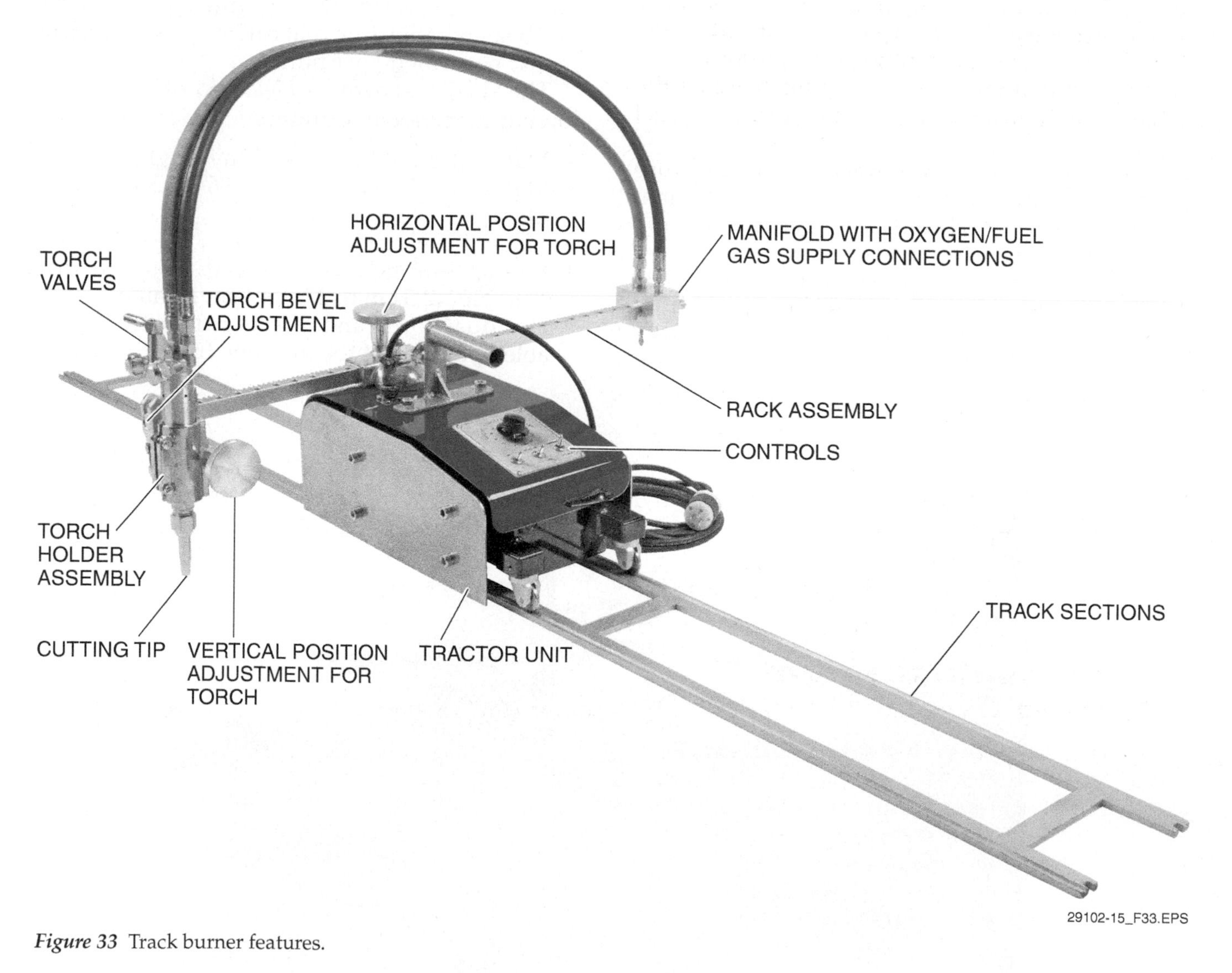

Figure 33 Track burner features.

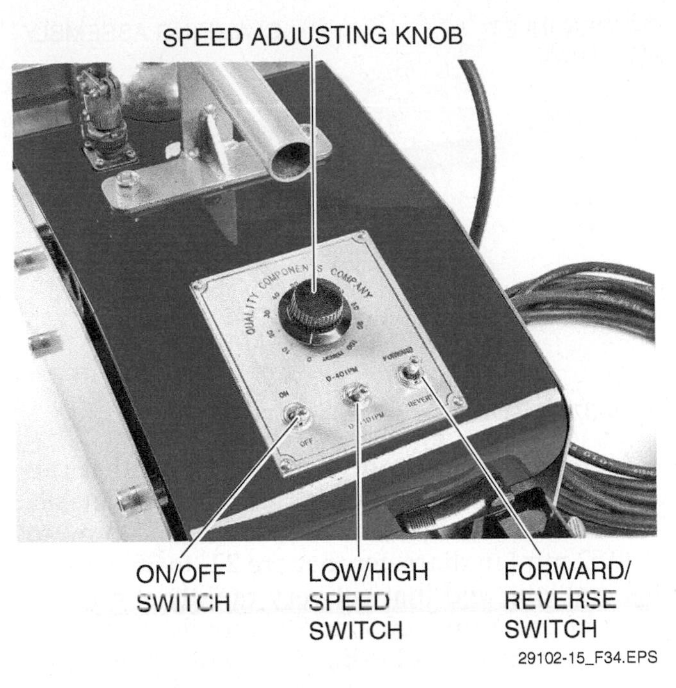

Figure 34 Track burner controls.

and *36*). The stainless steel band track cutter uses a chain and motor sprocket drive to rotate the machine cutting torch around the pipe a full 360 degrees. The all-aluminum ring gear type of cutter/beveler is positioned on the pipe, and then the saddle is clamped in place. In operation, the ring gear and the cutting torch rotate at different rates around the saddle for a full 360-degree cut.

2.5.3 Exothermic Oxygen Lances

Exothermic (combustible) oxygen lances are a special oxyfuel cutting tool usually used in heavy industrial applications and demolition work. The lance is a steel pipe that contains magnesium- and aluminum-cored powder or rods (fuel). In operation, the lance is clamped into a holder (*Figure 37*) that seals the lance to a fitting that supplies oxygen to it through a hose at pressures of 75 to 80 psi (517 to 552 kPa). With the oxygen turned on, the end of the lance is ignited with an acetylene torch or flare. As long as the oxygen is applied, the lance will burn and consume itself. The oxygen-fed flame of the burning magnesium, aluminum, and steel pipe creates temperatures approaching 10,000°F (5,538°C). At this temperature, the lance will rapidly cut or pierce any material, including steel, metal alloys, and cast iron, even under water. The lances for the holder shown in *Figure 37* are 10' (≈3 m) long and range in size from ⅜" (9.5 mm) to 1" (25.4 mm) in diameter. The larger sizes can be coupled to obtain a longer lance.

Figure 35 Band-track pipe cutter/beveler.

Shop-Made Straight-Line Cutting Guide

A simple solution for straight-line cutting is to clamp a piece of angle iron to the work and use a band clamp around the cutting torch tip to maintain the cutting tip distance from the work. When the cut is started, the band clamp rests on the top of the vertical leg of the angle iron, and the torch is drawn along the length of the angle iron at the correct cutting speed.

Figure 36 Ring gear pipe cutter/beveler.

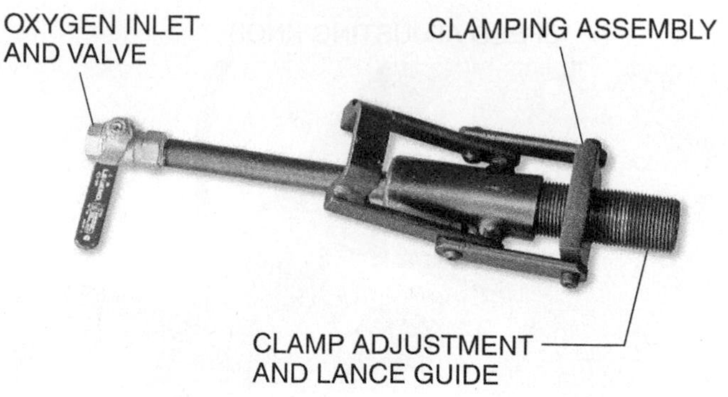

Figure 37 Oxygen lance holder.

A small pistol-grip heat-shielded unit that can be used with an electric welder is also available. This small unit uses lances from ¼" (6.35 mm) to ⅜" (9.52 mm) in diameter that are 22" to 36" (55.9 to 91.4 cm) long and that cut very rapidly at a maximum burning time of 60 to 70 seconds. The small unit is primarily used to burn out large frozen pins and frozen headless bolts or rivets. Like a large lance, it can be used to cut any material, including concrete-lined pipe. Both units are relatively inexpensive and can be set up in the field with only an oxygen cylinder, hose, and ignition device.

Additional Resources

ANSI Z49.1, Safety in Welding, Cutting, and Allied Processes. Miami, FL: American Welding Society.

Uniweld Products, Inc. Numerous videos are available at **www.uniweld.com/en/uniweld-videos**. Last accessed: November 30, 2014.

The Harris Products Group, a division of Lincoln Electric. Numerous videos are available at **www.harrisproductsgroup.com/en/Expert-Advice/videos.aspx**. Last accessed: November 30, 2014.

2.0.0 Section Review

1. A lighter-than-air compound of carbon and hydrogen that is formed by dissolving calcium carbide in water is ____ .
   a. argon
   b. acetylene
   c. propane
   d. propylene

2. Safety devices used on regulators, hoses, and torches to prevent reverse gas flow and protect against fires and explosions are ____ .
   a. check valves and flashback arrestors
   b. orifice plates and torch tips
   c. HEPA filters and flashback valves
   d. distribution manifolds and friction strikers

3. What basic type of oxyfuel cutting torch is designed for use with fuel that is supplied through a regulator from a pressurized fuel storage cylinder?
   a. Motor-controlled torch
   b. Vacuum-injector torch
   c. Neutral-flame torch
   d. Positive-pressure torch

4. The most common material used for marking cutting lines on metal is a form of heat-resistant talc called ____ .
   a. soapstone
   b. Masonite
   c. graphite
   d. flint

5. A type of oxyfuel cutting tool that uses a steel pipe containing magnesium- and aluminum-cored powder or rods for heavy industrial cutting applications and demolition work is a(n) ____ .
   a. circle cutting accessory
   b. plasma machine torch
   c. exothermic oxygen lance
   d. ring gear beveler

SECTION THREE

3.0.0 OXYFUEL EQUIPMENT SETUP AND SHUTDOWN

Objective

Explain how to set up, light, and shut down oxyfuel equipment.

a. Explain how to properly prepare a torch set for operation.
b. Explain how to leak test oxyfuel equipment.
c. Explain how to light the torch and adjust for the proper flame.
d. Explain how to properly shut down oxyfuel cutting equipment.

Performance Tasks

1. Set up oxyfuel equipment.
2. Light and adjust an oxyfuel torch.
3. Shut down oxyfuel cutting equipment.
4. Disassemble oxyfuel cutting equipment.
5. Change empty gas cylinders.

Trade Terms

Carburizing flame: A flame burning with an excess amount of fuel; also called a reducing flame.

Neutral flame: A flame burning with correct proportions of fuel gas and oxygen.

Oxidizing flame: A flame burning with an excess amount of oxygen.

Operators should be trained and tested in the correct methods of safely preparing, starting, testing for leaks, and shutting down an oxyfuel cutting station. This part of the module examines procedures for setting up oxyfuel equipment, leak testing the equipment, lighting and adjusting the torch, and shutting down the equipment after use.

3.1.0 Setting Up Oxyfuel Equipment

When setting up oxyfuel equipment, follow procedures to ensure that the equipment operates properly and safely. The following sections explain the procedures for setting up oxyfuel equipment.

3.1.1 Transporting and Securing Cylinders

Cylinders should be transported to the workstation in an upright position on an appropriate hand truck or bottle cart. Once the cylinders are at the workstation, they must be secured with chain to a fixed support so that they cannot be knocked over accidentally. Leaving the cylinders in a proper cylinder cart is common; removal of the cylinders from the cart is not required. Then the protective cap from each cylinder can be removed and the outlet nozzles inspected to make sure that the seat and threads are not damaged. Place the protective caps where they will not be lost and where they will be readily available for reinstallation.

WARNING!

Always handle cylinders with care. They are under high pressure and should never be dropped, knocked over, rolled, or exposed to heat in excess of 140°F (60°C). When moving cylinders, always be certain that the valve caps are in place. Use a cylinder cage to lift cylinders. Never use a sling or electromagnet for cylinder lifting.

3.1.2 Cracking Cylinder Valves

To crack open a cylinder valve, start by ensuring that the cylinder is fully secured. Then crack open the cylinder valve momentarily to remove any dirt from the valve opening (*Figure 38*).

WARNING!

Operators should always stand with the valve stem between themselves and the regulator when opening valves to avoid injury from dirt that may be lodged in the valve. If a cloth is used during the cleaning process, it must not have any oil or grease on it. Oil or grease mixed with compressed oxygen can cause an explosion.

Hoisting Cylinders

Never attempt to lift a cylinder using the holes in a safety cap. Always use a lifting cage. Make sure that the cylinder is secured in the cage. Cages designed for storing and lifting cylinders are available in various sizes. The model shown here includes a partition to separate oxygen cylinders from fuel gas cylinders.

29102-15_SA09.EPS

CAUTION

Use care not to over-tighten connections. The brass connections used will strip if over-tightened. Repair or replace equipment that does not seal properly. If the torch valves leak, try lightly tightening the packing nut at the valves. If a cutting attachment or torch tip leaks, disassemble and check the sealing surfaces and/or O-rings for damage. If the equipment or torch tip is new and the sealing surfaces do not appear damaged, over-tighten the connection slightly to seat the sealing surfaces; then loosen and retighten normally. If the leaks persist or the equipment is visibly damaged, replace the torch, cutting attachment (if used), or cutting tip as necessary.

29102-15_F38.EPS

Figure 38 Cracking a cylinder valve.

3.1.3 Attaching Regulators

To attach the regulators, first check that the regulator is closed (adjustment screw is backed out and loose/turns with no resistance).

Check the regulator fittings to ensure that they are free of oil and grease (*Figure 39*).

Connect and tighten the oxygen regulator to the oxygen cylinder using a torch wrench (*Figure 40*).

WARNING!

Do not work with a regulator that shows signs of damage, such as cracked gauges, bent thumbscrews, or worn threads. Set it aside for repairs. Operators should never attempt to repair regulators. When tightening the connections, always use a torch wrench.

CHECK THAT FITTINGS ARE CLEAN

29102-15_F39.EPS

Figure 39 Checking connection fittings.

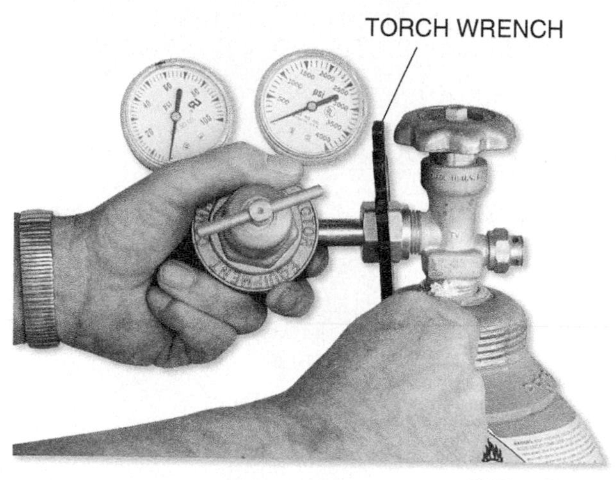

29102-15_F40.EPS

Figure 40 Tightening regulator connection.

Connect and tighten the fuel gas regulator to the fuel gas cylinder. Remember that most fuel gas fittings have left-hand threads. Next, clean the outlet connection of the regulator. Crack the cylinder valve slightly and turn the pressure adjustment screw clockwise until you feel some resistance and the regulator begins to allow gas to pass through and expel any debris. Shut the cylinder valve and close the regulator (*Figure 41*).

3.1.4 Installing Flashback Arrestors or Check Valves

The installation of flashback arrestors or check valves is important and easy to accomplish if they are not already installed. Attach a flashback arrestor and/or check valve to the hose connection on the oxygen regulator or torch handle (*Figure 42*) and tighten with a torch wrench. Then attach and tighten a flashback arrestor and/or check valve to the fuel gas regulator or torch handle.

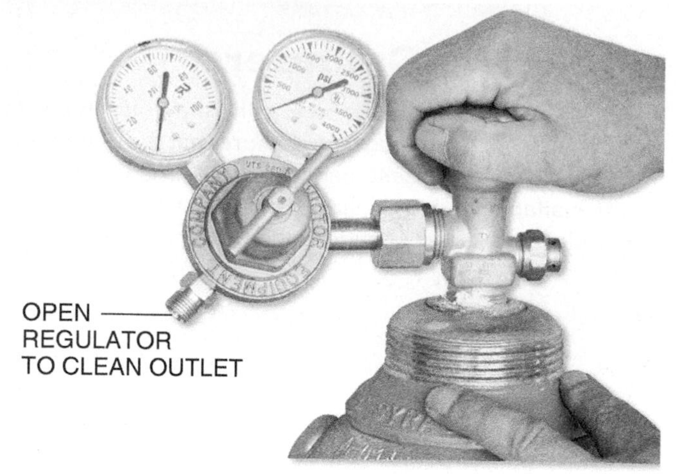

29102-15_F41.EPS

Figure 41 Cleaning the regulator.

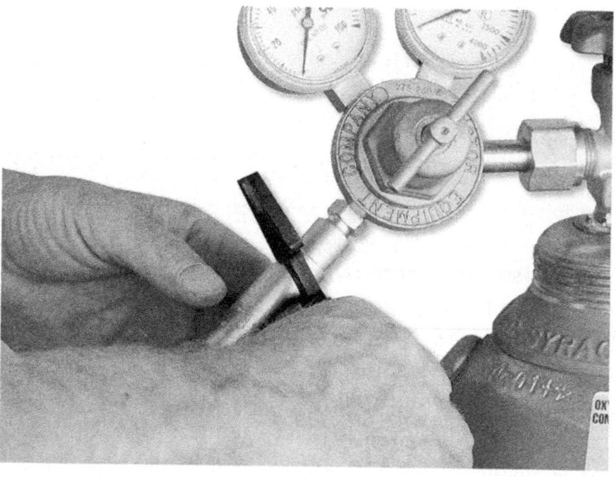

29102-15_F42.EPS

Figure 42 Attaching a flashback arrestor.

Once again, remember that fuel gas fittings have left-hand threads.

> **WARNING!**
>
> At least one flashback arrestor must be used with each hose. The absence of a flashback arrestor could result in flashback. Flashback arrestors can be attached either to the regulator, the torch, or both; however, flashback arrestors installed at the torch handle are preferred if only one is being used.

3.1.5 Connecting Hoses to Regulators

New hoses contain talc and possibly loose bits of rubber. These materials must be blown out of the hoses before the torch is connected. If they are not blown out, they will clog the tiny torch needle valves or tip openings.

To connect the hoses to the regulators, first inspect both the oxygen and fuel gas hoses for any damage, burns, cuts, or fraying. Replace any damaged hoses. Then connect the oxygen hose to the oxygen regulator flashback arrestor or check valve (*Figure 43*), and connect the fuel gas hose to the fuel gas regulator flashback arrestor and/or check valve. Complete the installation by opening the cylinder valves and regulators and purging the hoses until they are clear.

3.1.6 Attaching Hoses to the Torch

To attach the hoses to the torch, first attach flashback arrestors to the oxygen and fuel gas hose connections on the torch body—unless the torch has built-in flashback arrestors and check valves. Attach and tighten the oxygen hose to the oxygen fitting on the flashback arrestor or torch (*Figure 44*). Then attach and tighten the hose to the fuel gas fitting on the flashback arrestor or torch.

3.1.7 Connecting Cutting Attachments (Combination Torch Only)

If cutting attachments are being connected to a combination torch, be sure to check the torch manufacturer's instructions for the correct installation method. Then connect the attachment and tighten by hand as required.

3.1.8 Installing Cutting Tips

Before installing a cutting tip in a cutting torch, first identify the thickness of the material to be cut. Then identify the proper size cutting tip from the manufacturer's recommended tip size chart for the fuel being used.

> **WARNING!**
>
> If acetylene fuel is being used, make sure that the maximum fuel flow rate per hour of the tip does not exceed one-seventh of the fuel cylinder capacity. If a purplish flame is observed when the torch is operating, the fuel rate is too high and acetone is being withdrawn from the acetylene cylinder along with the acetylene gas.

Once the cutting tip has been selected, inspect the cutting tip sealing surfaces and orifices for damage or plugged holes. If the sealing surfaces are damaged, discard the tip. If the orifices are plugged, clean them with a tip cleaner or drill.

Check the torch manufacturer's instructions for the correct method of installing cutting tips. Then install the cutting tip and secure it with a torch wrench or by hand as required (*Figure 45*).

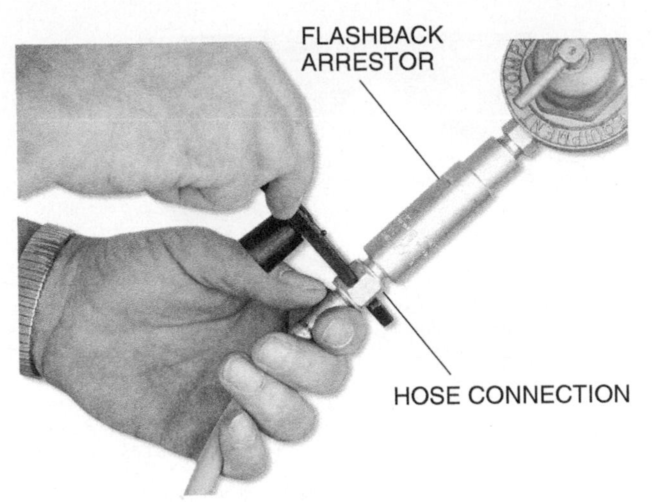

Figure 43 Connecting hose to regulator flashback arrestor.

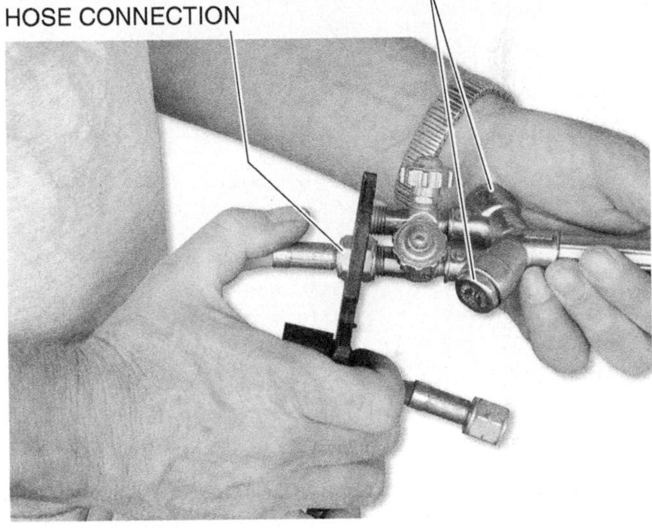

Figure 44 Connecting hoses to torch body.

3.1.9 Closing Torch Valves and Loosening Regulator Adjusting Screws

Closing the torch valves and loosening the regulator adjusting screws (*Figure 46*) are done before opening either cylinder valve. First, check the fuel and oxygen valves on the torch to be sure they are closed. Then check both the oxygen and fuel gas regulator adjusting screws to be sure they are loose (backed out).

> **CAUTION**
>
> Loosening regulator adjusting screws closes the regulators and prevents damage to the regulator diaphragms when the cylinder valves are opened.

29102-15_F45.EPS

Figure 45 Installing a cutting tip.

29102-15_F46.EPS

Figure 46 Torch valves and regulator adjusting screws.

3.1.10 Opening Cylinder Valves

To open cylinder valves (*Figure 47*), stand on the opposite side of the cylinder from the regulator and crack open the oxygen cylinder valve until the pressure on the regulator gauge rises and stops. The pressure in the cylinder and the pressure at the inlet of the regulator are now equal. Now open the oxygen cylinder valve all the way.

> **WARNING!**
> When opening a cylinder valve, keep the cylinder valve stem between you and the regulator. Never stand directly in front of or behind a regulator. The regulator adjusting screw can blow out, causing serious injury. Always open the cylinder valve gradually. Quick openings can damage a regulator or gauge or even cause a gauge to explode.

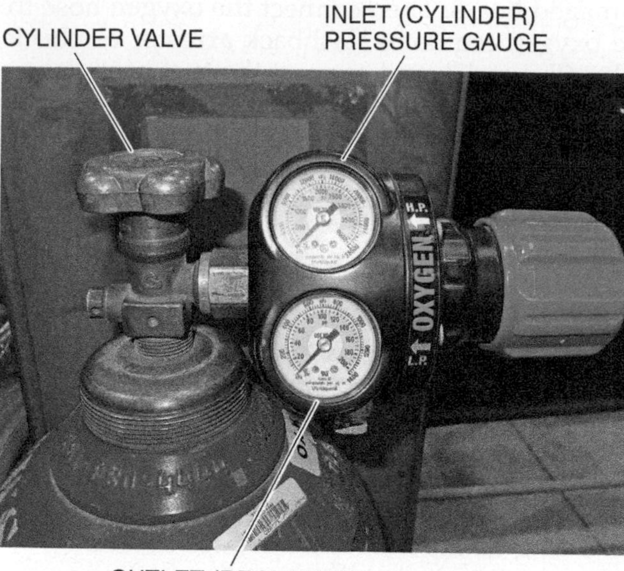

29102-15_F47.EPS

Figure 47 Cylinder valve and gauges.

Oxygen cylinder valves must be opened all the way until the valve seats at the top. Seating the valve at the fully open position prevents high-pressure leaks at the valve stem.

Once the oxygen cylinder valve is open, slowly open the fuel gas cylinder valve until the cylinder pressure gauge indicates the cylinder pressure, but no more than one and a half turns. This allows it to be quickly closed in case of a fire. This is especially important with acetylene.

3.1.11 Purging the Torch and Setting the Working Pressures

After the oxygen and fuel gas tank valves have been opened, the torch valves are opened to purge the torch and set the working pressures on the regulators.

Fully open the oxygen valve on the torch. Then depress and hold the cutting oxygen lever. Turn the oxygen regulator adjusting screw clockwise until the working pressure gauge shows the correct working pressure with the gas flowing. Allow the gas to flow for five to ten seconds to purge the torch and hoses of air or fuel gas. Then release the cutting lever and close the oxygen valve.

Note that, on single-stage regulators, the working pressure shown on the gauge will likely rise when the cutting lever is released. You will see the pressure gauge fall somewhat each time the lever is opened; this is normal operation. Set the pressure on the oxygen regulator while the gas is flowing to ensure it will be as desired when working.

Open the fuel valve on the torch about one eighth of a turn. Turn the fuel regulator adjusting screw clockwise until the working pressure gauge shows the correct fuel gas working pressure with the gas flowing. Allow the gas to flow for five to ten seconds to purge the hoses and torch of air. Then close the torch fuel valve. If acetylene is used, check that the acetylene static pressure does not rise above 15 psig (103 kPa). If it does, immediately open the torch fuel valve and reduce the regulator output pressure as needed. Because of its instability, acetylene cannot be used at pressures of more than 15 psig (103kPa) when in gaseous form. At higher pressures, acetylene gas breaks down chemically, producing heat and pressure that could result in a violent explosion inside the hose.

> **WARNING!**
>
> The working pressure gauge readings on single-stage regulators will rise after the torch valves are turned off. This is normal. However, if acetylene is being used as the fuel gas, make sure that the static pressure does not rise above 15 psig (103 kPa). Make sure that equipment is purged and leak tested in a well-ventilated area to avoid creating an explosive concentration of gases.

3.2.0 Testing for Leaks

Equipment must be tested for leaks immediately after it is set up and periodically thereafter. The torch should be checked for leaks before each use. Leaks could cause a fire or explosion. To test for leaks, apply a commercially prepared leak-testing formula (*Figure 48*) or a solution of detergent and water to each potential leak point. If bubbles form, a leak is present. Be aware though, that solutions made from common household soaps are not usually as effective as commercially prepared leak detection solutions.

> **WARNING!**
>
> If a detergent is used for leak testing, make sure the detergent contains no oil. In the presence of oxygen, oil can cause fires or explosions.

There are numerous leak points to test, including the following:

- Oxygen cylinder valve
- Fuel gas cylinder valve
- Oxygen regulator and regulator inlet and outlet connections
- Fuel gas regulator and regulator inlet and outlet connections

Portable Oxyacetylene Equipment

Most oxyfuel equipment used on large job sites and in shops is very heavy and is usually transported using a special hand truck. This type of equipment is typically used when extensive cutting must be accomplished. However, for small tasks in unusual locations, such as a commercial building rooftop, portable equipment that can be hand carried by one person is often used.

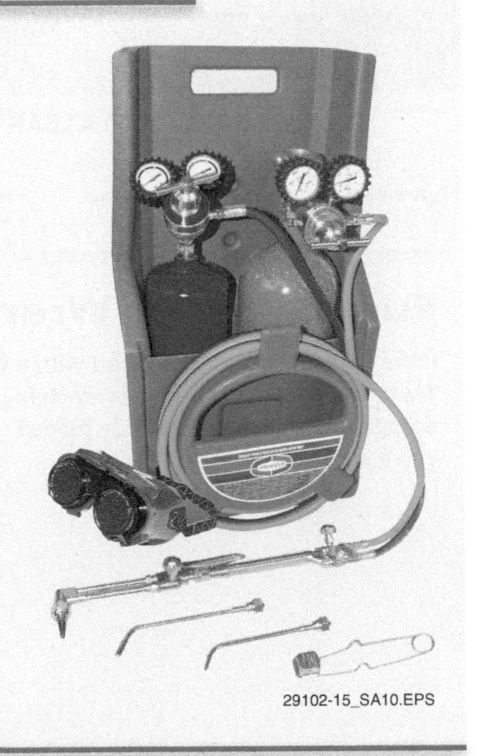

29102-15_SA10.EPS

APPLYING THE SOLUTION

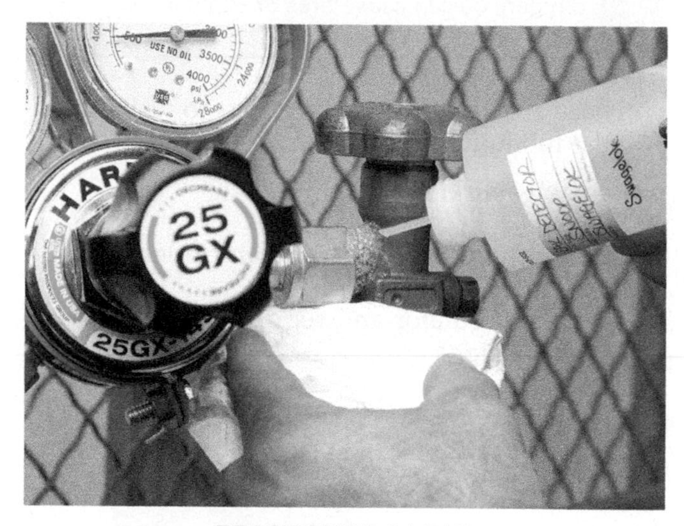

REACTION TO A LEAK

29102-15_F48.EPS

Figure 48 Leak detection fluid use.

- Hose connections at the regulators, check valves/ flashback arrestors, and torch
- Torch valves and cutting oxygen lever valve
- Cutting attachment connection (if used)
- Cutting tip

If there is a leak at the fuel gas cylinder valve stem, attempt to stop it by tightening the packing gland at the base of the stem. If this does not stop the leak, mark and remove the cylinder from service and notify the supplier. For other leaks, tighten the connections slightly with a wrench. If this does not stop the leak, turn off the gas pressure, open all connections, and inspect the fitting for damage.

WARNING!

Do not use Teflon® tape or pipe dope on these fittings since they do not provide a good seal. Make sure that equipment is purged and leak tested in a well-ventilated area to avoid creating an explosive concentration of gases.

3.2.1 Initial and Periodic Leak Testing

Initial and periodic leak testing is performed during initial equipment setup and periodically thereafter. First, set the equipment to the correct working pressures with the torch valves turned off. Then, using a leak-test solution, check for leaks at the cylinder valves, regulator relief ports, and regulator gauge connections (*Figure 49*). Also, check for leaks at hose connections, regulator connections, and check valve/flame arrestor connections up to the torch.

Fuel Cylinder Wrench

If the fuel cylinder is equipped with a valve requiring a T-wrench, always leave the wrench in place on the valve so that the fuel can be quickly turned off. This type of valve is obsolete but still in use.

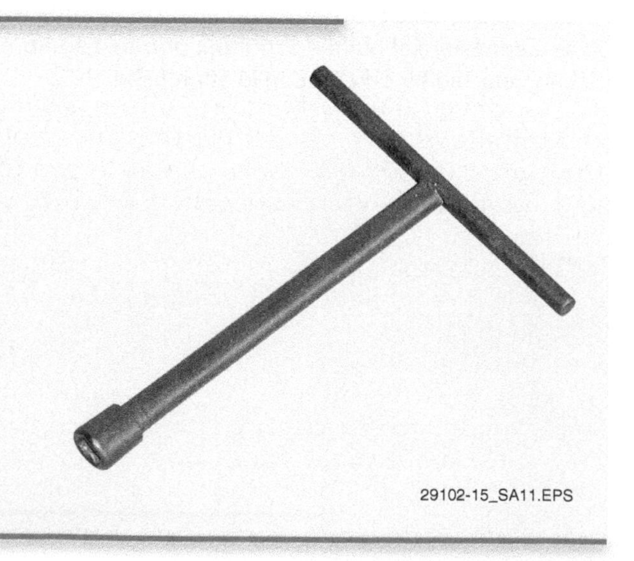

29102-15_SA11.EPS

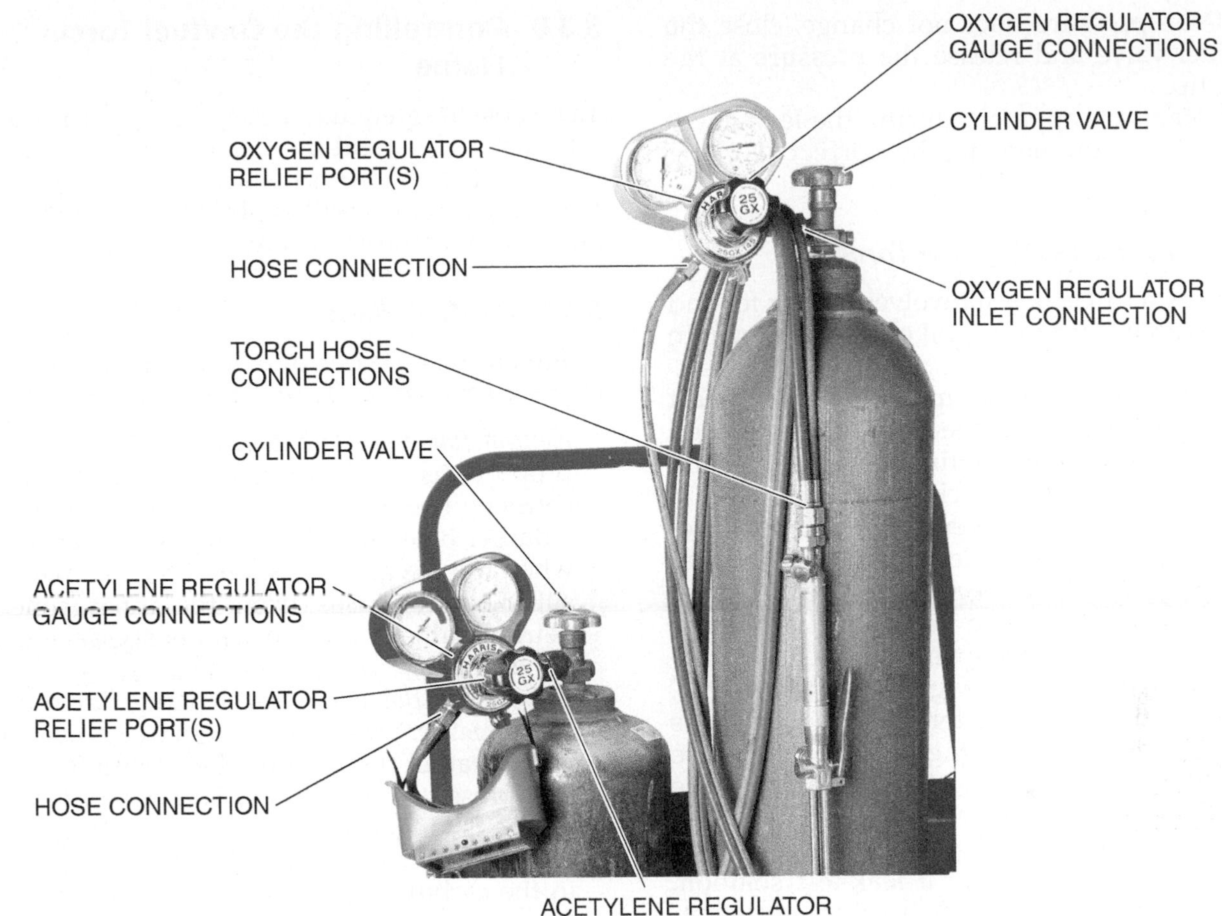

Figure 49 Typical initial and periodic leak-test points.

3.2.2 Leak-Down Testing of Regulators, Hoses, and Torch

Before the torch is ignited for use, the regulators, hoses, and torch should be quickly tested for leaks. First, set the equipment to the correct working pressures with the torch valves turned off. Then loosen both regulator adjusting screws. Check the working pressure gauges after a minute or two to see if the pressure drops. If the pressure drops, check the hose connection and regulators for leaks; otherwise, proceed with the test.

Place a thumb or finger over the cutting tip orifices and press tightly to block them (*Figure 50*). Turn on the torch oxygen valve and then depress and hold the cutting oxygen lever down. After the gauge pressure drops slightly, observe the oxygen working pressure gauge for a minute to see if the pressure continues to drop. If the pressure keeps dropping, perform the leak test described in the following section to determine the source of the leak. If the pressure does not change, close the torch oxygen valve and release the pressure at the cutting tip.

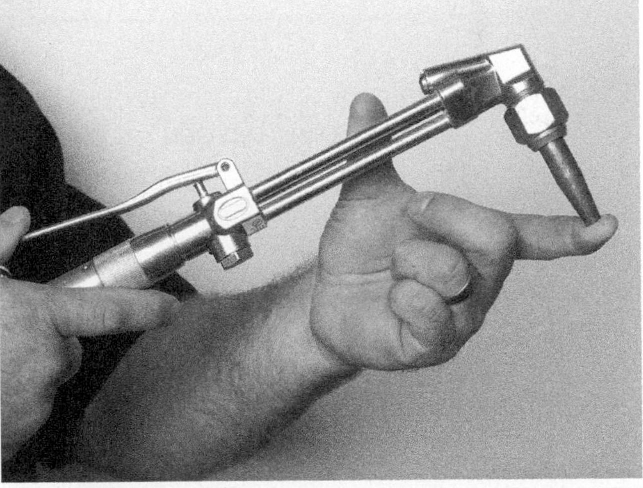

Figure 50 Blocking cutting tip for a leak test.

Next, block the tip again and turn on the torch fuel valve. After the gauge pressure drops slightly, carefully observe the fuel working pressure gauge for a minute. If the pressure continues to drop, perform the leak test described in the following section to determine the source of the leak. If the pressure does not change, close the torch fuel valve and release the pressure at the cutting tip.

If no leaks are apparent during the leak-down test, set the equipment to the correct working pressures.

3.2.3 Full Leak Testing of a Torch

Performing a full leak test involves testing for and isolating torch leaks in several places, as shown in *Figure 51*.

Start by setting the equipment to the correct working pressures with the torch valves turned off. Then, place a thumb or finger over the cutting tip orifices and press to block the orifices.

Turn on the torch oxygen valve and then depress the cutting oxygen lever. With the cutting tip blocked, check for leaks using a leak-test solution at the torch oxygen valve, cutting oxygen lever valve, cutting attachment connection to the handle (if used), preheat oxygen valve (if present), and cutting tip seal at the torch head. If no leaks are found, release the cutting oxygen lever, close the torch oxygen valve, and release the pressure at the cutting tip.

Next, with the cutting tip again blocked, open the torch fuel valve. Using a leak-test solution, check for leaks at the torch fuel valve, cutting attachment connection to the handle (if used), and cutting tip seal at the torch head. If no leaks are detected, close the torch fuel valve and release the pressure at the cutting tip.

3.3.0 Controlling the Oxyfuel Torch Flame

To be able to safely use a cutting torch, the operator must understand the flame and be able to adjust it and react to unsatisfactory conditions. The following sections will explain the oxyfuel flame and how to control it safely.

3.3.1 Oxyfuel Flames

There are three types of oxyfuel flames: **neutral flame**, **carburizing flame**, and **oxidizing flame**.

- *Neutral flame* – A neutral flame burns proper proportions of oxygen and fuel gas. The inner cones will be light blue in color, surrounded by a darker blue outer flame envelope that results when the oxygen in the air combines with the super-heated gases from the inner cone. A neutral flame is used for all but special cutting applications.
- *Carburizing flame* – A carburizing flame has a white feather created by excess fuel. The length of the feather depends on the amount of excess fuel present in the flame. The outer flame envelope is longer than that of the neutral flame, and it is much brighter in color. The excess fuel in the carburizing flame (especially acetylene) produces large amounts of carbon. The carbon will combine with red-hot or molten metal, making the metal hard and brittle. The carburizing flame is cooler than a neutral flame and is never used for cutting. It is used for some special heating applications.

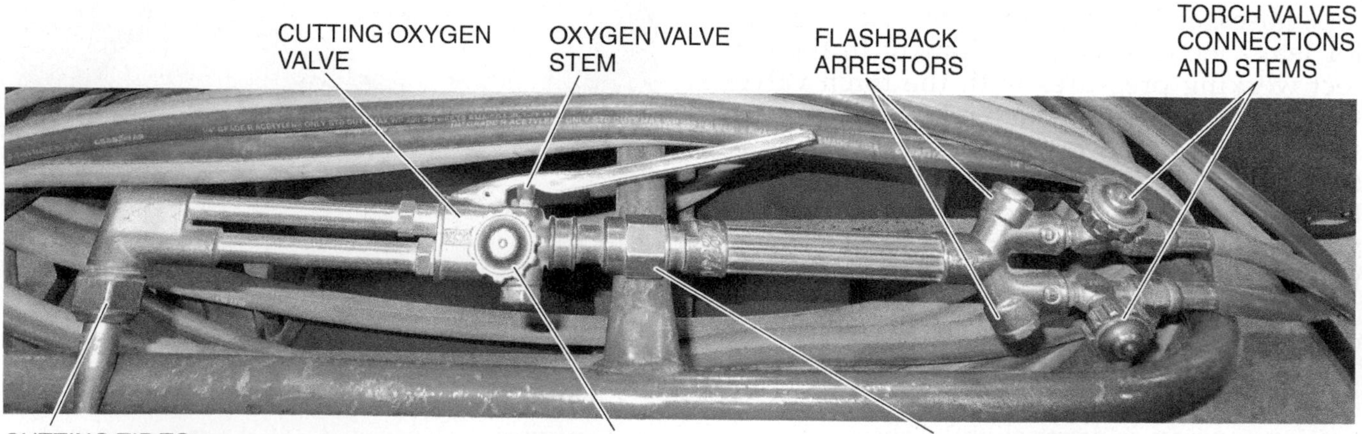

Figure 51 Torch leak-test points.

- *Oxidizing flame* – An oxidizing flame has an excess of oxygen. The inner cones are shorter, much bluer in color, and more pointed than a neutral flame. The outer flame envelope is very short and often fans out at the ends. An oxidizing flame is the hottest flame. A slightly oxidizing flame is recommended with some special fuel gases, but in most cases it is not used. The excess oxygen in the flame can combine with many metals, forming a hard, brittle, low-strength oxide. However, the preheat flames of a properly adjusted cutting torch will be slightly oxidizing when the cutting oxygen is shut off.

Figure 52 shows the various flames that occur at a cutting tip for both acetylene and LP gas.

Always Leak Test a Cutting Torch

Always take the time to perform a leak test on regulators, hoses, and the torch before using them the first time each day. A leak test should also be performed on a torch after tips have been changed or after converting the torch from one operation to another. Leaks, especially fuel gas leaks, can cause a fire or explosion after a cutting operation begins. A dangerous fire occurring at or near the torch may not be immediately noticed by the operator due to limited visibility through the tinted lenses.

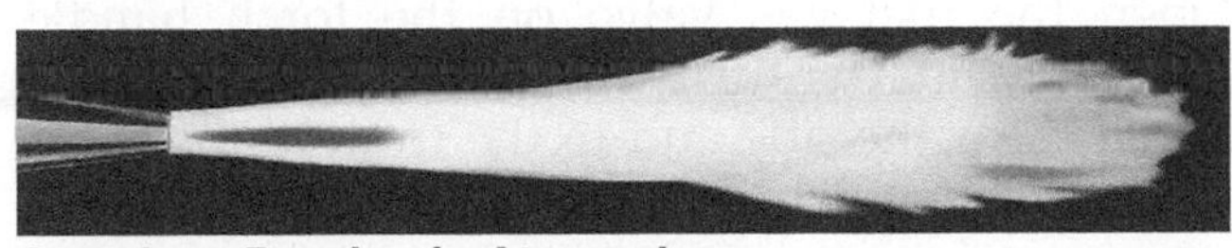

Acetylene Burning in Atmosphere
Open fuel gas valve until smoke clears from flame.

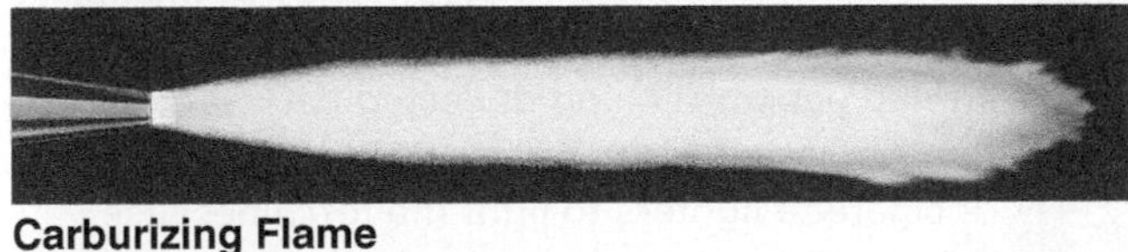

Carburizing Flame
(Excess acetylene with oxygen)
Preheat flames require more oxygen.

Neutral Flame
(Acetylene with oxygen) Temperature 5589°F (3087°C).
Proper preheat adjustment when cutting.

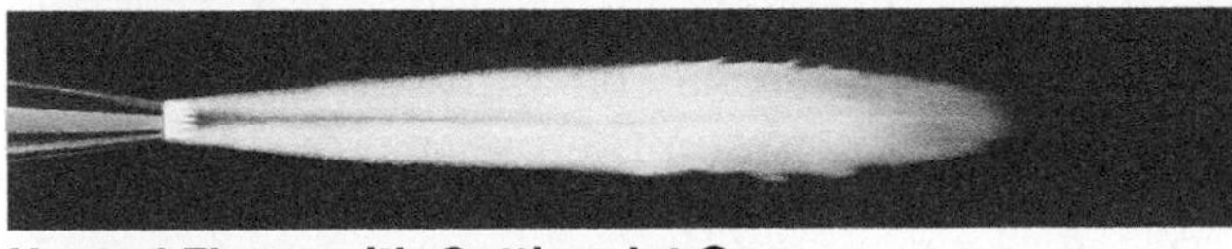

Neutral Flame with Cutting Jet Open
Cutting jet must be straight and clean.
If it flares, the pressure is too high for the tip size.

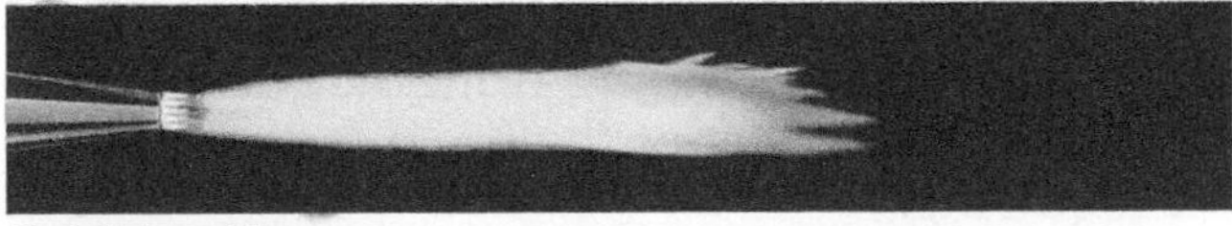

Oxidizing Flame
(Acetylene with excess oxygen) Not recommended for average cutting. However, if the preheat flame is adjusted for neutral with the cutting oxygen on, then this flame is normal after the cutting oxygen is off.

OXYACETYLENE FLAME

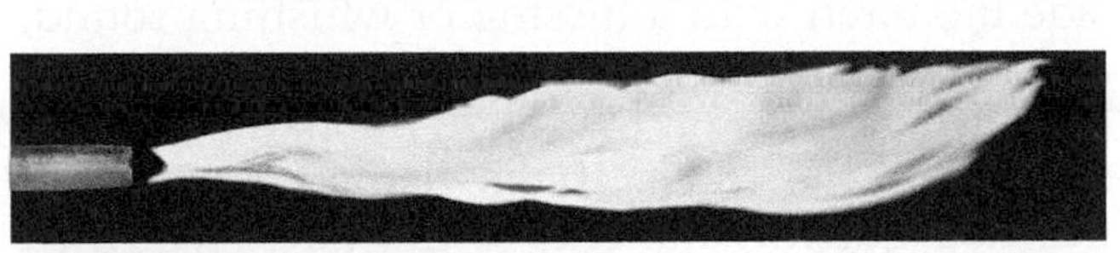

LP Gas Burning in Atmosphere
Open fuel gas valve until flame begins to leave tip end.

Reducing Flame
(Excess LP-gas with oxygen) Not hot enough for cutting.

Neutral Flame
(LP-gas with oxygen) For preheating prior to cutting.

Oxidizing Flame with Cutting Jet Open
Cutting jet stream must be straight and clean.

Oxidizing Flame without Cutting Jet Open
(LP-gas with excess oxygen) The highest temperature flame for fast starts and high cutting speeds.

OXYPROPANE FLAME

29102-15_F52.EPS

Figure 52 Acetylene and LP (propane) gas flames.

3.3.2 Backfires and Flashbacks

When the torch flame goes out with a loud pop or snap, a backfire has occurred. Backfires are usually caused when the tip or nozzle touches the work surface or when a bit of hot dross briefly interrupts the flame. When a backfire occurs, relight the torch immediately. Sometimes the torch even relights itself. If a backfire recurs without the tip making contact with the base metal, shut off the torch and find the cause. Possible causes are the following:

- Improper operating pressures
- A loose torch tip
- Dirt in the torch tip seat or a bad seat

When the flame goes out and burns back inside the torch with a hissing or whistling sound, a flashback is occurring. Immediately shut off the oxygen valve on the torch; the flame is burning inside the torch. If the flame is not extinguished quickly, the end of the torch will melt off. The flashback will stop as soon as the oxygen valve is closed. Therefore, quick action is crucial. Flashbacks can cause fires and explosions within the cutting rig and, therefore, are very dangerous. Flashbacks can be caused by the following:

- Equipment failure
- Overheated torch tip
- Dross or spatter hitting and sticking to the torch tip
- Oversized tip (tip is too large for the gas flow rate being used)

After a flashback has occurred, wait until the torch has cooled. Then, blow oxygen (not fuel gas) through the torch for several seconds to remove soot that may have built up in the torch during the flashback before relighting it. If the torch makes a hissing or whistling sound after it is reignited or if the flame does not appear to be normal, shut off the torch immediately and have the torch serviced by a qualified technician.

3.3.3 Igniting the Torch and Adjusting the Flame

After the cutting equipment has been properly set up and purged, the torch can be ignited and the flame adjusted for cutting. The procedure for igniting the torch starts with choosing the appropriate cutting torch tip according to the base metal thickness being cut and the fuel gas being used. Always inspect the cutting tip sealing surfaces and orifices prior to installation. Attach the tip to the cutting torch by placing it on the end of the torch and tightening the nut. Some manufacturers recommend tightening the nut with a torch wrench, while others recommend tightening the nut by hand. Check the manufacturer's documentation for the equipment in use to ensure the tip is being installed correctly.

NOTE

Refer to the manufacturer's charts. Depending on the tip selected, the oxygen and fuel gas pressure may have to be adjusted.

Prior to opening the oxygen and fuel gas valves, be sure to put on the proper PPE. Also ensure that you are not depressing the oxygen cutting lever. If present, close the preheat oxygen valve and open the torch oxygen valve fully. Open the fuel gas valve on the torch handle about one-quarter turn. Then, holding the friction lighter near the side and to the front of the torch tip, ignite the torch.

WARNING!

Hold the friction lighter near the side of the tip, rather than directly in front of it, to prevent the ignited gas from being deflected backwards. Always use a friction lighter. Never use matches or cigarette lighters to light the torch because this could result in severe burns and/or could cause the lighter to explode. Always point the torch away from yourself, other people, equipment, and flammable material.

Once the torch is lit, adjust the torch fuel gas flame by adjusting the flow of fuel gas with the fuel gas valve. Increase the flow of fuel gas until the flame stops smoking or pulls slightly away from the tip. Decrease the flow until the flame returns to the tip. Open the preheat oxygen valve (if present) or the oxygen torch valve very slowly and adjust the torch flame to a neutral flame. Then press the cutting oxygen lever all the way down and observe the flame. It should have a long, thin, high-pressure oxygen cutting jet up to 8" (≈20 cm) long, extending from the cutting oxygen hole in the center of the tip. If it does not, do the following:

- Check that the working pressures are set as recommended on the manufacturer's chart.
- Clean the cutting tip. If this does not clear up the problem, change the cutting tip.

With the cutting oxygen on, observe the preheat flame. If it has changed slightly to a carburizing flame, increase the preheat oxygen until the flame is neutral. After this adjustment, the preheat flame will change slightly to an acceptable oxidizing flame when the cutting oxygen is shut off.

3.3.4 Shutting Off the Torch

Shutting off the torch itself is done by releasing the cutting oxygen lever and then closing the torch or preheat oxygen valves. After that, quickly close the torch fuel gas valve to extinguish the flame.

> **WARNING!**
>
> Always turn off the oxygen flow first to prevent a possible flashback into the torch.

3.4.0 Shutting Down Oxyfuel Cutting Equipment

When a cutting job is completed and the oxyfuel equipment is no longer needed, it must be shut down. *Figure 53* identifies the order in which various pieces of the oxyfuel cutting equipment are shut down.

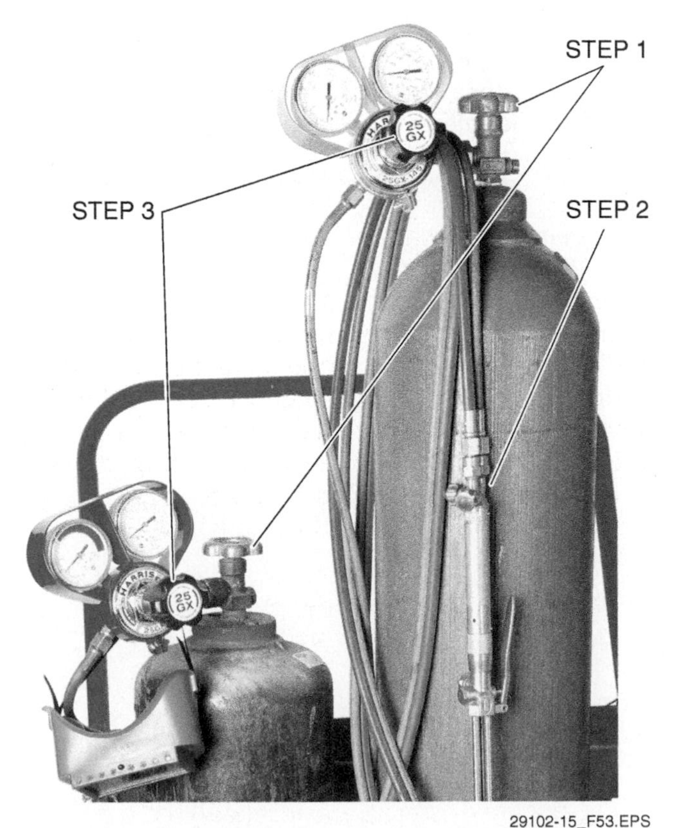

Figure 53 Shutting down oxyfuel cutting equipment.

Step 1 To begin the shutdown, first close the fuel gas and oxygen cylinder valves. Leave the regulators at their present setting.

Step 2 Open the fuel gas valve on the torch to allow all remaining gas to escape, and then close it. Next, open the oxygen valve on the torch to allow all remaining gas to escape, and then close it. These actions relieve the gas pressure in the hose and regulators, all the way back to the cylinder valve. Do not proceed to the next step until all pressure is released and all regulator gauges — both inlet and outlet — read zero.

Step 3 Turn the fuel gas and oxygen regulator adjusting screws counterclockwise to back them out, until they are loose.

Step 4 Coil and secure the hose and torch to prevent damage.

3.4.1 Disassembling Oxyfuel Equipment

In some situations, it may be necessary to disassemble the oxyfuel equipment after it has been used. Before any disassembly takes place, make sure that the equipment has been properly shut down. This includes checking that the cylinder valves are closed and all pressure gauges read zero.

Remove both hoses from the torch assembly and then detach the hoses from the regulators. Remove both regulators from the cylinders and reinstall the protective caps on the cylinders. The cylinders should now be returned to their proper storage place.

> **WARNING!**
>
> Always transport and store gas cylinders in the upright position. Be sure they are properly secured (chained) and capped. Regardless of whether the cylinders are empty or full, never store fuel gas cylinders and oxygen cylinders together without providing the required separation distance or fire-rated partition.

Obtaining Maximum Fuel Flow

Increasing the fuel flow until the flame pulls away from the tip and then decreasing the flow until the flame returns to the tip sets the maximum fuel flow for the tip size in use.

3.4.2 *Changing Cylinders*

Empty is a relative term when discussing gas cylinders. These cylinders should never be completely emptied because reverse flow could occur. Oxygen tanks, for example, should never get below the required working pressure, or about 25 psi (172 kPa). Once the cylinder pressure drops near the working pressure value, the torch will stop performing properly and the cylinder will have to be replaced. As a result, some residual pressure is typically present. Follow these procedures to change a cylinder:

> **WARNING!**
>
> When moving cylinders, always be certain that they are in the upright position and the valve caps are secured in place. Never use a sling or electromagnet to lift cylinders. To lift cylinders, use a cylinder cage.

Step 1 Begin by making sure that the equipment has been properly shut down. This includes checking that the cylinder valves are closed and all pressure gauges read zero.

Step 2 Remove the regulator from the empty cylinder and replace the protective cap on the cylinder.

Step 3 Mark MT (empty) and the date (or the accepted site notation for indicating an empty cylinder) near the top of the cylinder using soapstone (*Figure 54*).

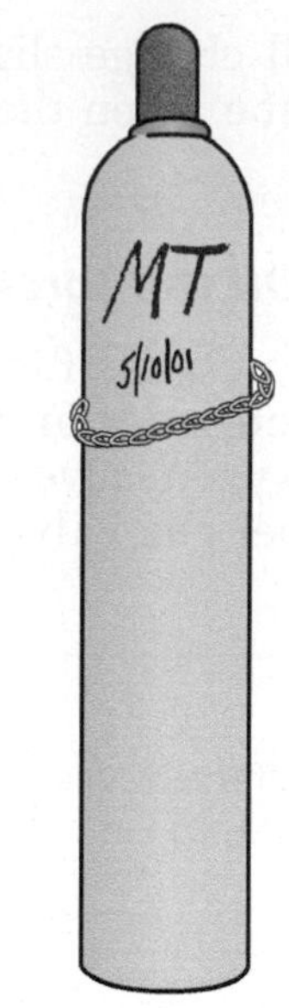

Figure 54 Typical empty cylinder marking.

Step 4 Transport the empty cylinder from the workstation to the storage area. Place the empty cylinder in the empty cylinder section of the storage area for the type of gas it contained.

> **Marking and Tagging Cylinders**
>
> Do not use permanent markers on cylinders; use soapstone or another temporary marker. If a cylinder is defective, place a warning tag on it.

Additional Resources

ANSI Z49.1, Safety in Welding, Cutting, and Allied Processes. Miami, FL: American Welding Society.

Uniweld Products, Inc. Numerous videos are available at **www.uniweld.com/en/uniweld-videos**. Last accessed: November 30, 2014.

The Harris Products Group, a division of Lincoln Electric. Numerous videos are available at **www.harrisproductsgroup.com/en/Expert-Advice/videos.aspx**. Last accessed: November 30, 2014.

3.0.0 Section Review

1. Gas cylinders can be lifted to height ____ .
   a. with a strong electromagnet
   b. using a cylinder cage
   c. with a sling routed through the openings in the cap
   d. with a cable routed through the openings in the cap

2. Immediately after setup and periodically thereafter, oxyfuel equipment must be tested for ____ .
   a. purging
   b. carburizing
   c. flashbacks
   d. leaks

3. An oxidizing torch flame is one that has a(n) ____ .
   a. excess of fuel gas
   b. proper oxygen/fuel gas mix
   c. excess of oxygen
   d. abnormally low temperature

4. When oxyfuel cutting equipment is being shut down, how should the fuel gas and oxygen regulator adjusting screws be positioned after bleeding the remaining gas pressure in the hoses?
   a. Both screws tight
   b. Fuel gas screw tight; oxygen screw loose
   c. Fuel gas screw loose; oxygen screw tight
   d. Both screws loose

SECTION FOUR

4.0.0 PERFORMING CUTTING PROCEDURES

Objective

Explain how to perform various oxyfuel cutting procedures.

a. Identify the appearance of both good and inferior cuts and their causes.
b. Explain how to cut both thick and thin steel.
c. Explain how to bevel, wash, and gouge.
d. Explain how to make straight and bevel cuts with portable oxyfuel cutting machines.

Performance Tasks

6. Cut shapes from various thicknesses of steel, emphasizing:
   - Straight line cutting
   - Square shape cutting
   - Piercing
   - Beveling
   - Cutting slots
7. Perform washing.
8. Perform gouging.
9. Use a track burner to cut straight lines and bevels.

Trade Terms

Drag lines: The lines on the edge of the material that result from the travel of the cutting oxygen stream into, through, and out of the metal.

The following sections explain how to recognize good and bad cuts, how to prepare for cutting operations, and how to perform straight-line cutting, piercing, bevel cutting, washing, and gouging.

4.1.0 Preparing for Oxyfuel Cutting with a Hand Cutting Torch

Before metal can be cut, the equipment must be set up and the metal prepared. One important step is to properly lay out the cut by marking it with soapstone or punch marks. The few minutes this takes will result in a quality job, reflecting craftsmanship and pride. The following procedures describe how to prepare to make a cut.

Prepare the metal to be cut by cleaning any rust, scale, or other foreign matter from the surface. If possible, position the work so that it can be cut comfortably. Mark the lines to be cut with soapstone or a scriber. Then select the correct cutting torch tip according to the thickness of the metal to be cut, the type of cut to be made, the amount of preheat needed, and the type of fuel gas to be used. Ignite the torch and use the procedures outlined in the following sections for performing specific types of cutting operations.

4.1.1 Inspecting the Cut

Before attempting to make a cut, operators must be able to recognize good and bad cuts and know what causes bad cuts. This is explained in the following list and illustrated in *Figure 55*:

- A good cut features a square top edge that is sharp and straight, not ragged. The bottom edge can have some dross adhering to it but not an excessive amount. What dross there is should be easily removable with a chipping hammer. The **drag lines** should be near vertical and not very pronounced.
- When preheat is insufficient, bad gouging results at the bottom of the cut because of slow travel speed.
- Too much preheat will result in the top surface melting over the cut, an irregular cut edge, and an excessive amount of dross.
- When the cutting oxygen pressure is too low, the top edge will melt over because of the resulting slow cutting speed.
- Using cutting oxygen pressure that is too high will cause the operator to lose control of the cut, resulting in an uneven kerf.
- A travel speed that is too slow results in bad gouging at the bottom of the cut and irregular drag lines.
- When the travel speed is too fast, there will be gouging at the bottom of the cut, a pronounced break in the drag line, and an irregular kerf.
- A torch that is held or moved unsteadily across the metal being cut can result in a wavy and irregular kerf.
- When a cut is lost and then not restarted carefully, bad gouges will result at the point where the cut is restarted.

A square kerf face with minimal notching not exceeding 1⁄16" (1.6 mm) deep is expected and, in fact, required in the Performance Accreditation Tasks for this module.

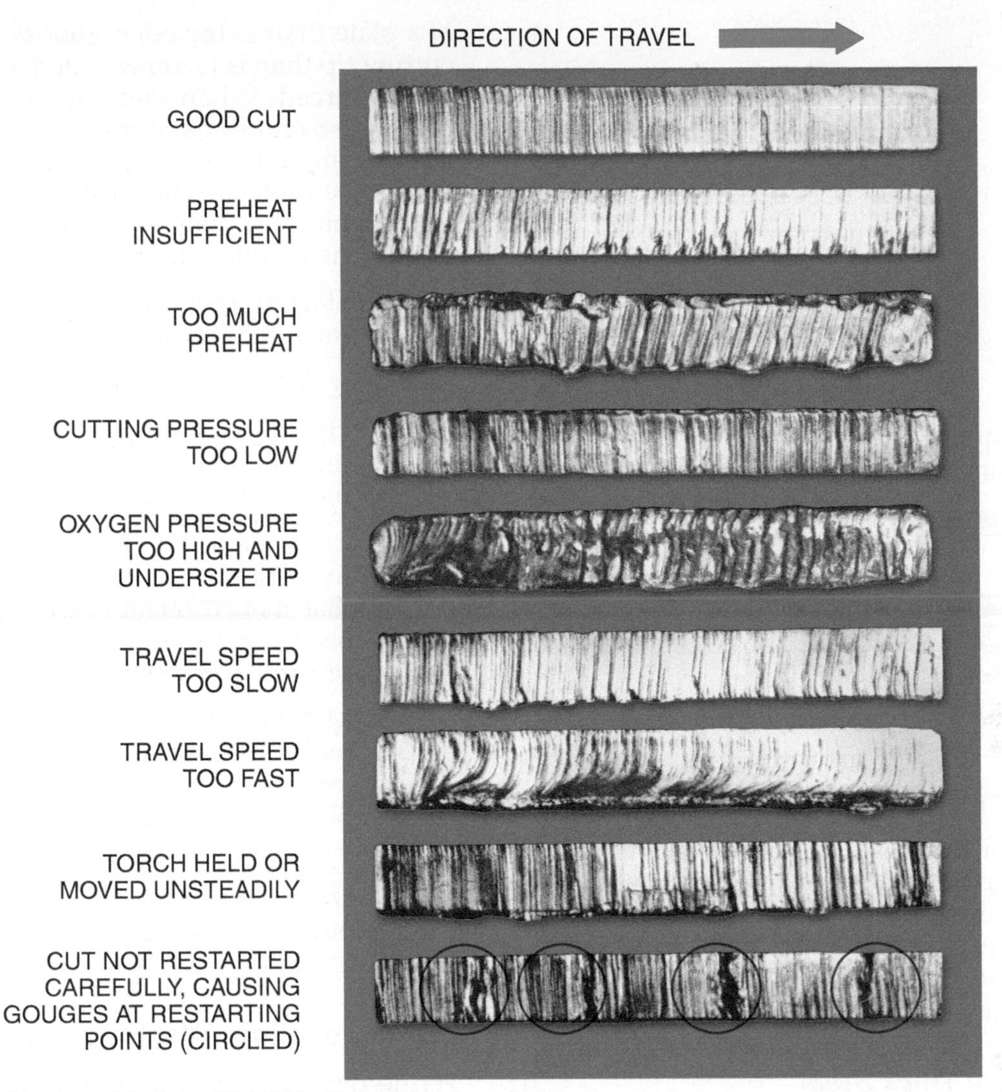

Figure 55 Examples of good and bad cuts.

The tasks in the sections that follow are designed to develop skills with a cutting torch. Each task should be practiced until there is thorough familiarity with the procedure. After each task is completed, it should be taken to the instructor for evaluation. Do not proceed to the next task until the instructor says to continue.

4.2.0 Cutting Steel

The effectiveness of cutting steel with an oxyfuel cutting outfit depends on factors such as the thickness of the steel, the cutting tip that is being used, and the skill of the operator.

4.2.1 Cutting Thin Steel

Thin steel is considered material 3⁄16" (≈5 mm) thick or less. A major concern when cutting thin steel is distortion caused by the heat of the torch and the cutting process. To minimize distortion, move as quickly as possible without losing the cut.

To begin the process for cutting thin steel, first prepare the metal surface. Then light the torch and hold it so that the tip is pointing in the direction the torch is traveling at a 15- to 20-degree angle. Make sure that a preheat orifice and the cutting orifice are centered on the line of travel next to the metal (*Figure 56*).

> **CAUTION**
>
> Holding the tip upright (perpendicular to the metal) when cutting thin steel will overheat the metal, causing distortion. Maintain the 15 to 20-degree push angle as shown.

Preheat the metal to a dull red. Use care not to overheat thin steel because this will cause distortion. The edge of the tip can be lightly rested on

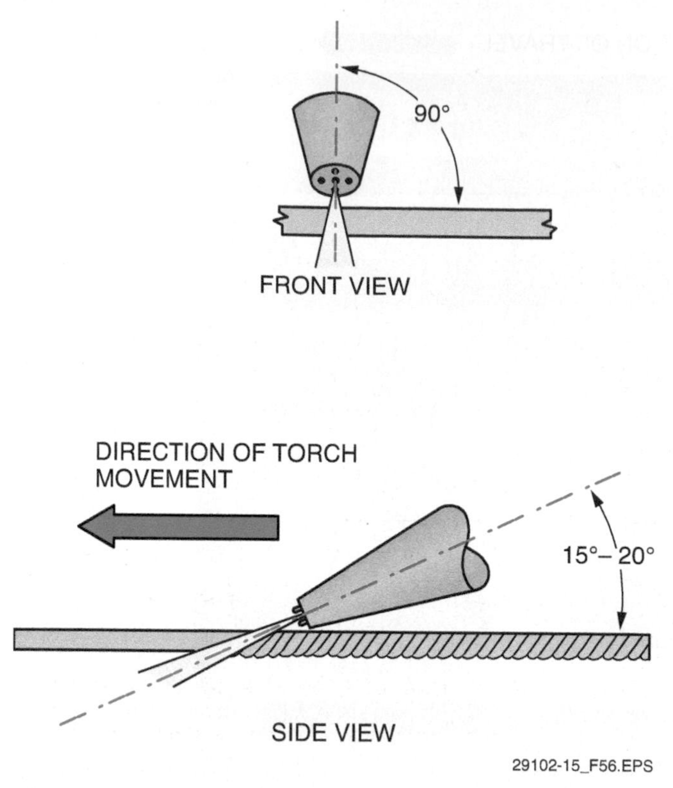

Figure 56 Cutting thin steel.

the surface of the metal being cut and then slid along the surface when making the cut in thin metal. Press the cutting oxygen lever to start the cut, and then move quickly along the line. To minimize distortion, move as quickly as possible without losing the cut.

4.2.2 Cutting Thick Steel

Most oxyfuel cutting is done on steel that is more than 3⁄16" (≈5 mm) thick. Whenever heat is applied to metal, distortion is a problem, but as the steel gets thicker, it becomes less of a problem.

To cut thick steel with a cutting torch, start by preparing the metal surface. Then light the torch and adjust the torch flame. Follow the number sequence shown in *Figure 57* to perform the cut.

The torch can be moved from either right to left or left to right. Choose the direction that allows the best visibility of the cut. When cutting begins, the tips of the preheat flame should be held 1⁄16" to 1⁄8" (1.6 to 3.2 mm) above the workpiece. For steel up to 3⁄8" (10 mm) thick, the first and third procedures can usually be omitted.

4.2.3 Piercing a Plate

Before holes or slots can be cut in a plate, the plate must be pierced. Piercing puts a small hole through the metal where the cut can be started. Because more preheat is necessary on the surface of a plate than at the edge, choose the next-larger cutting tip than is recommended for the thickness to be pierced. When piercing steel that is more than 3" (≈8 cm) thick, it may help to first preheat the bottom side of the plate directly under the spot to be pierced. The following steps describe how to pierce a plate for cutting. *Figure 58* provides a visual reference.

Step 1 Start by preparing the metal surface and the torch for cutting.

Step 2 Ignite the torch and adjust the flame.

Step 3 Hold the torch tip 1⁄4" to 5⁄16" (6.4 mm to 7.9 mm) above the spot to be pierced until the surface is a bright cherry red.

Step 4 Raise the tip about 1⁄2" (12.7 mm) above the metal surface and tilt the torch slightly so that molten metal does not blow directly back into the tip as the oxygen lever is depressed. Depress the oxygen lever.

Step 5 Maintain the tipped position until a hole burns through the plate. Then rotate the torch back to the vertical position (perpendicular to the plate).

Step 6 Lower the torch back to the initial distance from the plate and continue to cut outward from the original hole to the line to be cut. Then follow the line.

4.3.0 Beveling, Washing, and Gouging

While oxyfuel cutting equipment is commonly associated with cutting through metal plate, it is also well suited for cutting angles in the edge of steel plate, removing bolts and rivets, and cutting groves in metal surfaces.

4.3.1 Cutting Bevels

Bevel cutting is often performed to prepare the edge of steel plate for welding. The procedure for bevel cutting is illustrated in *Figure 59*.

Step 1 Prepare the metal surface and the torch.

Step 2 Ignite the torch and adjust the flame.

Step 3 Hold the torch so that the tip faces the metal at the desired bevel angle. Using a piece of angle iron as a cutting guide as shown in *Figure 59* will result in a 45-degree bevel angle. Angle iron can be used as a guide for any angle, as long as the operator consciously maintains the torch at the proper bevel angle.

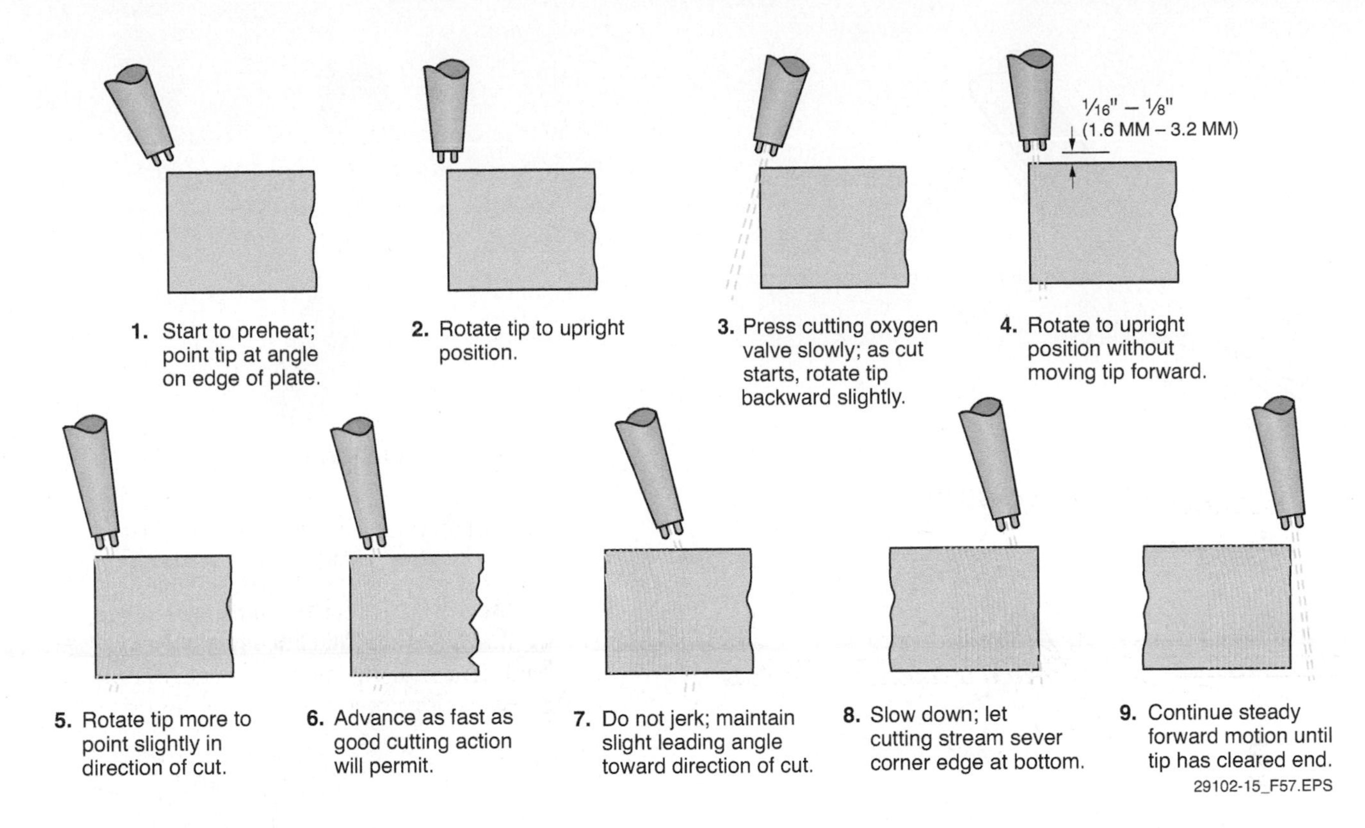

1⁄16" – 1⁄8"
(1.6 MM – 3.2 MM)
1. Start to preheat; point tip at angle on edge of plate.
2. Rotate tip to upright position.
3. Press cutting oxygen valve slowly; as cut starts, rotate tip backward slightly.
4. Rotate to upright position without moving tip forward.
5. Rotate tip more to point slightly in direction of cut.
6. Advance as fast as good cutting action will permit.
7. Do not jerk; maintain slight leading angle toward direction of cut.
8. Slow down; let cutting stream sever corner edge at bottom.
9. Continue steady forward motion until tip has cleared end.
29102-15_F57.EPS

Figure 57 Flame cutting with a hand torch.

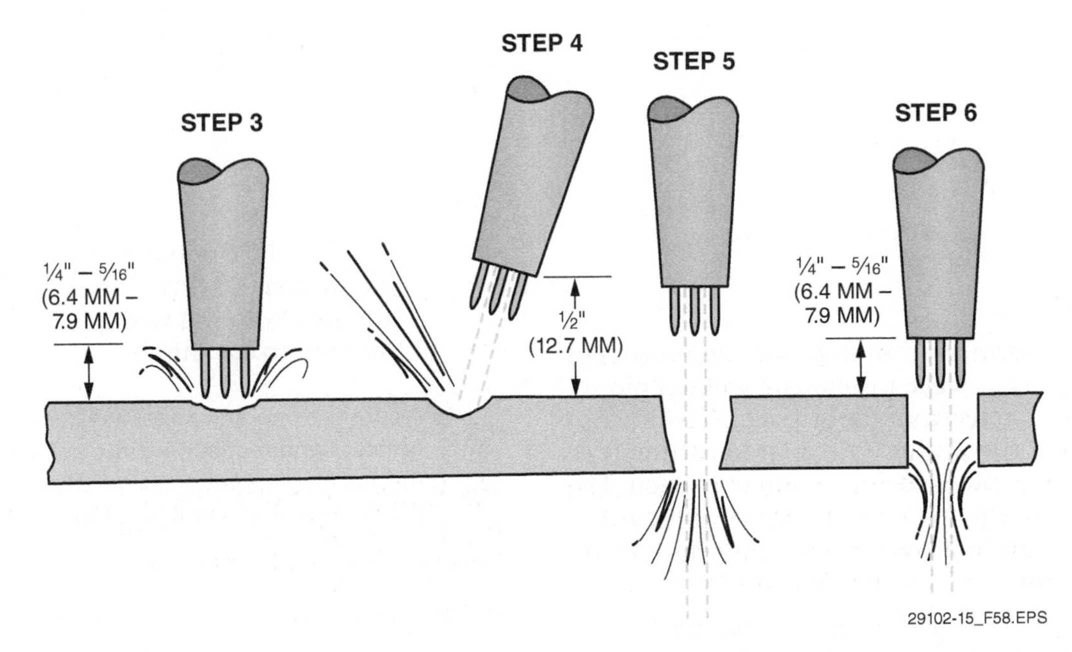

Figure 58 Steps for piercing steel.

Step 4 Preheat the edge to a bright cherry red.

Step 5 Press the cutting oxygen lever to start the cut.

Step 6 As cutting begins, move the torch tip at a steady rate along the line to be cut. Pay particular attention to the torch angle to ensure it creates a uniform bevel along the entire length of the cut.

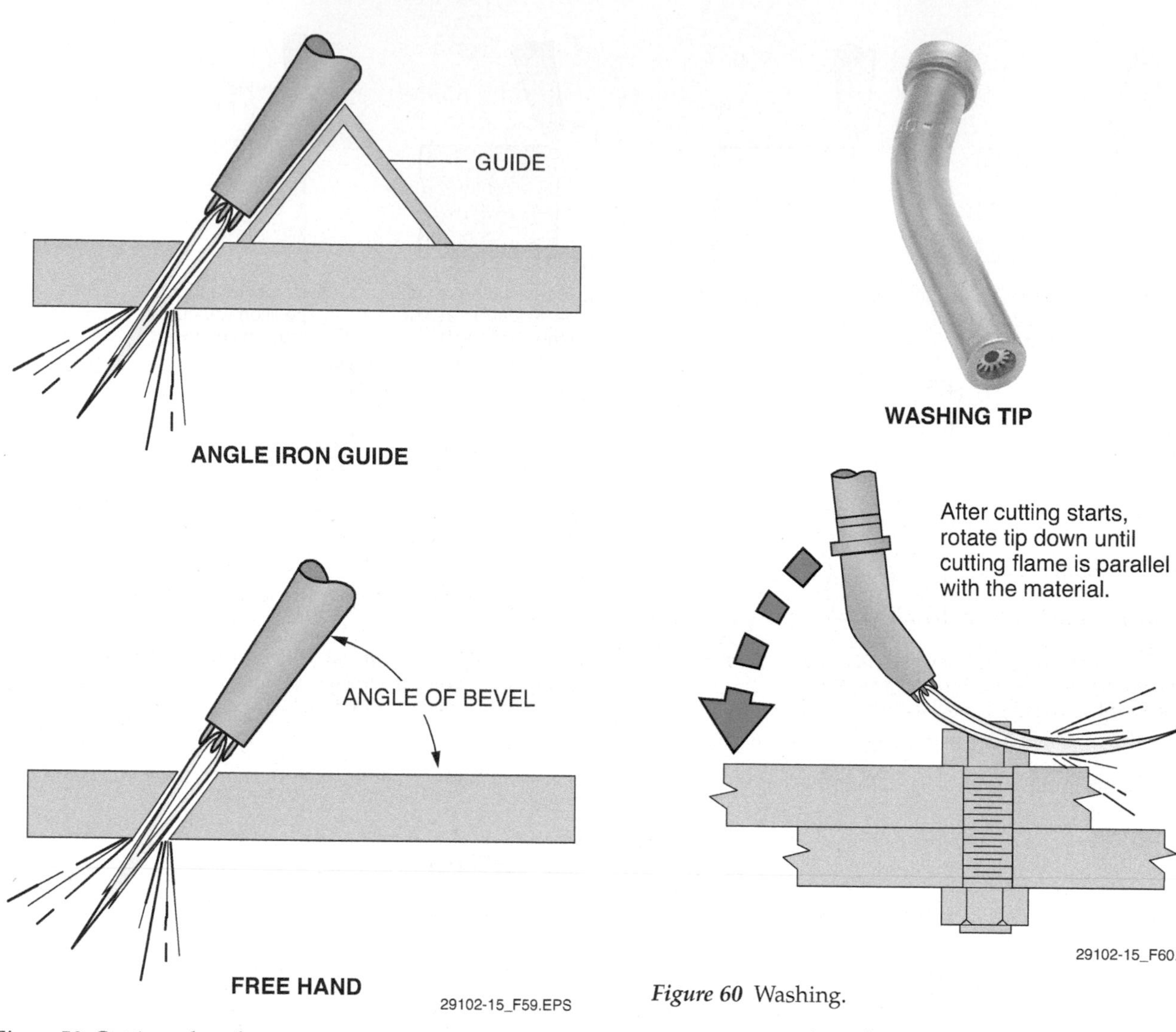

Figure 59 Cutting a bevel.

Figure 60 Washing.

4.3.2 Washing

Washing is a term used to describe the process of cutting out bolts, rivets, previously welded pieces, or other projections from the surface. Washing operations use a special tip with a large cutting hole that produces a low-velocity stream of oxygen. The low-velocity oxygen stream helps prevent cutting into the surrounding base metal. *Figure 60* is a simplified illustration of a washing procedure.

Step 1 Prepare the metal surface and torch.

Step 2 Ignite the torch and adjust the flame.

Step 3 Preheat the metal to be cut until it is a bright cherry red.

Step 4 Rotate the cutting torch tip to roughly a 55-degree angle to the metal surface.

Step 5 At the top of the material, press the cutting oxygen lever to begin cutting the material to be removed. Continue moving back and forth across the material while rotating the tip to a position parallel with the material. Move the tip back and forth and down to the workpiece surface. Take care not to cut into it.

CAUTION

As the surrounding metal heats up, there is a greater danger of cutting into it. Try to complete the washing operation as quickly as possible. If the surrounding metal gets too hot, stop and let it cool down.

4.3.3 Gouging

Gouging (*Figure 61*) is the process of cutting a groove into a surface. Gouging operations use a special curved tip that produces a low-velocity stream of oxygen that curves up, allowing the operator to control the depth and width of the groove. It is an effective means to gouge out cracks or weld defects for welding. Gouging tips can also be used to remove steel backing from welds or to wash off bolt or rivet heads. However, gouging tips are not as effective as washing tips for removing the shank of a bolt or rivet.

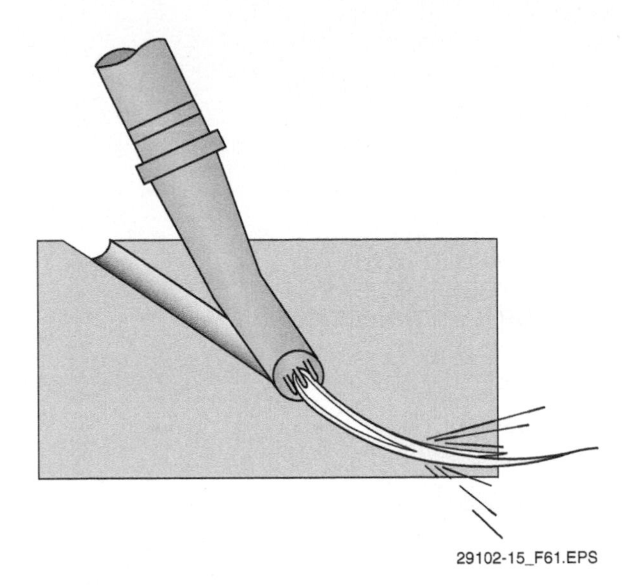

Figure 61 Gouging.

The travel speed and torch angle are very important when gouging. If the travel speed or torch angle is incorrect, the gouge will be irregular and there will be a buildup of dross inside the gouge. Practice until the gouge is clean and even, with a consistent depth.

Step 1 Prepare the metal surface and torch.

Step 2 Ignite the torch and adjust the flame.

Step 3 Holding the torch so that the preheat holes are pointed directly at the metal, preheat the surface until it becomes a bright cherry red.

Step 4 When the steel has been heated to a bright cherry red, slowly roll the torch away from the metal so that the holes are at an angle that will cut the gouge to the correct depth. While rolling the torch away, depress the cutting oxygen lever gradually.

Step 5 Continue to move the cutting torch along the line of the gouge while rocking it back and forth to create a gouge of the required depth and width.

4.4.0 Operating Oxyfuel Track Burners

Oxyfuel track burners, such as the one shown in *Figure 62*, provide a convenient way for operators to make straight cuts, curved cuts, and beveled cuts in the field. They can also enhance precision and uniformity in cuts. When a number of cuts are needed, track burners significantly increase productivity. Most models are portable, allowing them to be set up on the job site.

4.4.1 Torch Adjustment

The rack assembly on the track burner permits the torch holder assembly to move toward or away from the tractor unit. The torch holder allows vertical positioning of the torch. The torch bevel adjustment allows torch positioning at any commonly required angle. After adjusting the torch to the desired position, tighten all clamping screws to prevent the torch from making any unexpected movements.

4.4.2 Straight-Line Cutting

The following provides some basic information regarding the set-up of track burners. Although most track burners have a great deal in common, be sure to follow the manufacturer's operating procedures for the system in use.

To perform straight-line cutting with an oxyfuel track burner, first place the machine track on the workpiece and line it up before placing the machine on the track. Be sure the track is long enough for the cut to be made. If not, install additional track. Connect track sections carefully and ensure it is properly supported. When properly connected, the machine should travel smoothly from one track section to the next. If the cut is long, the track may have to be clamped at both ends beyond the cut to keep the track from moving during the cutting process.

WARNING!

Many cutting machines are not designed to detect the end of their track or workpiece. Take care that an unattended machine does not fall from an elevated workpiece while in operation.

Once the track has been positioned, place the machine on the track. Be sure that the supply gas hoses and the power lines are long enough and free to move with the machine so that it can complete the cut properly. Move the machine to the approximate point where the cut will start. Then set the Low/High speed switch to the desired cutting speed or speed range. Set the On/Off switch to the Off position. Next, plug the power cord into an appropriate power supply outlet. Ensure that all clamping screws are properly tightened. Now ignite and properly adjust the torch, and preheat the start of the cut. Set the Forward/Reverse switch to the desired direction of travel. Simultaneously turn on the cutting oxygen and rotate the cutting speed control knob to the desired rate of travel. When the cut is completed, stop the machine and shut off the torch.

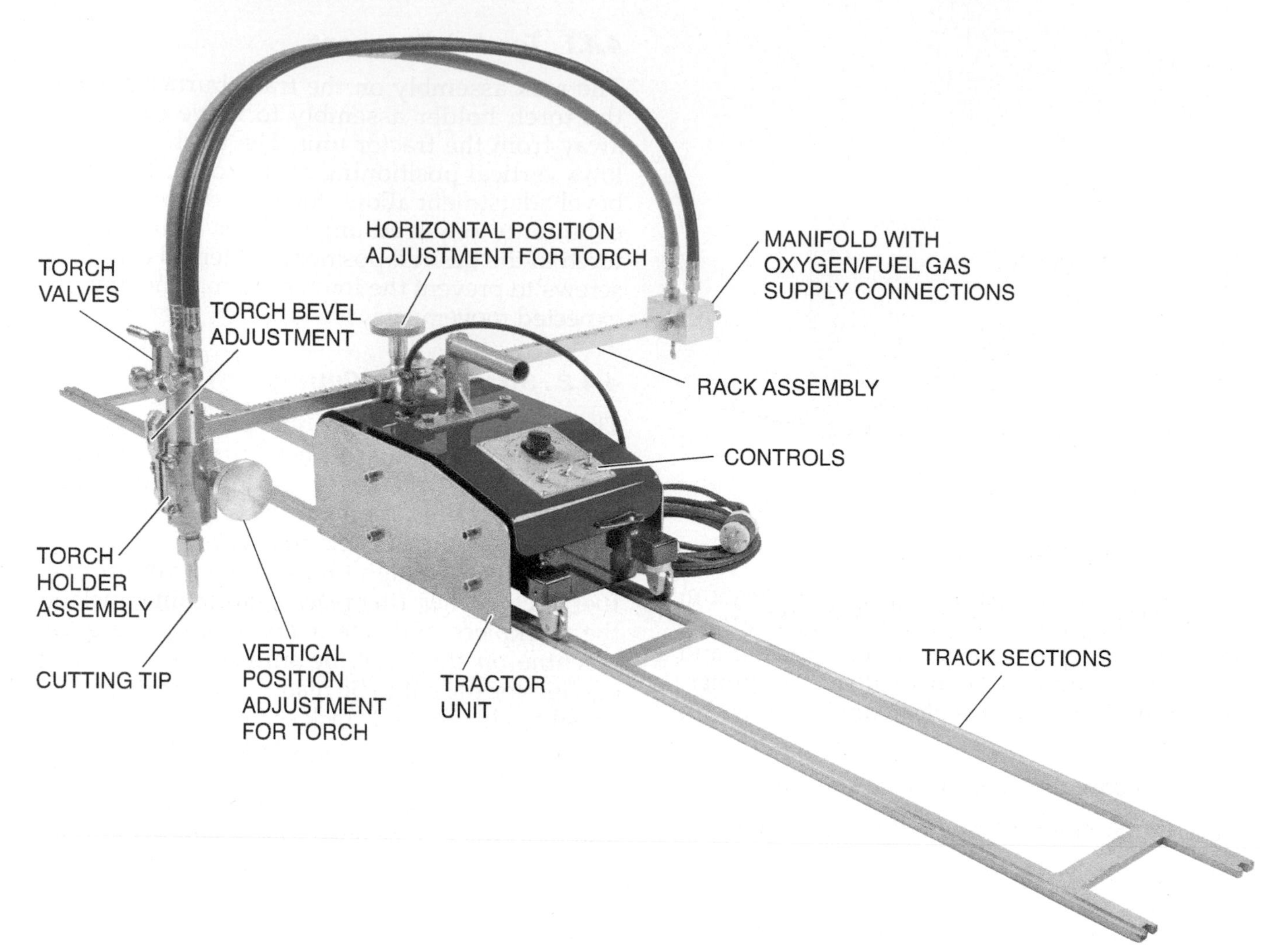

Figure 62 Portable oxyfuel track burner.

4.4.3 Bevel Cutting

Bevel cutting can also be performed with a portable oxyfuel track burner. As before, first place the machine track on the workpiece and line it up before placing the machine on the track. The track must be long enough for the cut to be made. If it is not, install additional track. Connect track sections carefully. Extend the track on both sides of the cut and support the track. When properly connected, the machine should travel smoothly from one track section to the next. If the cut is long, the track may have to be clamped at both ends beyond the cut to keep the track from moving during the cut.

With the track properly positioned, place the machine on the track. Make sure the supply gas hoses and the power lines are long enough and free to move with the machine so that it can complete the cut properly. Loosen the bevel adjusting knob, set the torch angle to the desired bevel angle, and then tighten the bevel adjusting knob.

Move the machine to the approximate point where the cut will start. Then set the Low/High speed switch to the desired cutting speed. Set the On/Off switch to the Off position. Next, plug the power cord into a 115 alternating current (AC), 60 Hertz (Hz) power outlet. Ensure that all clamping screws are properly tightened. Now ignite and properly adjust the torch, and preheat the start of the cut. Set the Forward/Reverse switch to the desired direction of travel. Simultaneously turn on the cutting oxygen and rotate the cutting speed control knob to the desired rate. When the cut is completed, stop the machine and shut off the torch.

Additional Resources

ANSI Z49.1, Safety in Welding, Cutting, and Allied Processes. Miami, FL: American Welding Society.

Uniweld Products, Inc. Numerous videos are available at **www.uniweld.com/en/uniweld-videos**. Last accessed: November 30, 2014.

The Harris Products Group, a division of Lincoln Electric. Numerous videos are available at **www.harrisproductsgroup.com/en/Expert-Advice/videos.aspx**. Last accessed: November 30, 2014.

4.0.0 Section Review

1. An oxyfuel cut that has gouging at the bottom, a pronounced break in the drag line, and an irregular kerf is most likely to occur when the ____ .
   a. travel speed is too fast
   b. preheat is insufficient
   c. cutting oxygen pressure is too low
   d. cutting oxygen pressure is too high

2. The process of burning a small hole through the metal where an oxyfuel cut can be started is called ____ .
   a. gouging
   b. beveling
   c. piercing
   d. washing

3. A washing operation uses a special torch tip with a large cutting hole that produces a ____ .
   a. high-velocity stream of oxygen
   b. grooved cut in the base metal surface
   c. cleansing layer of dross in the cut
   d. low-velocity stream of oxygen

4. On a portable oxyfuel track burner, the cutting torch can be angled in a plane that is perpendicular to the track by using the torch ____ .
   a. clamping screws
   b. bevel adjustment
   c. kerf positioner
   d. holder assembly

Summary

Oxyfuel cutting has many uses on many different job sites. It can be used to cut metal plate and shapes to size, prepare joints for welding, clean metals or welds, and disassemble structures. Oxyfuel cutting equipment can range in size from small, portable sets to large, automated, fixed-position machines. In all cases, high pressures and flammable gases are involved. For that reason, there is always the danger of fire and explosion when using oxyfuel equipment. These risks can be minimized when the operator is well trained and knowledgeable. By understanding the safety precautions and equipment fundamentals presented in this module, operators will be better prepared to use oxyfuel cutting equipment and cutting techniques in their workplaces.

Review Questions

1. Oxyfuel cutting is best suited for use on ____.
   a. steel alloys
   b. ferrous metals
   c. nonferrous metals
   d. stainless steel

2. The stream of high-pressure cutting oxygen that is directed from an oxyfuel torch causes the metal to instantaneously ____ .
   a. solidify
   b. de-kerf
   c. magnitize
   d. oxidize

3. The recommended range of tinting for either the face shield or the safety glass lenses used during an oxyfuel cutting operation is ____ .
   a. 1 to 2
   b. 3 to 6
   c. 7 to 8
   d. 9 to 10

4. Most welding environment fires occur during ____ .
   a. oxyfuel gas equipment transporting
   b. carbon arc welding activities
   c. oxyfuel gas welding or cutting
   d. acetylene tank installation

5. When preparing a tank or vessel that might have contained flammable materials to be cut with an oxyfuel cutting torch, the best approach is to ____ .
   a. fill it with water
   b. purge it with oxygen
   c. fill it with air
   d. purge it with acetylene

6. When pure oxygen is combined with a fuel gas, it produces a ____ .
   a. high-pressure jet of cutting oxygen
   b. non-explosive, non-flammable vapor
   c. colorless, odorless, and tasteless gas
   d. high-temperature flame for cutting

7. The most common size of oxygen cylinder used in oxyfuel cutting applications is ____ .
   a. 85 cubic feet (2.4 cubic meters)
   b. 227 cubic feet (6.4 cubic meters)
   c. 350 cubic feet (9.9 cubic meters)
   d. 485 cubic feet (13.7 cubic meters)

8. If an acetylene cylinder is found lying on its side, what should be done before it is used?
   a. Contact the supplier for instructions.
   b. Release a small amount of the gas to atmosphere.
   c. Stand it upright and wait at least one hour.
   d. Add a liquid stabilizer to the gas.

9. The maximum hourly rate at which acetylene gas can be withdrawn from a cylinder is ____ .
   a. one half of the cylinder's capacity
   b. one third of the cylinder's capacity
   c. one fifth of the cylinder's capacity
   d. one seventh of the cylinder's capacity

10. The fuel gas that burns with the lowest flame temperature is ____ .
    a. propane
    b. propylene
    c. acetylene
    d. butane

11. Most gas pressure regulators contain two gauges—one that indicates the cylinder pressure and one that indicates the ____ .
    a. ideal cylinder pressure for the conditions
    b. pressure of the gas at the regulator outlet
    c. pressure in the accompanying cylinder
    d. maximum pressure for the cylinder

12. Oxyfuel cutting equipment that has left-hand threads and a V-notch in the nut are most likely to be ____ .
    a. oxygen fittings
    b. purging gas fittings
    c. aftermarket fittings
    d. fuel gas fittings

13. The type of cutting torch that uses a vacuum created by oxygen flow to draw in fuel from a very-low-pressure fuel source is a(n) ____ .
    a. siphon torch
    b. inert torch
    c. injector torch
    d. suspension torch

14. A friction lighter produces sparks when its steel surface is rubbed with a piece of ____ .
    a. flint
    b. soapstone
    c. graphite
    d. steatite

15. A computer-controlled plate cutting machine and an optical pattern-tracing machine are examples of ____ .
    a. portable oxyfuel cutters
    b. exothermic machines
    c. manual guide cutters
    d. fixed-location machines

16. Cracking the cylinder valves during the setup of oxyfuel equipment is a way of ____ .
    a. leak testing the valve regulators
    b. equalizing the cylinder pressures
    c. removing dirt from the valves
    d. venting residual gas from the cylinders

17. Before opening cylinder valves, verify that the adjusting screws on the oxygen and fuel gas regulators have been ____ .
    a. tightened
    b. closed
    c. loosened
    d. purged

18. Oxyfuel cutting equipment is typically leak tested using a(n) ____ .
    a. supervisor's sense of smell
    b. solution that produces bubbles
    c. lit match or a candle
    d. ultrasonic detector

19. An oxyfuel cutting flame that has an excess of fuel is called a(n) ____ .
    a. hot lean flame
    b. neutral flame
    c. oxidizing flame
    d. carburizing flame

20. When disassembling oxyfuel equipment, verify that all pressure gauges are ____ .
    a. reading zero
    b. open and showing atmospheric pressure
    c. chained securely to each other
    d. marked MT for storage

21. Gas cylinders should never be completely emptied because of the risk of ____ .
    a. reverse flow
    b. valve cracking
    c. disproportional mixing
    d. cylinder implosion

22. When a cut has been made with oxyfuel cutting equipment, the drag lines of the cut should be close to ____ .
    a. thirty degrees with minimal notching
    b. forty-five degrees with a wavy kerf
    c. horizontal and notched at the kerf
    d. vertical and not very pronounced

23. A major concern when cutting thin steel with an oxyfuel cutting torch is ____ .
    a. flashback
    b. sparking
    c. distortion
    d. backfire

24. An oxyfuel cutting process that is often used for cutting off bolts, rivets, and other projections is called ____ .
    a. beveling
    b. gouging
    c. piercing
    d. washing

25. What part of an oxyfuel track burner allows the vertical positioning of the torch to be controlled?
    a. The clamping screws
    b. The torch holder
    c. The kerf regulator
    d. The bevel adjustment

Trade Terms Quiz

Fill in the blank with the correct term that you learned from your study of this module.

1. The gap produced by a cutting process is called a(n) _______ .
2. A loud snap or pop that can be heard as a torch flame is extinguished is called a(n) _______ .
3. The soft, white material that is commonly used to mark metal is _______ .
4. Metals that contain iron are called _______ .
5. Creating a groove in the surface of a workpiece is a process called _______ .
6. A flame that is burning with too much fuel is called a(n) _______ .
7. A flame that is burning with too much oxygen is called a(n) _______ .
8. When correct portions of fuel gas and oxygen are fed to a flame, the flame is said to be a(n) _______ .
9. When a thermal process is used for cutting, the material expelled from the kerf is called _______ .
10. Cutting off projections such as bolts, rivets, and previous welded pieces is a process referred to as _______ .
11. The name used to describe what occurs when an oxyfuel cutting torch penetrates a metal plate is _______ .
12. When a flame burns back into the tip of a torch and causes a high-pitched whistling sound, the condition is called _______ .
13. The lines on the edge of a cut that result from the cutting oxygen streaming into, through, and out of the metal are called _______ .

Trade Terms

Backfire
Carburizing flame
Drag lines
Dross
Ferrous metals
Flashback
Gouging
Kerf
Neutral flame
Oxidizing flame
Pierce
Soapstone
Washing

Appendix

PERFORMANCE ACCREDITATION TASKS

The American Welding Society (AWS) School Excelling through National Skills Standards Education (SENSE) program is a comprehensive set of minimum Standards and Guidelines for Welding Education programs. The following performance accreditation is aligned with and designed around the SENSE program.

The Performance Accreditation Tasks (PATs) correspond to and support the learning objectives in *AWS EG2.0, Guide for the Training and Qualification of Welding Personnel: Entry-Level Welder*.

Note that in order to satisfy all learning objectives in *AWS EG2.0*, the instructor must also use the PATs contained in the second level of the NCCER Welding curriculum.

PATs 1 and 2 correspond to *AWS EG2.0, Module 8 – Thermal Cutting Processes, Unit 1 – Manual OFC Principles*, Key Indicators 5, 6, and 7.

PAT 3 corresponds to *AWS EG2.0, Module 8 – Thermal Cutting Processes, Unit 1 – Manual OFC Principles*, Key Indicators 3 and 4.

PATs provide specific acceptable criteria for performance and help to ensure a true competency-based welding program for students.

The following tasks are designed to test your competency with an oxyfuel cutting torch. Do not perform these cutting tasks until directed to do so by your instructor.

SETTING UP, IGNITING, ADJUSTING, AND SHUTTING DOWN OXYFUEL EQUIPMENT

Using oxyfuel equipment that has been completely disassembled, demonstrate how to:

- Set up oxyfuel equipment
- Ignite and adjust the flame
  - Carburizing
  - Neutral
  - Oxidizing
- Shut off the torch
- Shut down the oxyfuel equipment

Criteria for Acceptance:

- Set up the oxyfuel equipment in the correct sequence ____________
- Demonstrate that there are no leaks ____________
- Properly adjust all three flames ____________
- Shut off the torch in the correct sequence ____________
- Shut down the oxyfuel equipment ____________

CUTTING A SHAPE

Using a carbon steel plate, lay out and cut the shape and holes shown in the figure. If available, use a machine track cutter to straight cut the longer dimension.

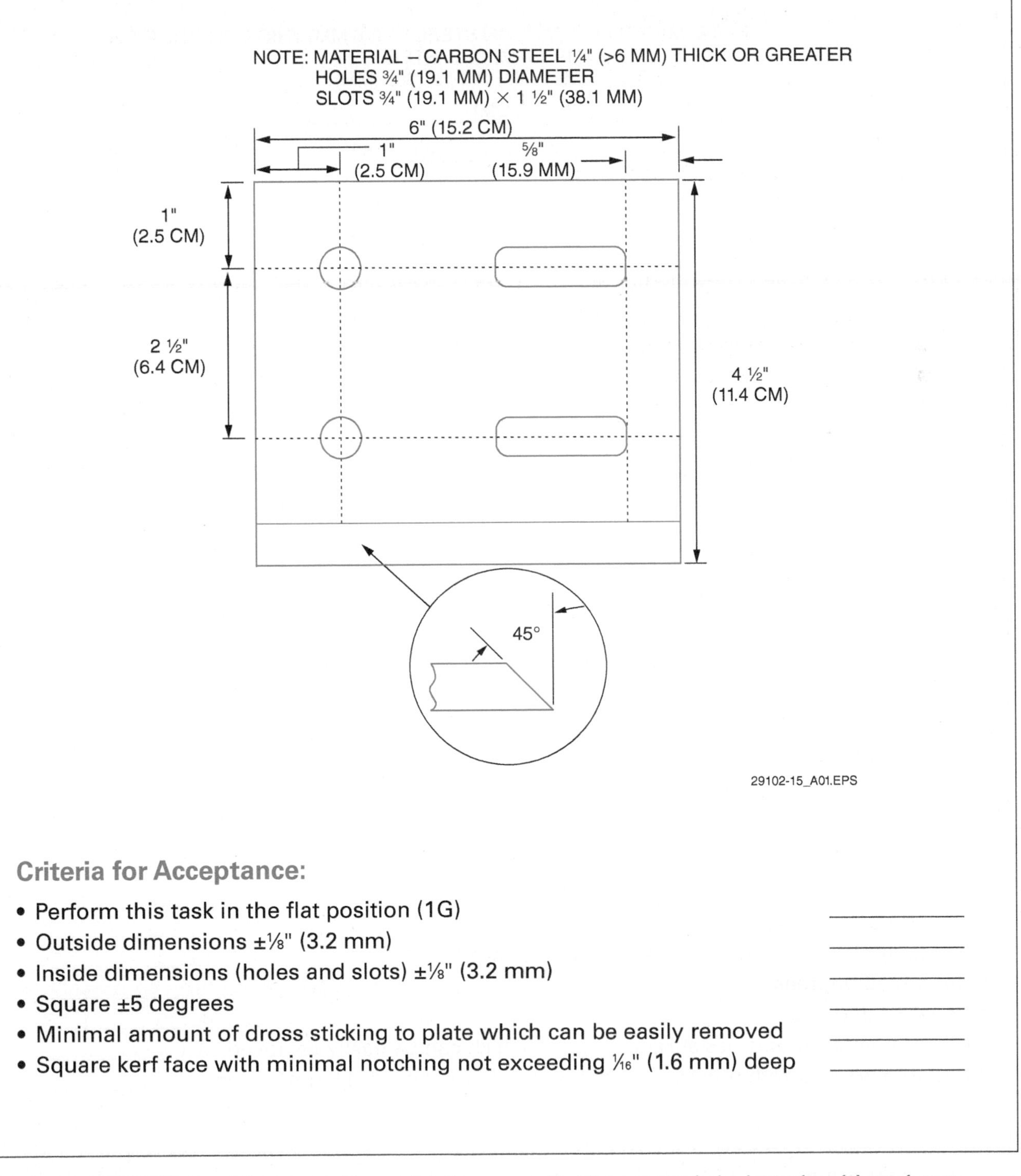

Criteria for Acceptance:

- Perform this task in the flat position (1G) __________
- Outside dimensions ±⅛" (3.2 mm) __________
- Inside dimensions (holes and slots) ±⅛" (3.2 mm) __________
- Square ±5 degrees __________
- Minimal amount of dross sticking to plate which can be easily removed __________
- Square kerf face with minimal notching not exceeding ¹⁄₁₆" (1.6 mm) deep __________

CUTTING A SHAPE

Using a carbon steel plate, lay out and cut the shape and holes shown in the figure. If available, use a machine track cutter to bevel and straight cut the longer dimension.

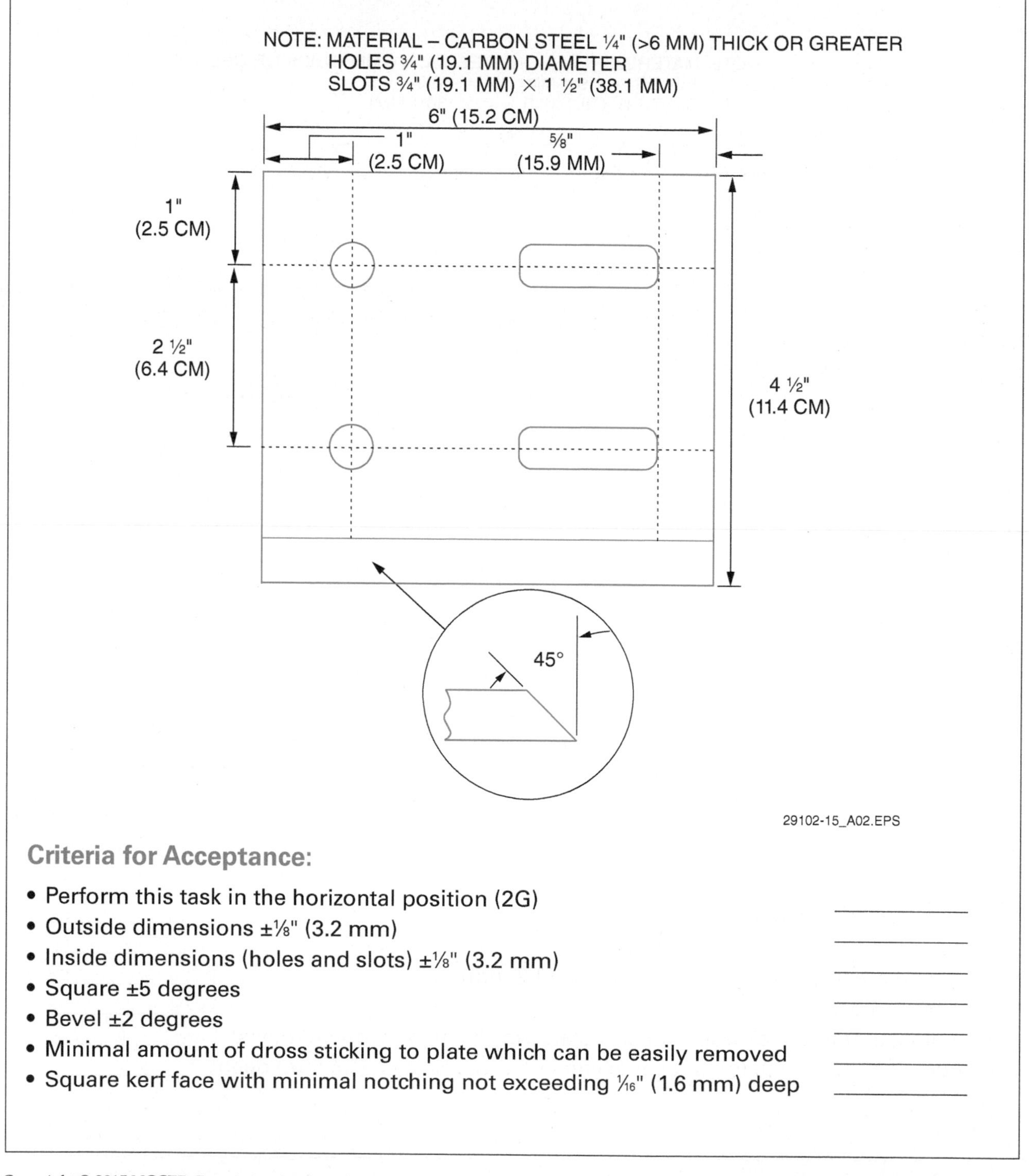

29102-15_A02.EPS

Criteria for Acceptance:

- Perform this task in the horizontal position (2G) ________
- Outside dimensions ±⅛" (3.2 mm) ________
- Inside dimensions (holes and slots) ±⅛" (3.2 mm) ________
- Square ±5 degrees ________
- Bevel ±2 degrees ________
- Minimal amount of dross sticking to plate which can be easily removed ________
- Square kerf face with minimal notching not exceeding 1⁄16" (1.6 mm) deep ________

Trade Terms Introduced in This Module

Backfire: A loud snap or pop as a torch flame is extinguished.

Carburizing flame: A flame burning with an excess amount of fuel; also called a reducing flame.

Drag lines: The lines on the edge of the material that result from the travel of the cutting oxygen stream into, through, and out of the metal.

Dross: The material (oxidized and molten metal) that is expelled from the kerf when cutting using a thermal process. It is sometimes called slag.

Ferrous metals: Metals containing iron.

Flashback: The flame burning back into the tip, torch, hose, or regulator, causing a high-pitched whistling or hissing sound.

Gouging: The process of cutting a groove into a surface.

Kerf: The gap produced by a cutting process.

Neutral flame: A flame burning with correct proportions of fuel gas and oxygen.

Oxidizing flame: A flame burning with an excess amount of oxygen.

Pierce: To penetrate through metal plate with an oxyfuel cutting torch.

Soapstone: Soft, white stone used to mark metal.

Washing: A term used to describe the process of cutting out bolts, rivets, previously welded pieces, or other projections from the metal surface.

Additional Resources

This module presents thorough resources for task training. The following resource material is suggested for further study.

ANSI Z49.1, Safety in Welding, Cutting, and Allied Processes. Miami, FL: American Welding Society.

Plasma Cutters Handbook: Choosing Plasma Cutters, Shop Safety, Basic Operation, Cutting Procedures, ANSI Z49.1, Safety in Welding, Cutting, and Allied Processes. Miami, FL: American Welding Society.

The Harris Products Group, a division of Lincoln Electric. Numerous videos are available at **www.harrisproductsgroup.com/en/Expert-Advice/videos.aspx**. Last accessed: November 30, 2014.

Uniweld Products, Inc. Numerous videos are available at **www.uniweld.com/en/uniweld-videos**. Last accessed: November 30, 2014.

Figure Credits

The Lincoln Electric Company, Cleveland, OH, USA, Module Opener, Figures 1, 2, 5, 14, 32–34, 48–50, 53, 60 (photo), 62, SA04

Topaz Publications, Inc., Figures 3, 7–9, 11, 13, 15, 16, 18, 27, 28, 29A, 30, 37–45, SA02, SA03, SA06–SA08, SA11

Courtesy of Uniweld Products, Figure 20, SA10

Victor Technologies, Figure 21

Vestil Manufacturing, Figure 29B

Koike Aronson, Inc. – Worldwide manufacturer of cutting, welding and positioning equipment, Figure 31

Courtesy of H & M Pipe Beveling Machine Company, Inc., Figures 35, 36

Zachry Industrial, Inc., Figures 46, 47, 51

Courtesy of Smith Equipment, Figure 52

© American Welding Society (AWS) *Welding Handbook* 1991, Welding Processes Volume No. 2, Edition No. 8, Miami: American Welding Society, Figure 55

Xerafy, SA01

Courtesy of Saf-T-Cart, SA09

Section Review Answer Key

Answer	Section Reference	Objective
Section One		
1. b	1.1.0	1a
2. d	1.2.2	1b
Section Two		
1. b	2.1.2	2a
2. a	2.2.3	2b
3. d	2.3.1	2c
4. a	2.4.4	2d
5. c	2.5.3	2e
Section Three		
1. b	3.1.1	3a
2. d	3.2.0	3b
3. c	3.3.1	3c
4. d	3.4.0	3d
Section Four		
1. a	4.1.1	4a
2. c	4.2.3	4b
3. d	4.3.2	4c
4. b	4.4.1	4d

This page is intentionally left blank.

NCCER CURRICULA — USER UPDATE

NCCER makes every effort to keep its textbooks up-to-date and free of technical errors. We appreciate your help in this process. If you find an error, a typographical mistake, or an inaccuracy in NCCER's curricula, please fill out this form (or a photocopy), or complete the online form at **www.nccer.org/olf**. Be sure to include the exact module ID number, page number, a detailed description, and your recommended correction. Your input will be brought to the attention of the Authoring Team. Thank you for your assistance.

Instructors – If you have an idea for improving this textbook, or have found that additional materials were necessary to teach this module effectively, please let us know so that we may present your suggestions to the Authoring Team.

NCCER Product Development and Revision

13614 Progress Blvd., Alachua, FL 32615

Email: curriculum@nccer.org
Online: www.nccer.org/olf

❑ Trainee Guide ❑ Lesson Plans ❑ Exam ❑ PowerPoints Other ____________

Craft / Level: Copyright Date:

Module ID Number / Title:

Section Number(s):

Description:

Recommended Correction:

Your Name:

Address:

Email: Phone:

This page is intentionally left blank.

Ladders and Scaffolds

OVERVIEW

Ladders and scaffolds are some of the most important tools on a job site. Used properly, they make a pipefitter's job much easier. Carelessness, however, can be fatal. Common accidents like falling, being struck by falling objects, and electrocution can be avoided if safety precautions are followed. Considerations for the type of ladder or scaffolding being used, as well as environmental factors, personal positioning, fall arrest systems, and proper assembly and care of all equipment contribute to safe operations.

Module 08105

Trainees with successful module completions may be eligible for credentialing through the NCCER Registry. To learn more, go to www.nccer.org or contact us at 1.888.622.3720. Our website has information on the latest product releases and training, as well as online versions of our *Cornerstone* magazine and Pearson's product catalog.

Your feedback is welcome. You may email your comments to curriculum@nccer.org, send general comments and inquiries to info@nccer.org, or fill in the User Update form at the back of this module.

This information is general in nature and intended for training purposes only. Actual performance of activities described in this manual requires compliance with all applicable operating, service, maintenance, and safety procedures under the direction of qualified personnel. References in this manual to patented or proprietary devices do not constitute a recommendation of their use.

08105 V4.0

08105
Ladders and Scaffolds

Objectives

Successful completion of this module prepares trainees to:

1. Identify various types of ladders and describe how to safely use them.
   a. Identify common ladders and basic safety guidelines.
   b. Explain how to use stepladders.
   c. Explain how to use straight ladders and extension ladders.
2. Identify and describe how to use scaffolding.
   a. Identify common scaffolds and explain how to use them safely.
   b. Explain how to use and care for tubular buck scaffolds.
   c. Explain how to use and care for pole scaffolds.
   d. Explain how to use and care for rolling scaffolds.
3. Identify scaffold hazards and state guidelines for safe use.
   a. Identify specific scaffold hazards.
   b. State specific scaffold safety guidelines.

Performance Tasks

Under the supervision of your instructor, you should be able to do 2 of the following tasks:

1. Select, inspect, and use stepladders, straight ladders, and extension ladders.
2. Demonstrate 3-point contact and the 4-to-1 rule.
3. Erect, inspect, and disassemble tubular buck scaffolding.

Trade Terms

4-to-1 rule
Base plates
Casters
Coupling pin
Cross braces
Duty rating
Guardrails
Harness
Hinge pin
Leveling jack
Scaffold
Scaffolding
Scaffold floor
Toeboard
Vertical upright

Industry Recognized Credentials

If you are training through an NCCER-accredited sponsor, you may be eligible for credentials from NCCER's Registry. The ID number for this module is 08105. Note that this module may have been used in other NCCER curricula and may apply to other level completions. Contact NCCER's Registry at 888.622.3720 or go to www.nccer.org for more information.

Contents

Figures

This page is intentionally left blank.

SECTION ONE

1.0.0 LADDERS

Objective

Identify and describe how to safely use ladders.

a. Identify common ladders and basic safety guidelines.
b. Explain how to use stepladders.
c. Explain how to use straight ladders and extension ladders.

Performance Tasks

1. Select, inspect, and use stepladders, straight ladders, and extension ladders.
2. Demonstrate 3-point contact and the 4-to-1 rule.

Trade Terms

4-to-1 rule: The safety rule for straight ladders and extension ladders, which states that a ladder should be placed so that the distance between the base of the ladder and the supporting wall is one-fourth the working distance of the ladder.

Duty rating: American National Standards Institute (ANSI) rating assigned to ladders. It indicates the type of use the ladder is designed for (industrial, commercial, or household) and the maximum working load limit (weight capacity) of the ladder. The working load limit is the maximum combined weight of the user, tools, and any materials bearing down on the rungs of a ladder.

Harness: A device that straps securely around the body and is connected to a lifeline. It is part of a personal fall arrest system (PFAS).

Scaffold: An elevated work platform for workers and materials.

Scaffolding: A temporary, built-up framework or suspended platform or work area designed to support workers, materials, and equipment at elevated or otherwise inaccessible job sites.

Ladders and scaffolds are common sights on construction jobs. They allow employees to work safely at elevated levels. Ladders can be used for short distances, while a **scaffold** offers a work platform for laborers and materials when the work site is many stories above ground (*Figure 1*). **Scaffolding** is a temporary, built-up framework or suspended platform; it is designed to support workers, materials, and equipment at elevated or otherwise inaccessible job sites.

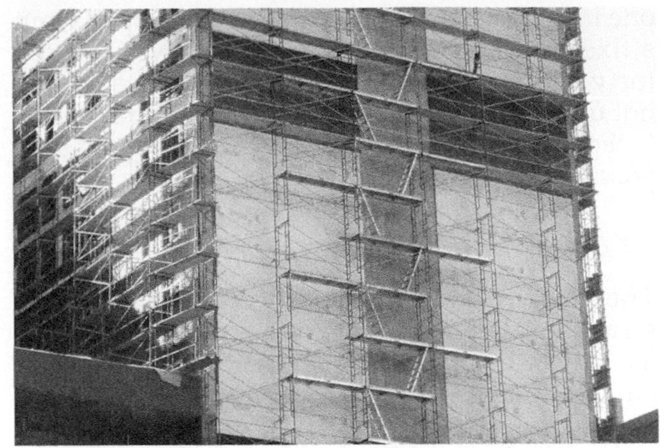

Figure 1 Scaffolding can be many stories above the ground.

Safety is the top priority on a job; one careless mistake could mean the difference between life and death. Any time work is done above ground, there is a chance that an incident or accident may occur. The unsafe use of ladders and scaffolds has always been a problem in the workplace. The most common hazards associated with ladders and scaffolding are falling, being struck by falling objects, and electrocution.

Inspecting ladders and scaffolds and being familiar with their safe operation can greatly reduce the risk of incidents and accidents. It is your responsibility to learn how to set up, use, and maintain equipment. Workers must remember to always wear a safety harness when working from a ladder or scaffold that is more than six feet in elevation. As part of a personal fall arrest system (PFAS), the safety **harness** straps securely around a worker's body and is connected to a lifeline. In

Case History

Bad Weather and Unsafe Conditions Can Kill

A laborer was working on the third level of a tubular welded frame scaffolding which was covered with ice and snow. The planking on the scaffolding wasn't sturdy and the scaffolding didn't have a guardrail. There was also no access ladder for the various scaffolding levels. The worker slipped and fell approximately 20 feet to the pavement below. He died of a head injury.

The Bottom Line: Make sure that all scaffolding has solid planking and guardrails.

Source: The Occupational Safety and Health Administration (OSHA)

one instance, a worker died when he slipped from a fixed ladder attached to a water tower and fell forty feet to the ground. He died because he was not using the proper fall protection equipment.

Following proper safety procedures could save your life and the lives of your co-workers.

1.1.0 Common Ladders

The basic types of ladders include stepladders, straight ladders, and extension ladders and each was designed for specific purposes and circumstances (*Figure 2*). Aluminum ladders are corrosion-resistant and can be used in situations where they might be exposed to the elements. They are also lightweight and can be used where they need to be frequently lifted and moved. Wooden ladders, which are heavier and sturdier than fiberglass or aluminum ladders, can be used when heavy loads must be moved up and down. Fiberglass ladders are nonconductive and also highly durable, so they are useful in situations involving electrical work or where some amount of rough treatment is unavoidable. Both fiberglass and aluminum are easier to clean than wood.

Selecting the right ladder is essential: consider its features and whether it is best for the job. Always consider the highest duty rating and weight limit needed, as well as the height requirements. A **duty rating** is assigned by the American National Standards Institute (ANSI) and it describes what a particular type of ladder is supposed to be used for as well as how much weight it can hold. A ladder that is too long or too short will not allow the work surface to be reached easily, safely, or comfortably.

The following safety guidelines must be followed when using any type of ladder:

- Choose the right ladder for the job.
  - Never use stepladders for straight-ladder work.
  - Do not use metal, metal-reinforced, or wet ladders where direct contact with a live power source is possible. Use extreme caution when working near electrical wires, service, and equipment.
  - Never use makeshift substitutes for ladders.
- Inspect the ladder before each use.
  - Test all working parts for proper attachment and operation.
  - Check the entire ladder for loose nails, screws, brackets, or other hardware. If any hardware problems are found, tighten the loose parts or have the ladder repaired before use.
  - Never use a ladder with broken or missing rungs or steps, broken or split side rails, or other faulty construction.
  - Keep steps, rungs, and soles of your shoes free of grease, oil, paint, and other slippery substances.
  - When defective ladders are discovered, first remove them from service and clearly mark them with Do Not Use. Then report the de-

Figure 2 Types of ladders.

fect to the appropriate person on your work site or to your supervisor.

- Set it up properly and securely.
  - Place the ladder feet on a firm, suitable surface and keep the area around the bottom and the top of the ladder clear.
  - The distance between the foot of a ladder and the base of the structure it is leaning against must be one-fourth of the distance between the ground and the point where the ladder touches the structure. Stated another way, there should be a **4-to-1** ratio between the distances. For example, if the height of the wall shown in *Figure 3* is 16 ft. (4.9 m), the base of the ladder should be 4 ft. (1.2 m) from the base of the wall. If you are going to step off a ladder onto a platform or roof, the top of the ladder should extend at least 3 ft. (0.9 m) above the point where the ladder touches the platform, roof, side rails, etc.
  - Always use appropriate safety feet or non-slip bases. If the ladder has to be placed on a slippery surface, take additional precautions.
  - Use a stabilizer leg (*Figure 4*) if the ladder is on uneven ground. Do not use wood blocks or bricks.
  - If outrigger safety legs are provided on a ladder, they are used whenever the ladder is used.
  - Place the ladder so that it leaves 6 inches of clearance in back of the ladder and 30 inches of clearance in front of the ladder.
  - Place the ladder so that it leans against a solid, immovable surface. Never place a ladder against a window, door, doorway, sash, box, or loose or movable wall.
  - If a ladder must be placed in front of a door that opens toward the ladder, the door must be locked or blocked open so that it cannot strike the ladder.
  - Do not place a ladder in a doorway, passageway, driveway, or other area where it is in the way of any other work unless it is protected by barricades or guards.
  - Move the ladder in line with the work to be done. Never lean sideways away from the

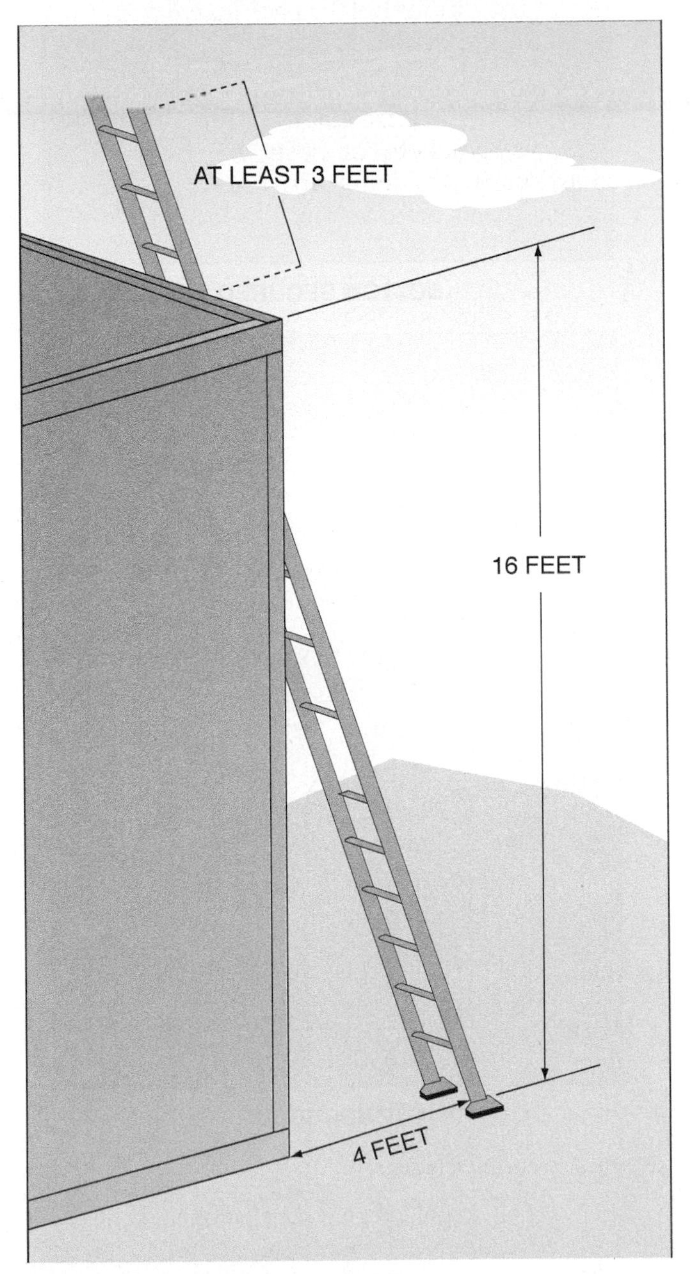

Figure 3 Proper ladder positioning.

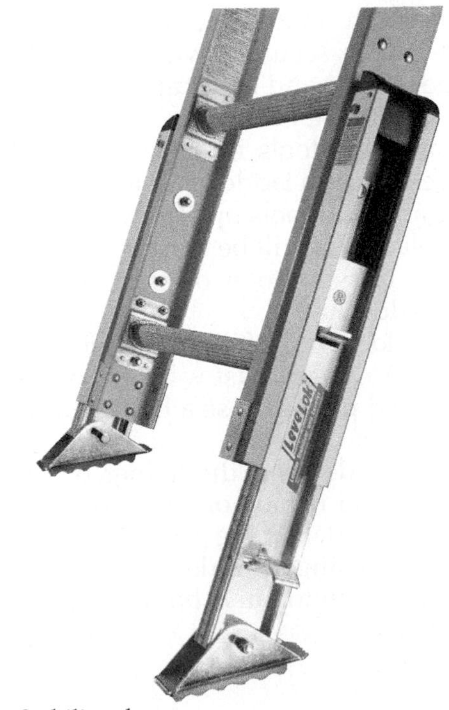

Figure 4 Stabilizer leg.

ladder in order to reach the work area.
  - Never try to move a ladder while someone is on it.
  - Block, tie, or otherwise secure portable ladders that are in use to prevent them from being displaced.
  - Do not use ladders during high winds. If you must use a ladder in windy conditions, make sure you securely lash the ladder to prevent slippage.
- Follow manufacturers' recommended guidelines, as well as OSHA regulations, when using a ladder.
  - Maintain three points of contact at all times, as shown in *Figure 5*. This can be two feet and one hand, or one foot and two hands.
  - Never go up or down a ladder while facing away from it. Face the ladder at all times.
  - Always keep your body weight in the center of the ladder between the rails.
  - Do not stand or sit on the ladder top or the top step of the ladder.
  - Never use a ladder in the horizontal position as a platform, walk board, or scaffold.
  - Most ladders are intended to carry one person at a time. Do not overload.
  - Double stepladders are designed with steps on the front and back so they can be climbed from either side. These ladders are usually identified with a sticker that indicates that both sides may be used to climb. When two people use the ladder, the maximum load limit must not be exceeded.
  - Keep both hands free so that you can hold the ladder securely while climbing. Use a rope to raise and lower any tools and materials you might need.
  - Don't carry tools in your hands while you are climbing a ladder. Use a hand line or tagline and pull tools up once you have reached the place you will be working.
  - Never rest any tools or materials on the top of a ladder.
  - Use ladders only for short periods of elevated work. If you must work from a ladder for extended periods, use a lifeline fastened to a safety belt.
  - Climb or descend the ladder one rung at a time. Never run up or slide down a ladder.
  - Lay the ladder on the ground when you have finished using it, unless it is anchored securely at the top and bottom where it is being used.

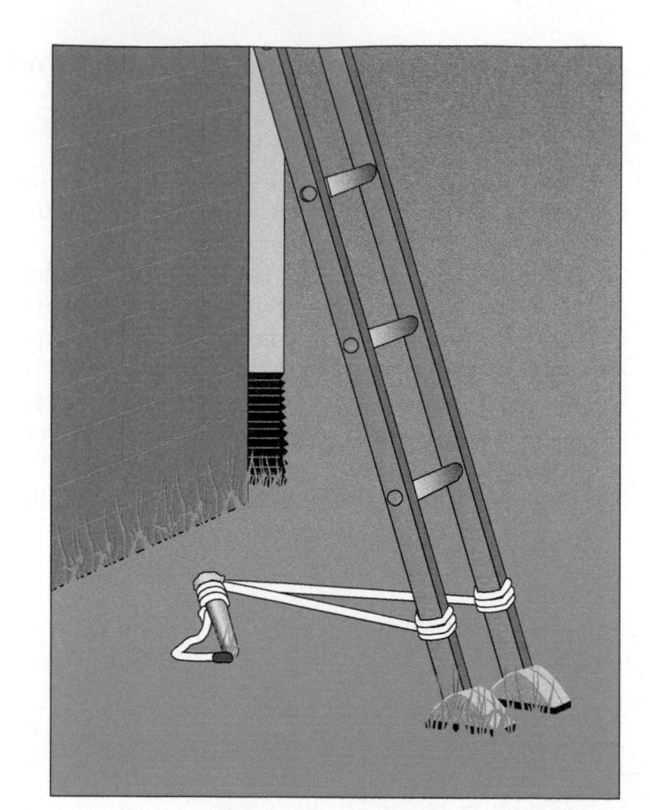

Figure 5 Securing a ladder.

- Maintain the ladder and store it appropriately.
  - Store a ladder in a cool, dry, ventilated place and provide easy access for inspection.
  - Provide sufficient support to prevent sagging when laying a ladder flat.

WARNING!

If you aren't sure about whether a stepladder can be climbed from either side, ask your supervisor for help before you use the ladder. You may prevent an accident.

  - Do not change or alter any type of ladder.
  - Do not paint a wooden ladder, because paint can hide defects and flaws.

1.2.0 Stepladders

Stepladders are self-supporting ladders made of two sections hinged at the top (*Figure 6*). The section of a stepladder used for climbing consists of rails and rungs like those on straight ladders. The other section consists of rails and braces. Spreaders are the hinged arms between the sections that keep the ladder stable and prevent it from folding while in use. A stepladder may have a pail shelf to hold paint or tools. A good rule is to never allow your belt buckle to pass beyond the stepladder rail.

Figure 6 Typical stepladder.

Stepladders, straight ladders, and extension ladders should be inspected in similar fashion. Pay special attention to the hinges and spreaders to be sure they are in good repair. Be sure the rungs are clean. A stepladder's rungs are usually flat, so oil, grease, or dirt can easily build up on them and make them slippery.

When using stepladders:

- Be sure that all four feet are on a hard, even surface when the stepladder is positioned. If they are not, the ladder can rock from side to side or corner to corner as it is climbed.
- Never stand on the top step or the top of a stepladder. Putting your weight this high will make the ladder unstable. The top of the ladder is made to support the hinges, not to be used as a step.
- Make sure the spreaders are locked in the fully open position when the ladder is in position.
- If the ladder is designed to be used from only one side, never use the braces for climbing even though they may look like rungs. They are not designed to support bodyweight.
- Some stepladders are designed to be used from either side, but others are not. Be sure you know if the ladder you are using can be climbed from either side before you use it.

Figure 7 shows some common ladder safety precautions.

NOTE

According to OSHA standards, industrial stepladders are designed for heavy-duty work and cannot exceed 20 feet in height.

DOs

DON'Ts

- Be sure your ladder has been properly set up and is used in accordance with safety instructions and warnings.
- Wear shoes with non-slip soles.

- DON'T stand above the highest safe standing level.
- DON'T stand above the second step from the top of a stepladder and the 4th rung from the top of an extension ladder. A person standing higher may lose their balance and fall.

- Keep your body centered on the ladder. Hold the ladder with one hand while working with the other. Never let your belt buckle pass beyond either ladder rail.

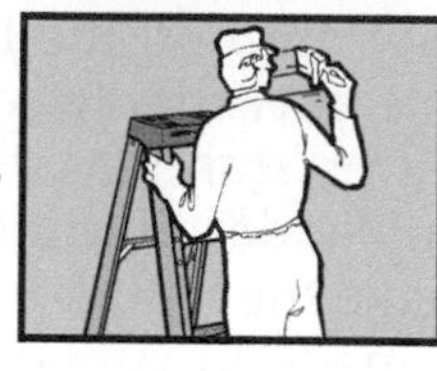

- DON'T climb a closed stepladder. It may slip out from under you.
- DON'T climb on the back of a stepladder. It is not designed to hold a person.

- Move materials with extreme caution. Be careful pushing or pulling anything while on a ladder. You may lose your balance or tip the ladder.

- DON'T stand or sit on a step-ladder top or pail shelf. They are not designed to carry your weight.
- DON'T climb a ladder if you are not physically and mentally up to the task.

- Get help with a ladder that is too heavy to handle alone. If possible, have another person hold the ladder when you are working on it.

- DON'T exceed the Duty Rating, which is the maximum load capacity of the ladder. Do not permit more than one person on a single-sided stepladder or on any extension ladder.

- Climb facing the ladder. Center your body between the rails. Maintain a firm grip.
- Always move one step at a time, firmly setting one foot before moving the other.

- DON'T place the base of an extension ladder too close to the building as it may tip over backward.
- DON'T place the base of an extension ladder too far away from the building, as it may slip out at the bottom. **Please refer to the 4 to 1 Ratio Box.**

- Haul materials up on a line rather than carry them up an extension ladder.
- Use extra caution when carrying anything on a ladder.

- DON'T over-reach, lean to one side, or try to move a ladder while on it. You could lose your balance or tip the ladder. **Climb down and then reposition the ladder closer to your work!**

Read ladder labels for additional information.

4 TO 1 Ratio

Place an extension ladder at a 75-1/2° angle. The set-back ("S") needs to be 1 ft. for each 4 ft. of length ("L") to the upper support point.

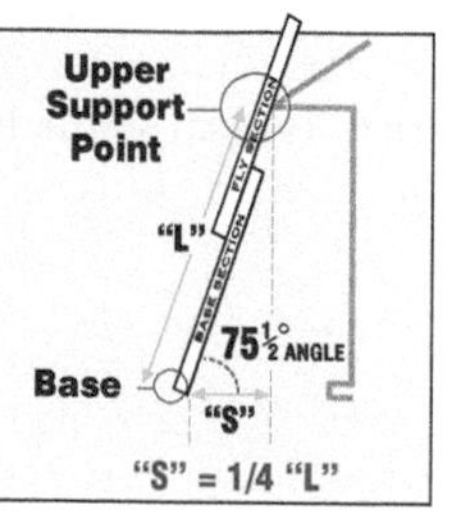

Figure 7 Ladder safety.

Case History

Even Falls from Short Distances Can Be Fatal

An employee was working alone from a 6-foot fiberglass stepladder when one of her co-workers heard a loud noise and went to check on her. The employee had fallen from the ladder, resulting in a head injury from which she later died.

Bottom Line: Even falls from short distances can be fatal.

Source: Occupational Safety & Health Administration website.

1.3.0 Straight and Extension Ladders

Straight ladders consist of two rails, rungs between the rails, and safety feet on the bottom of the rails (*Figure 8*). The straight ladders used in construction are generally made of wood or fiberglass.

WARNING!

Metal ladders conduct electricity and should not be used around electrical equipment.

Straight ladders are typically 8 to 12 feet in length and are similar to one section of an extension ladder. The most important rule to remember when using a straight or extension ladder is to maintain the correct ladder angle. Always place the ladder so that the distance between the base of the ladder and the supporting wall is one-fourth the working distance of the ladder. This is known as the 4-to-1 rule. The working distance of the ladder is the distance between the base of the ladder and the point where the ladder makes contact with the supporting wall. *Figure 9* shows the correct ladder angle. If the base of the ladder is too far from the wall, the ladder can slip out from under you. If the base of the ladder is too close to the wall, it is more likely to tip over backwards.

Setup is the next step after inspecting a ladder. In addition to the general guidelines for all ladders, use these to safely set up a straight ladder:

- Place the straight ladder at the proper angle before using it. A ladder placed at an improper angle will be unstable and could cause you to fall.

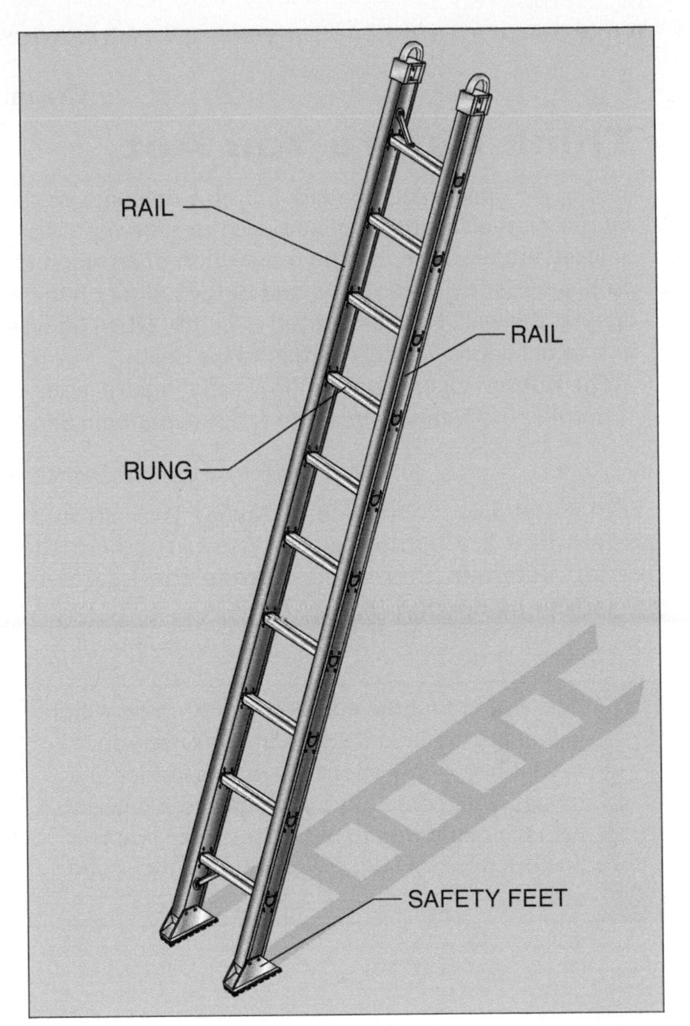

Figure 8 Straight ladder.

- Straight ladders should be used only on stable and level surfaces unless they are secured at both the bottom and the top to prevent any accidental movement (*Figure 10*).
- The distance between the foot of a ladder and the base of the structure it is leaning against must be one-quarter the distance between the ground and the point where the ladder touches the structure.

Once you have inspected and set up the ladder, you can begin working on it. In addition to the guidelines for all ladders, the use of straight ladders calls for additional safeguards:

- Make sure the ladder you are about to climb or descend is properly secured.
- Make sure the ladder's feet are solidly positioned on firm, level ground.
- Ensure that the top of the ladder is firmly positioned and in no danger of shifting once you begin your climb.
- Keep both hands on the rails when climbing a straight ladder.

Case History

Think Before You Act

During the construction of a building, a masonry worker was instructed by his foreman to prepare a batch of mortar on the second level and use the stairway to carry it to the third level. This worker decided it would be quicker and easier to use the top section of an extension ladder (without safety feet) instead of the stairway. He set up the ladder by placing one end of the ladder on the wet concrete floor and leaning the other end of the ladder against the wall. He then started to climb. When he was halfway up, the ladder slipped on the wet floor, causing him to fall approximately 12 feet to his death.

The Bottom Line: Always follow safety instructions.

Source: The National Institute for Occupational Safety and Health (NIOSH)

An extension ladder is actually two straight ladders that are connected so you can adjust the overlap between them and change the length of the ladder as needed (*Figure 11*).

WARNING!

Remember that the addition of your own weight will affect the ladder's steadiness once you mount it. It is important to test the ladder first by applying some of your weight to it without actually beginning to climb. This way you will be sure that the ladder remains steady as you ascend.

Extension ladders are positioned and secured following the same rules as straight ladders. There are, however, some safety rules that are unique to extension ladders:

- Inspect the rope used to raise the moveable section of an extension ladder before each use for frayed and worn spots. Replace damaged ropes.
- Inspect rung locks closely for damage. Do not use the ladder if the rung locks do not work correctly.
- Make sure the section locking mechanism is fully hooked over the desired rung.

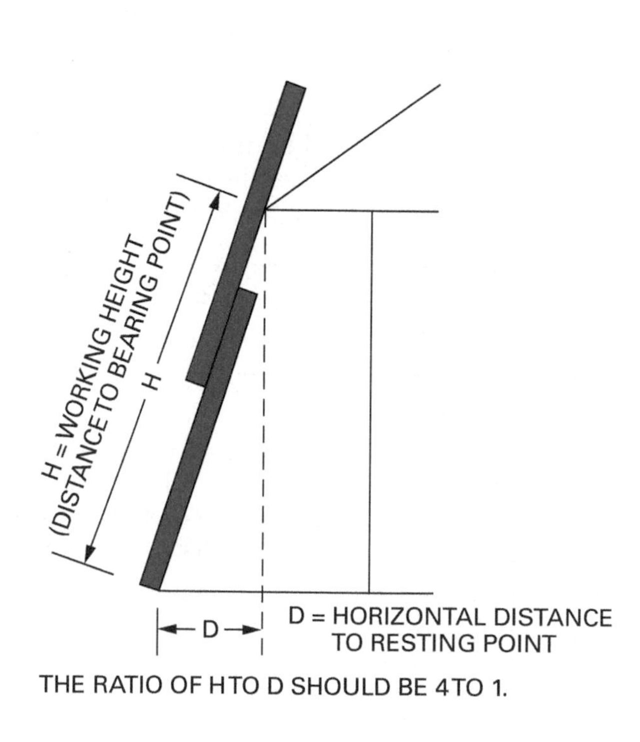

Figure 9 Proper positioning.

Figure 10 Ladder safety feet.

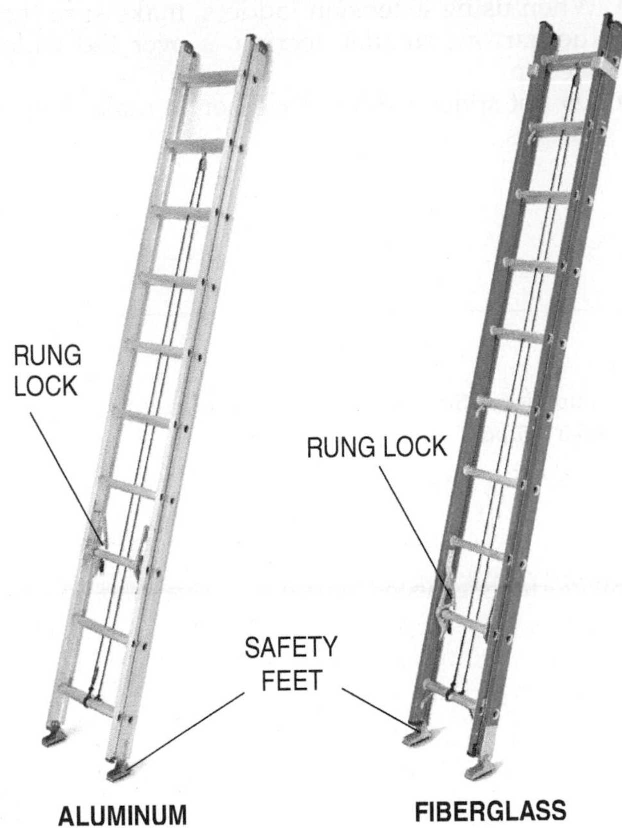

Figure 11 Extension ladders.

- Make sure that all ropes used for raising and lowering the extension are clear and untangled.
- Make sure the extension ladder overlaps between the two sections (*Figure 12*).
  - For ladders up to 36 feet long, the overlap must be at least 3 feet.
  - For ladders 36 to 48 feet long, the overlap must be at least 4 feet.
  - For ladders 48 to 60 feet long, the overlap must be at least 5 feet.
- Never stand above the highest safe standing level on a ladder. On an extension ladder, this is the fourth rung from the top. If you stand higher, you may lose your balance and fall.
  - Some ladders have colored rungs to show where you should not stand.
- Always adjust the ladder from the bottom, never from the top. The rung locks cannot always be seen from the top.
- Rungs must be on 12-inch centers.
- Make sure that the rungs have been treated or constructed to prevent slipping.
- Two-section ladders must not exceed 48 feet, and three-section ladders must not exceed 60 feet.

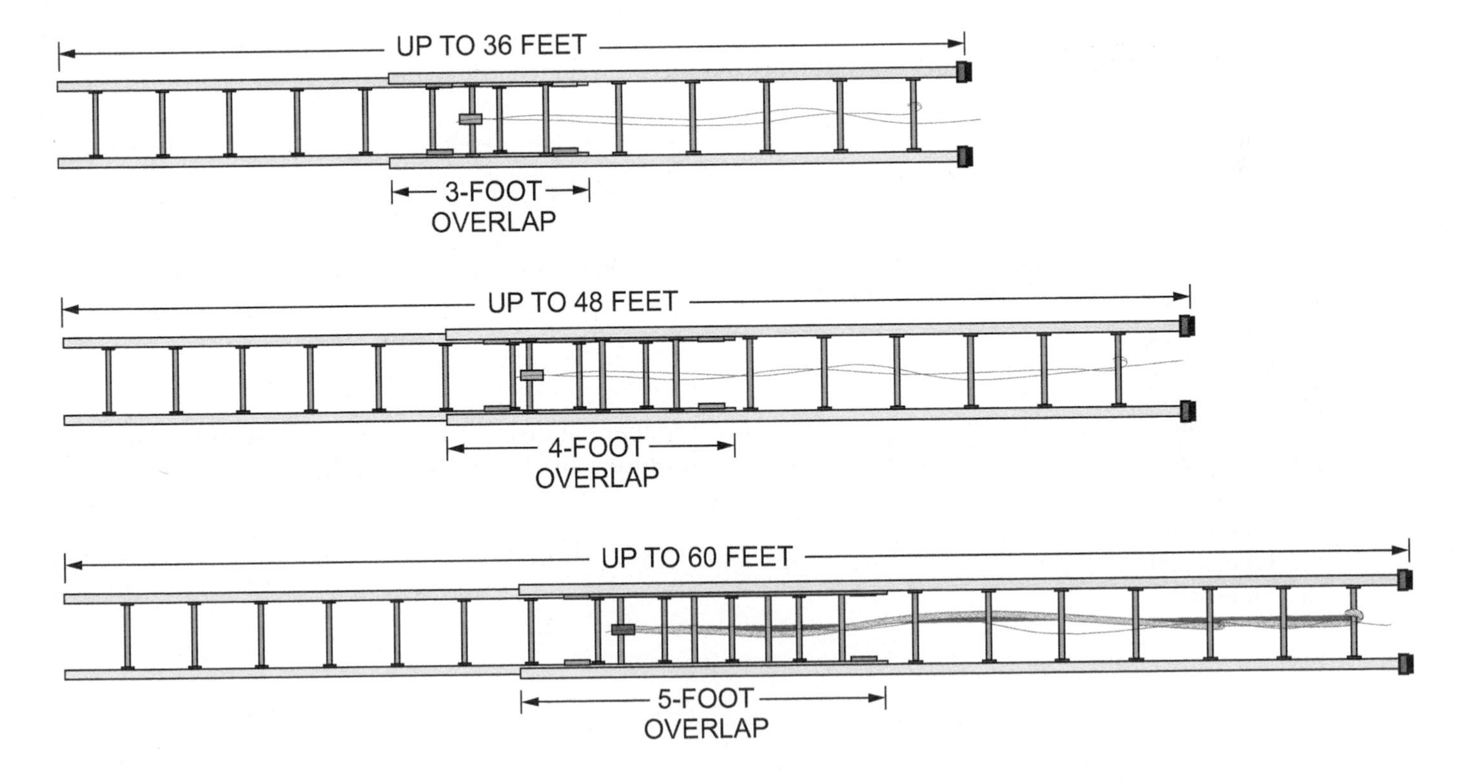

Figure 12 Overlap lengths for extension ladders.

- If you are going to step off the top of a ladder onto a roof or platform, the side rails of the ladder must extend over the top of the roof or platform by at least 36 inches. If this is not possible, grab rails must be installed on the top of the ladder to provide a secure grip for personnel.
- When using extension ladders, make sure that the narrow, smaller section is over the wider section.
- Do not splice ladders together to make longer ladders.

Additional Resource

Fall Protection in Construction. US Department of Labor: Occupational Safety and Health Administration. 2015. Available at: **https://www.osha.gov/Publications/OSHA3146.pdf**.

1.0.0 Section Review

1. Which of the following offers a temporary, suspended platform for elevated work?
   a. Ladders
   b. Stationary Swingboards
   c. Scaffolding
   d. Walk boards

2. _____ are actually two straight ladders.
   a. Extension ladders
   b. Step ladders
   c. Scaffolds
   d. Walk boards

3. Which type of ladder is useful in electrical work, because it is nonconductive?
   a. Aluminum
   b. Metal alloy
   c. Copper-plated
   d. Fiberglass

4. True or false? When using a stepladder, it is okay to stand on the top step, so long as the ladder has been properly positioned on a hard, even surface.
   a. True
   b. False

5. Always place straight ladders and extension ladders so that the distance between the base of the ladder and the supporting wall is _____ the working distance of the ladder.
   a. One-half
   b. One-third
   c. One-fourth
   d. One-fifth

SECTION TWO

2.0.0 SCAFFOLDING

Objective

a. Identify and describe how to use scaffolding.
b. Identify common scaffolds and explain how to use them safely.
c. Explain how to use and care for tubular buck scaffolds.
d. Explain how to use and care for pole scaffolds.
e. Explain how to use and care for rolling scaffolds.

Performance Task

3. Erect, inspect, and disassemble tubular buck scaffolding.

Trade Terms

Base plates: Flat discs or rectangles under scaffold legs that are used to evenly distribute the weight of the scaffold. Base plates come in different sizes for different scaffold heights.

Casters: Wheels that are attached to the bottoms of scaffold legs instead of base plates. Most casters come with brakes.

Coupling pin: A steel pin used to line up and join sections of a scaffold.

Cross braces: Steel pieces used to connect and support the vertical uprights of a scaffold.

Guardrails: Protective rails attached to a scaffold. The top rail is 42 inches above the scaffold floor, and the middle rail is halfway between the top rail and the toeboard.

Hinge pin: A small pin inserted through a leg and base plate or caster wheel and used to hold these parts together.

Leveling jack: A threaded, adjustable screw located between the legs and the base plates or caster wheels of a scaffold. It is used to raise or lower parts of a scaffold to level it on uneven surfaces.

Scaffold floor: A work area platform made of metal or wood.

Toeboard: A 4-inch railing attached around the scaffold floor to prevent tools and materials from falling off the scaffold.

Vertical upright: The end section of a scaffold, also known as a buck, made of welded steel. It supports other vertical uprights or upper end frames and the scaffold floor.

Scaffolds and scaffolding are common at construction sites because they make it easy to work in hard-to-reach areas (*Figure 13*). They increase the danger to workers, however, because of the elevation. With proper attention to safety procedures and OSHA regulations, the risks associated with using scaffolds can be greatly reduced.

2.1.0 Scaffold Types and Weight Classifications

Many different types of scaffolds are built to meet specific needs. The two types of scaffolds used most often by pipefitters are tubular buck scaffolds and pole scaffolds. *Figure 14* shows these types of scaffolds.

A scaffold is classified by the weight it is designed to support. The design capacity of a scaffold includes the total weight of all loads, which includes the working load and the weight of the

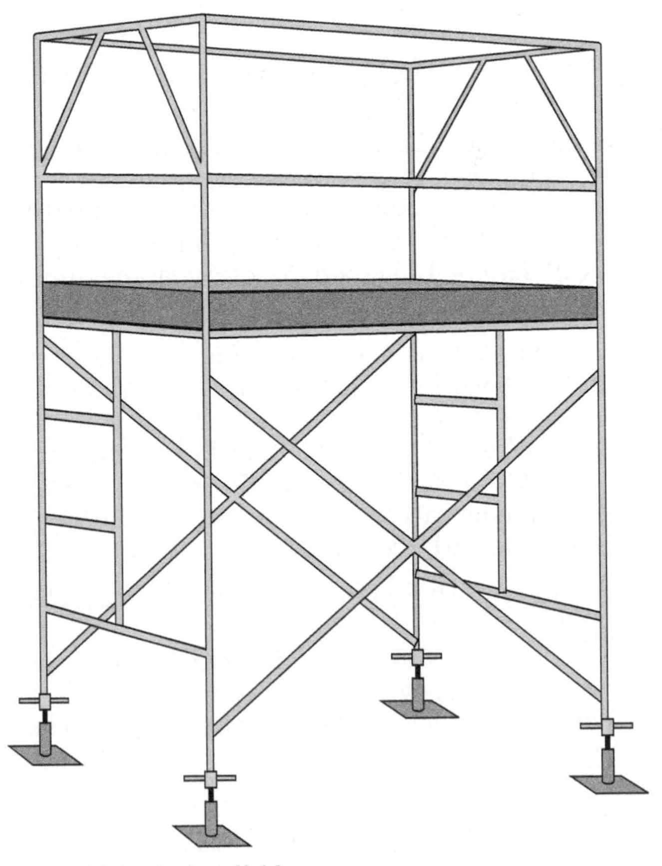

Figure 13 Typical scaffolding.

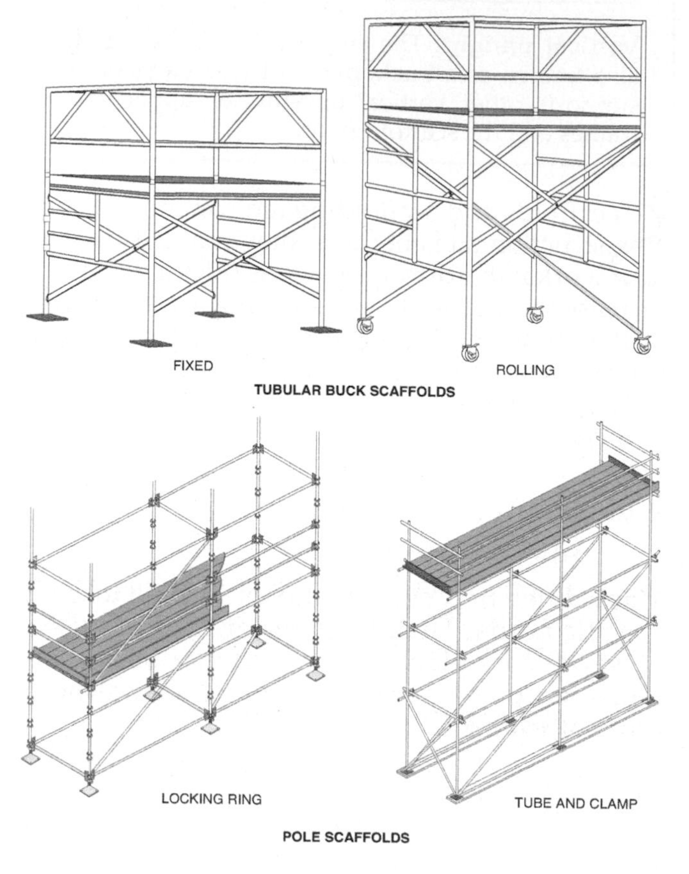

Figure 14 Scaffolds.

scaffold. The three basic scaffold classes are the following:

- *Heavy duty* – Designed to carry a maximum working load of 75 pounds per square foot
- *Medium duty* – Designed to carry a maximum working load of 50 pounds per square foot
- *Light duty* – Designed to carry a maximum working load of 25 pounds per square foot

The main part of the scaffolding is the working platform. A working platform must have a guardrail system that includes a top rail, midrail, toeboard, and screening. A **toeboard** is a 4-inch railing attached around the scaffold floor to prevent tools and materials from falling off the scaffold.

To be safe and effective:

- The top rail should be approximately 42 inches high
- The midrail should be located halfway between the toeboard and the top rail
- The toeboard should be a minimum of 4 inches high.

2.1.1 General Scaffold Safety Guidelines

Scaffolds are built and designed according to high safety standards, but normal wear and tear as well as accidental overstress can weaken a scaffold and make it unsafe. Be sure that the scaffolding you use has been tested using the American National Standards Institute (ANSI) standard. The ANSI standard for testing and rating scaffolding is *SC100*, which was written by the Scaffolding, Shoring, and Forming Institute (SSFI).

Most job sites require scaffold builders to be certified according to company policies and procedures. Always check with your immediate supervisor before erecting a scaffold; for safety purposes it is recommended that at least two people are involved in the process.

OSHA requires that the scaffolding supervisor be a competent person who has been specially trained to inspect scaffolding, oversee its assembly, and train workers in its safe and proper use. In addition, the scaffolding supervisor must inspect erected scaffolding at the start of each shift where it will be used and whenever the integrity of the scaffold is in question, such as after an accident.

The following safety guidelines must be followed when using any type of scaffold:

- Follow all state, local, and government codes, ordinances, and regulations pertaining to scaffolding.
- Ensure that the footing on which a scaffold is erected will withstand the maximum intended load without settling or displacement.
- Do not use unstable objects, such as barrels, boxes, loose bricks, or concrete blocks, to support scaffolding.
- Inspect all equipment before using it. Do not use any damaged equipment. Inspect erected scaffolds regularly to make sure they are in a safe working condition.

Use leveling jacks (*Figure 15*) instead of unstable objects to erect scaffolds on uneven grades. A **leveling jack** is a threaded, adjustable screw located between the legs and the base plates or caster wheels of a scaffold. It is used to raise or lower parts of a scaffold to level it on uneven surfaces. *Figure 16* shows a scaffold using leveling jacks.

WARNING!

Use extreme caution when working on scaffolds. Do not stand on the handrails and do not jump from platform to platform.

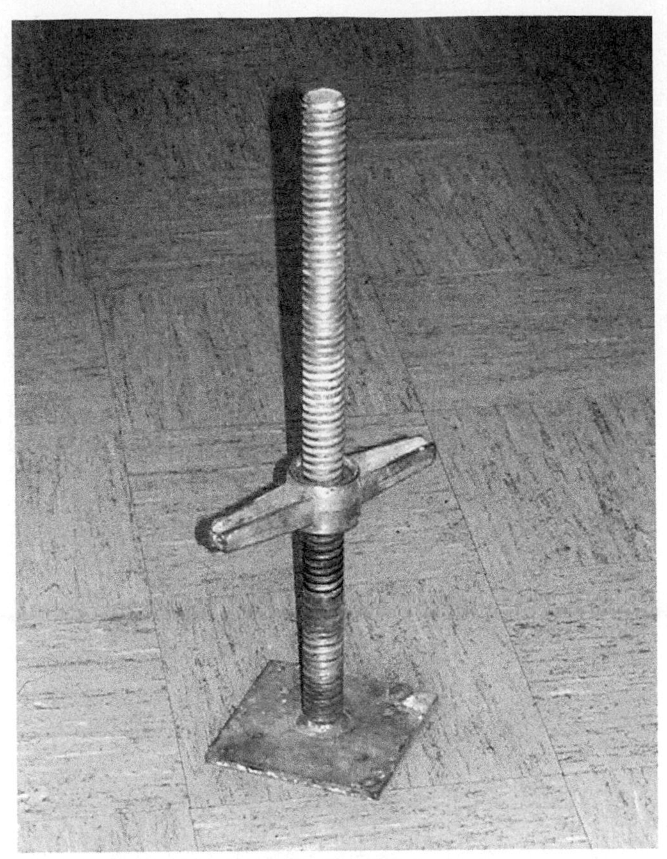
Figure 15 Typical adjustable leveling jack.

> **Did You Know?**
>
> **Scaffolding, Shoring, and Forming Institute**
>
> Manufacturers of scaffolding products take safety exceptionally serious. Many of these companies voluntarily belong to the Scaffolding, Shoring, and Forming Institute (SSFI), which promotes the technical advancement of scaffolding and shoring products. You can learn more about SSFI by visiting its website (www.ssfi.org).

2.2.0 Using and Caring for Tubular Buck Scaffolds

Erection methods for tubular buck scaffolds may differ by manufacturer, but most are erected in the same manner. The scaffold forms a rectangular framework made up of vertical uprights, or bucks, on the short sides of the scaffold and diagonal cross braces on the long sides of the scaffold. A **vertical upright** is the end section of a scaffold; **cross braces** are steel pieces used to connect and

Figure 16 Scaffold using leveling jacks.

support the vertical uprights. Access to the top of the scaffold is provided by ladders built into the vertical uprights or by a scaffold ladder attached to the scaffold. *Figure 17* shows basic tubular buck scaffold components.

2.2.1 Erecting Tubular Buck Scaffolds

A critical factor that must be considered before erecting a scaffold is its placement. You must fully examine the work area and consult with other workers in the area and then erect the scaffold so

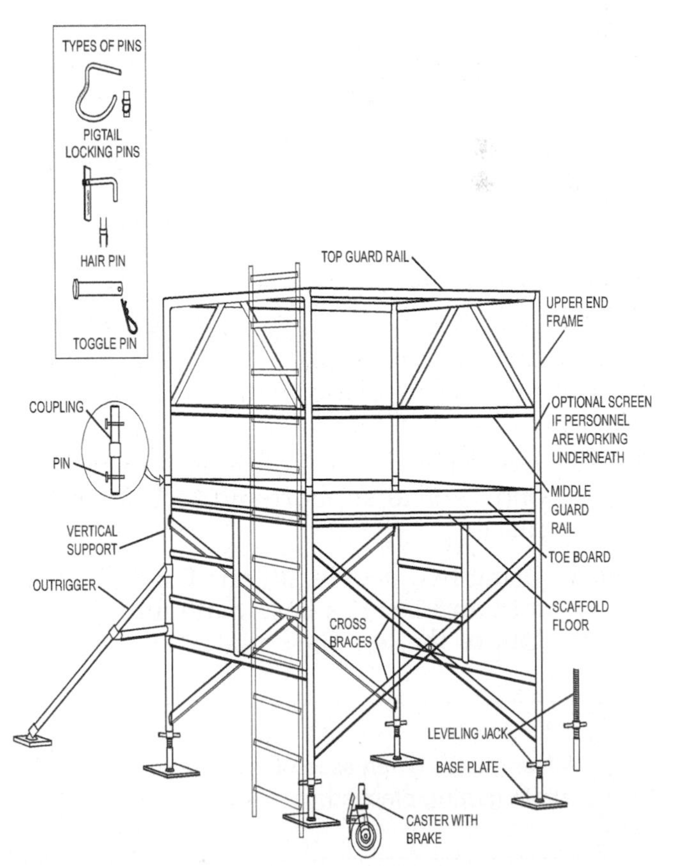

Figure 17 Basic tubular buck scaffold components.

that it will not interfere with the work being done in the general area and will not jeopardize safety.

To erect a tubular buck scaffold:

Step 1 Determine the height of the work to be performed.

Step 2 Identify the intended placement of the scaffold so that it will not interfere with your planned work activity or any other work being performed in the area.

Step 3 Calculate the number of vertical supports and cross braces needed to reach the work area.

> **NOTE**
> The size of the vertical supports will determine the number of supports needed to reach the work area.

Step 4 Obtain all the required components needed to erect the scaffold from the storeroom or supply house.

Step 5 Inspect all parts for damage and excessive rust, dirt, or grease.

> **NOTE**
> If a part is damaged or has excessive rust, do not use it. If there is excessive dirt or grease on the parts, clean them before using.

Step 6 Assemble two cross braces to one vertical support and lock them into place.

> **NOTE**
> For exterior installations, a mudsill must be placed under the scaffold.

Step 7 Assemble the cross braces into the second vertical support and lock them into place.

Step 8 Using two people, lift one leg of the scaffold and insert a base plate into the bottom of the leg. **Base plates** are flat discs or rectangles under scaffold legs that are used to evenly distribute the weight of the scaffold. They come in different sizes for different scaffold heights.

> **WARNING!**
> Wear gloves when assembling the scaffold to avoid getting pinched.

> **NOTE**
> If leveling jacks are needed, insert them between the legs and the base plates.

Step 9 Insert a hinge pin through the scaffold leg (or bottom part of the leveling jack) to secure the base plate. A **hinge pin** is a small pin inserted through a leg and base plate or caster wheel and used to hold these parts together. After inserting, lock it into place.

Step 10 Repeat Steps 8 and 9 to insert the base plates into the other scaffold legs.

Step 11 Ensure that the scaffold is plumb and level by adjusting the leveling jacks as necessary.

Step 12 Place scaffold planks between the vertical supports on the top rails to form the floor of the scaffold.

Step 13 Insert **coupling pins** into the tops of each of the vertical support legs and lock them into place. These are made of steel and used to line up and join sections of a scaffold.

Step 14 Place all items to complete the **guardrails** and the toeboards on the scaffold platform. The top rail of a guardrail is 42 inches above the **scaffold floor**, while the middle rail is halfway between the top rail and the toeboard. The scaffold floor is the work area platform made of metal or wood.

Step 15 Climb the ladder to the scaffold platform.

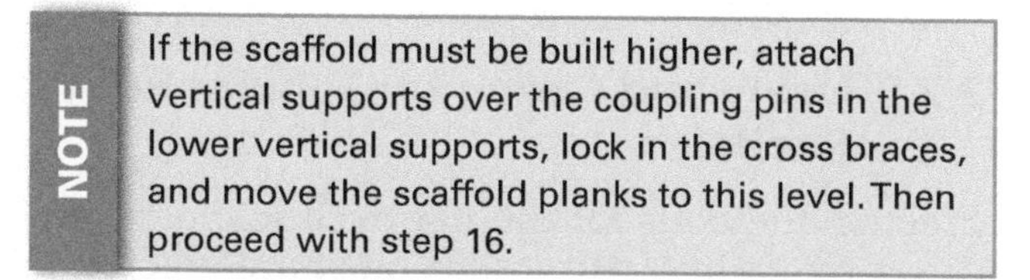

> **NOTE**
> If the scaffold must be built higher, attach vertical supports over the coupling pins in the lower vertical supports, lock in the cross braces, and move the scaffold planks to this level. Then proceed with step 16.

Step 16 Insert one upper end frame on the coupling pins of the lower vertical support. Lock it into place.

Step 17 Set the other upper end frame over the other vertical support and lock it into place.

Step 18 Secure the top guardrails to the scaffold.

Step 19 Secure the middle guardrails to the scaffold.

Step 20 Set the toeboards on all four sides of the scaffold.

2.2.2 Inspecting Tubular Buck Scaffolds

After erecting a tubular buck scaffold, it must be inspected before use.

To inspect a tubular buck scaffold:

Step 1 Check the base plates to ensure they are securely locked into place.

Step 2 Check the cross braces to ensure they are securely locked into the vertical supports.

Step 3 Check the top and middle guardrails to ensure that they are securely mounted.

Step 4 Check the toeboards to ensure they are properly installed.

Step 5 Check the coupling pins to ensure they are securely locked into place and cannot slip out.

Step 6 Check the scaffold floor to ensure it is securely locked into place and cannot slip out.

Step 7 Make sure the scaffold is level and standing on a firm surface that can support it.

2.2.3 Disassembling Tubular Buck Scaffolds

After removing all railings, the scaffold is disassembled from the top down.

To disassemble a tubular buck scaffold:

Step 1 Remove the toeboards from all four sides of the scaffold.

Step 2 Remove the middle guardrails from the scaffold.

Step 3 Remove the top guardrails from the scaffold.

Step 4 Remove the two upper end frames from the scaffold.

Step 5 Remove the four coupling rings from the tops of the two vertical supports.

Step 6 Remove the scaffold platform planks from the tops of the vertical supports.

Step 7 Remove the hinge pin from one of the base plates.

Step 8 Lift the leg of the scaffold and remove the base plate from that leg.

Step 9 Repeat Steps 7 and 8 to remove the other base plates.

Step 10 Disconnect the cross braces from one vertical support and remove that vertical support.

Step 11 Disconnect the cross braces from the remaining vertical support.

Step 12 Store all scaffold parts in their proper places.

2.3.0 Using and Caring for Pole Scaffolds

Erection methods for pole scaffolding also differ by manufacturer, but most are erected in basically the same manner. The pole scaffold consists of separate poles that serve as vertical posts, diagonal cross braces, and horizontal bearers or runners. The leveling jacks, guardrails, walk boards, and toeboards used for pole scaffolding are basically the same as those used for tubular buck scaffolding and all the safety rules that apply to tubular buck scaffolding also apply to pole scaffolds. The maximum interval for horizontal runners on a pole scaffold is 7 feet. Two types of pole scaffolds are locking ring scaffolds and tube and clamp scaffolds.

2.3.1 Locking Ring Scaffolds

Locking ring scaffolds are joined by locking rings and end connectors that are easily assembled and disassembled without the use of bolts and nuts or loose pins. *Figure 18* shows the components of a locking ring pole scaffold.

The vertical posts contain locking ring sets every 21 inches. The locking ring sets each contain two rings spaced $3\frac{1}{2}$ inches apart. The end connectors of the horizontal runners and vertical cross braces lock into these rings. To assemble the end connectors to the locking rings, hook the connector onto the rings and hammer the wedge underneath the bottom ring. To disassemble, pry the wedge out from underneath the bottom ring and unhook the end connector. The horizontal runners can be attached to the locking rings at any angle around the vertical post to allow for a wide variety of scaffolding configurations. *Figure 19* shows assembly and disassembly of the end connectors to the locking rings.

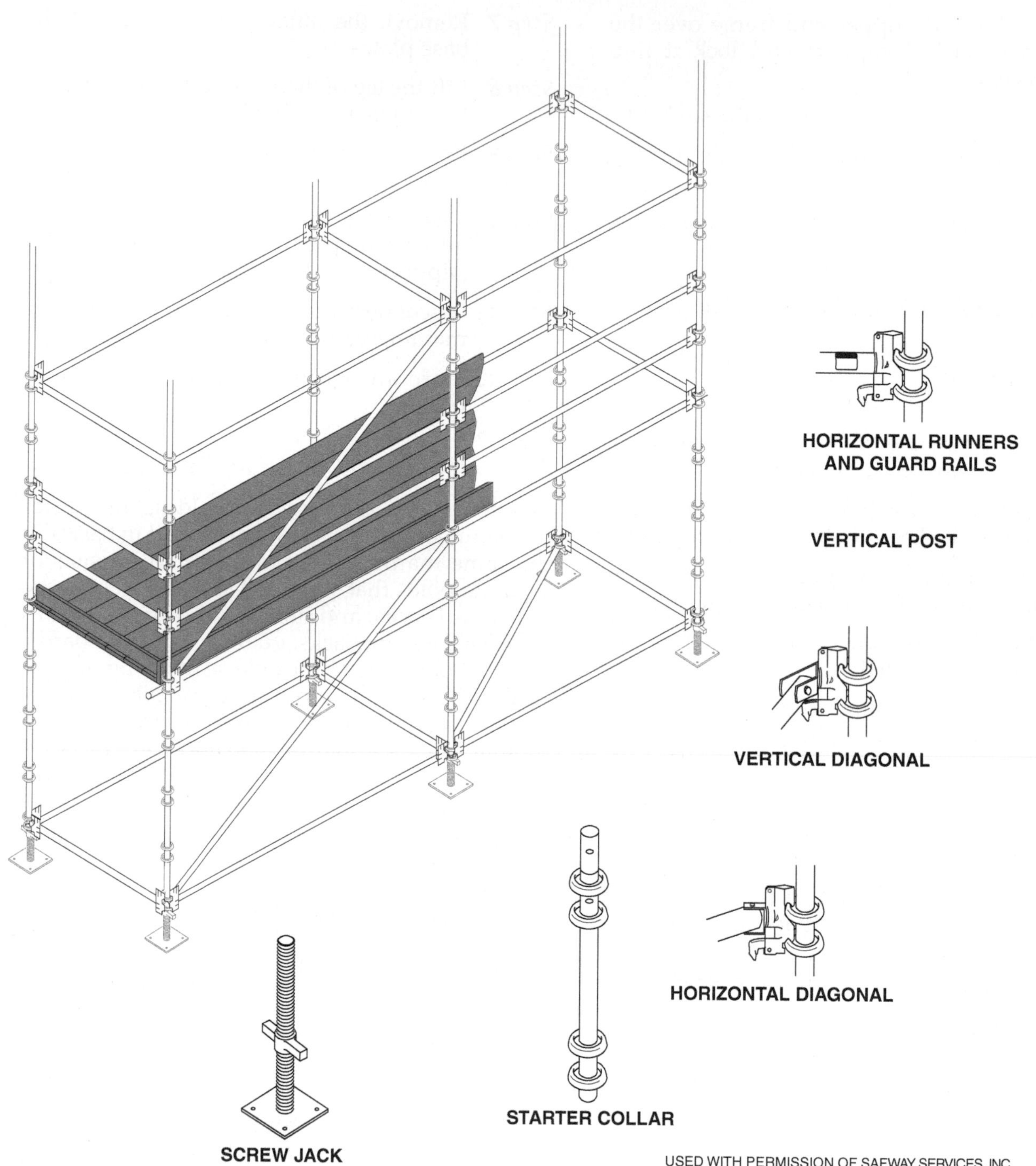

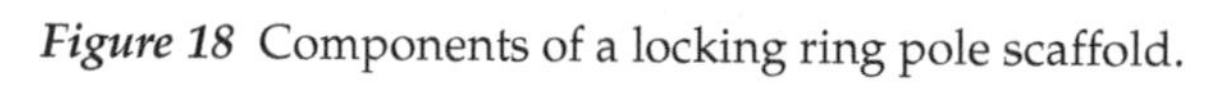

Figure 18 Components of a locking ring pole scaffold.

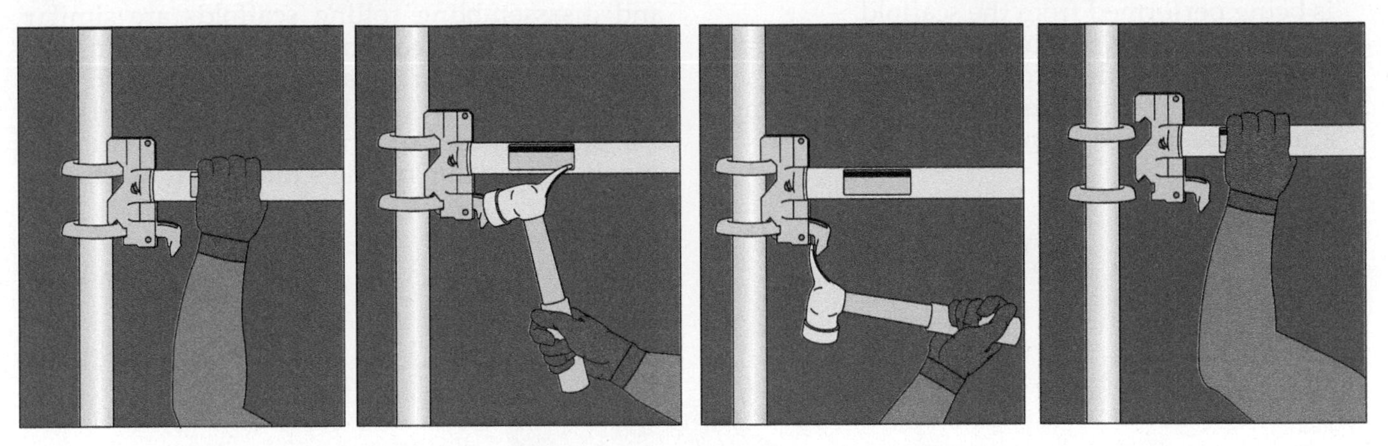

Figure 19 Assembly and disassembly of end connectors to locking rings.

2.3.2 Tube and Clamp Scaffolds

Like the locking ring scaffold, the tube and clamp scaffold is versatile and can be set up in many different configurations to suit different jobs. The vertical posts, horizontal runners, and diagonal cross braces are joined together with right-angle and swivel couplers (*Figure 20*). These couplers are secured to the poles with nuts and bolts. *Figure 21* shows the components of a tube and clamp pole scaffold.

Case History

Don't Become a Statistic

An employee was struck in the head by a scaffold rod that fell from above while he was working on the third level of scaffolding. He died later that day from a concussion. It was unclear whether he was wearing a hard hat at the time of the accident.

Bottom Line: Don't take chances. Wear your hard hat.

Source: Occupational Safety & Health Administration website.

2.4.0 Rolling Scaffolds

All the safety rules that apply to fixed scaffolds also apply to rolling scaffolds, but the additional safety rules that are specific to rolling scaffolds include the following:

- Do not use a working platform with a height that exceeds four times the smallest base dimension unless outriggers are used or the scaffold is guyed or braced against tipping.
- The scaffold as well as the caster wheels must be designed to support four times the intended maximum load.
- Use caster wheels with locking brakes to hold the scaffold in position.
- Brace rolling scaffolds properly with cross bracing and horizontal supports.
- Ensure that platforms are tightly planked for the full width of the scaffold and secured in place on each end.
- When moving a rolling scaffold, apply force as near to the base of the scaffold as possible to prevent tipping the scaffold over.
- Do not ride on a rolling scaffold.

Figure 20 Pole scaffolding swivel coupler.

- Lock the caster brakes at all times when work is being performed from the scaffold.
- Apply stationary scaffold rules for the guardrails and toeboards of rolling scaffolds.
- Do not extend the leveling jacks on a rolling scaffold more than 12 inches.

The procedures for assembling, inspecting, and disassembling rolling scaffolds are similar to those for the fixed scaffolds. Instead of installing base plates, you install caster wheels and lock them into place. As with the fixed scaffolds, actual assembly and securing procedures will vary slightly depending on the scaffold manufacturer.

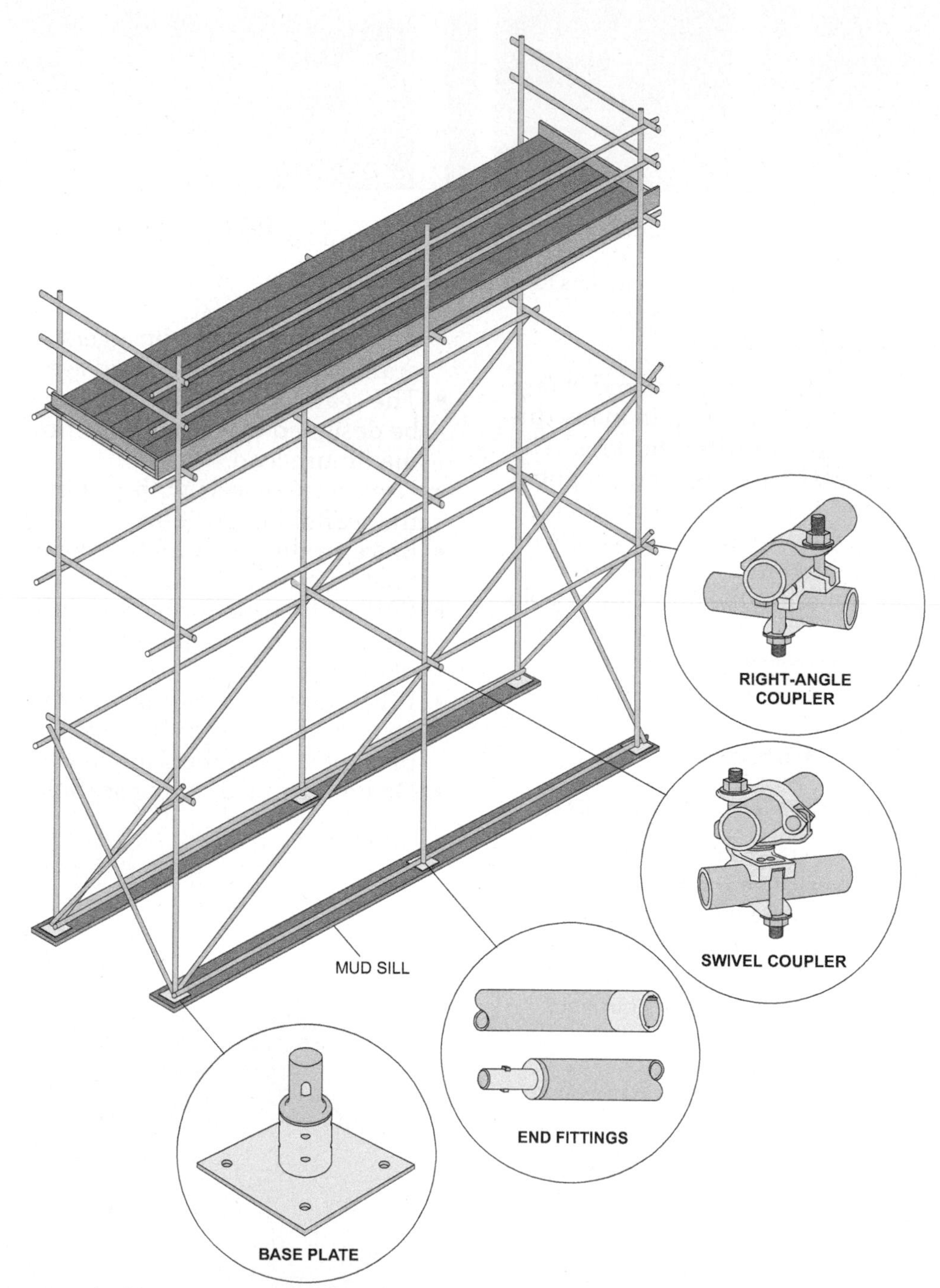

Figure 21 Components of tube and clamp pole scaffold.

Additional Resource

A Guide to Scaffold Use in the Construction Industry. US Department of Labor: Occupational Safety and Health Administration. 2002. Available at: **https://www.osha.gov/Publications/OSHA3150/osha3150.html**.

2.0.0 Section Review

1. On a rolling scaffolding, the scaffold as well as the caster wheels must be designed to support ____ times the intended maximum load.
   a. two
   b. three
   c. four
   d. five

2. True or false? All the safety rules that apply to tubular buck scaffolding also apply to pole scaffolds.
   a. True
   b. False

3. Vertical uprights are also known as _____.
   a. bucks
   b. bills
   c. does
   d. columns

4. The toeboard on the working platform of a scaffold should be a minimum of ____ inches high.
   a. 8
   b. 6
   c. 4
   d. 2

SECTION THREE

3.0.0 SCAFFOLD SAFETY

Objective

Identify scaffold hazards and state guidelines for safe use.

a. Identify specific scaffold hazards.
b. State specific scaffold safety guidelines.

With proper training, attention to one's surroundings, and effective communication and teamwork, scaffolds may be used safely. Improper or careless use of scaffolding can result in injury or death. Risks can be minimized by being aware of the hazards involved and following the proper safety procedures and guidelines.

3.1.0 Scaffold Hazards

The main hazards involved with the use of scaffolding are:

- Falls
- Being struck by falling objects
- Electric shocks

Falls can happen because fall protection has not been provided, is not used, or because it is installed or used improperly. Poorly planked scaffolding and working on scaffolding when conditions are dangerous, such as in high winds, ice, rain, and lightning, can lead to accidents. Falls also happen when scaffolding collapses because of improper construction.

Fall protection is required on any scaffolding 6 feet or more above ground level. Fall-protection devices consist of guardrail systems, personal fall-arrest systems, and/or safety nets. When using a fall arrest system, tie off above your head if possible as this will limit your freefall distance. Remember that the anchor point will change according to specifications and conditions.

A guardrail system must extend around all open sides of the scaffolding and is adequate fall protection for most scaffolding, but company policies must be followed at all times with many organizations requiring a harness and lanyard to be used. The side facing the work surface does not need to have a guardrail if it is located less than 14 inches away from the work surface. Any opening on a scaffolding platform must be protected by a guardrail system, including the access opening(s) and platforms that do not extend across the entire width of the scaffolding.

People who work on or pass under scaffolding may be hit by falling objects because tools, materials, debris, and scaffolding parts may fall to the surface below; for this reason, many tools are tied to a worker's tool belt. Those working on scaffolding may also be injured if there are others working above them or if the structure or workpiece extends above the work level of the scaffolding. Any worker who is exposed to the danger of falling objects is required to wear a hard hat. Depending on the situation, additional protection such as debris nets, screens or mesh, canopy structures, and toeboards may be needed. Barricades that prevent access under the scaffolding can also be used to protect workers and others.

WARNING!

Maintain awareness of the condition of anchorage of the scaffolding you are working on. Choose your anchor points wisely for fall protection – connect your lanyard to an anchor point that will protect you from scaffold collapse. Make an emergency exit plan before beginning work.

If the scaffolding you are working on shifts or begins to collapse, stop what you are doing and safely and quickly exit the scaffolding.

Because most scaffolding is made of metal, the chance of electric shock is always a hazard. Never assume that you can work around high-voltage wires just by avoiding contact since high voltages can arc through the air and cause electrocution without direct contact. When scaffolding must be erected close to power lines, the utility company must be called in to de-energize, move, and/or cover the lines with insulating protective barriers.

NOTE

Always refer to the competent person on site if you have any questions about the safety of scaffolding.

3.2.0 Scaffolding Safety Guidelines

Always use the right safety equipment, including a hard hat and personal fall-protection systems when working on, with, or near scaffolding. Follow the safety guidelines and OSHA regulations relevant to scaffold work.

Never work on scaffolding if you:

- Are subject to seizures

- Become dizzy or lightheaded when working at an elevation
- Take medication that might affect your stability and/or performance
- Are under the influence of drugs and/or alcohol

When working on scaffolding, always follow these guidelines:

- Erect and use scaffolding according to the manufacturer's instructions. They must also be erected and used in accordance with all local, state, federal, and OSHA requirements.
- Check all scaffolds for complete or incomplete inspection tags before mounting the scaffold. Do not mount a scaffold that is tagged as incomplete.
- Install guardrails and toeboards on all open sides and ends of platforms that are more than 6 feet high.
- OSHA requires that the scaffolding supervisor approve alterations to scaffolding plans.
- A professional engineer must design tube and coupler scaffolding that exceeds 125 feet in height and pole scaffolding that exceeds 60 feet in height.
- The front edge of scaffolding must not be more than 14 inches from the face of the work, unless guardrail systems are erected along the front edge and/or personal fall arrest systems are used to protect employees from falling. When outriggers are used, the front edge of the scaffolding must not be more than 3 inches from the face of the work.
- Outriggers or some other means of securing the scaffolding, such as guys, ties, and braces, must be used according to the scaffold manufacturer's recommendations or when the vertical height of the scaffolding is four times the minimum base dimension. Securing devices are set both vertically and horizontally. Vertical devices are set at no more than 20-foot intervals for scaffolds 3 feet (or less) wide, and at no more than 26-foot intervals for scaffolds greater than 3 feet wide. On completed scaffolds, the last securing device must be placed no greater than four times the minimum base dimension from the top of the scaffolding. The horizontal devices must be set no greater than 30 feet apart (measured from one end towards the other).
- Guardrails must be constructed from 2-inch by 4-inch lumber or OSHA-approved equivalent metal. It must be supported every 8 feet.
- The distance from the upper edge of the top rail to the scaffold platform must be 38 to 45 inches on scaffolds placed in service after January 1, 2000, while the distance must be 36 to 45 inches for scaffolds placed in service before January 1, 2000.
- Toeboards must be solid, or, if they have openings, the openings must be one inch or less at all points. They must be fastened at the outer edge of the platform with not more than ¼ inch clearance above the platform. Toeboards must be able to withstand - at all points - a force of at least 50 pounds in a horizontal or vertical direction without failure. There must be at least three-and-one-half inches from the top edge of the toeboard to the platform support.
- The space between the platform and the scaffolding upright support must not be more than 1 inch wide and there may be no more than 2 inches between planks.
- Ladders are not normally permitted on scaffolding except under the most exceptional of circumstances. You as a pipefitter should never make a decision to place a ladder on scaffolding except under the direction of the scaffolding supervisor.
- Install a screen between the toeboard and the guardrail when persons must work or pass underneath the scaffold. This screen must be 18-gauge US Standard 1-inch mesh or the equivalent.
- Scaffolds must be capable of supporting at least four times their maximum intended load weight.
- Provide an access ladder or equivalent safe access. Do not climb cross braces.
- Plumb and level all scaffolds as the erection proceeds. Do not force braces to fit, but level the scaffold until the proper fit can be made easily.
- Wooden scaffold boards must be No. 1 grade lumber and be equipped with cleats on each end. They must be secured to the scaffold with OSHA-certified, 9-gauge wire.
- Aluminum scaffold boards must be equipped with a turnkey latch to secure the board to the scaffolding.
- Never interchange parts of a scaffolding system made by different manufacturers.
- Attach a green, red, or yellow tag (*Figure 22*), as applicable, to any scaffolding that is assembled and erected to alert users of its current mechanical and/or safety status. Do not rely solely on the tag. Inspect all parts of scaffolding before each use.
- Have a safety plan in place before using scaffolding. At high levels there should be an individual tie-off to the main structure. If the

scaffolding begins to shift, exit the scaffolding immediately.

3.2.1 *Safety Guidelines for Built-Up Scaffolding*

Built-up scaffolding is, as its name implies, built from the ground up at a job site. Use the following guidelines when erecting and using tubular, built-up scaffolding:

- Inspect all scaffolding parts before assembly.
- Never use parts that are broken, damaged, or deteriorated. Be cautious of rusted materials.
- Follow the manufacturer's recommendations for the proper methods of erecting and using scaffolding.
- Do not interchange parts from different manufacturers.

Case History

Safety Equipment—It Might Have Saved His Life

A 32-year old journeyman pipefitter was wearing a body harness and lanyard when he climbed onto the roof of a building, but the lanyard wasn't attached to a lifeline or the scaffold. He fell 40 feet to the ground and died later from his injuries.

Bottom Line: Safety equipment works only when it's used properly.

Source: Alaska Health and Human Services website

- Do not force braces or other parts to fit. Adjust the level of the scaffolding until the connections can be made easily.
- Provide adequate sills or underpinnings for all scaffolding built on filled or soft ground and be sure to compensate for uneven ground by using adjusting screws or leveling jacks.
- Do not use boxes, concrete blocks, bricks, or other similar objects to support scaffolding.
- Keep scaffolding free of clutter and slippery materials.
- Be sure scaffolding is plumb and level at all times. Follow the prescribed spacing and positioning requirements for the parts of the scaffolding. Anchor or tie-in scaffolding to the building at prescribed intervals.
- Use ladders rather than cross braces to climb the scaffolding. Position ladders with caution to prevent the scaffolding tower from tipping.

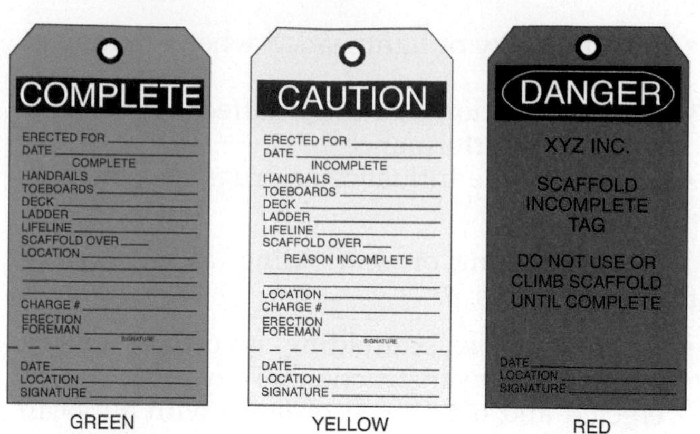

Figure 22 Typical scaffolding tags.

- Do not work on scaffolding that is more than 6 feet high without guardrails, midrails, and toeboards on open sides and ends.
- Lock the **casters** of mobile scaffolding when it is positioned for use. Casters are the wheels which are attached to the bottoms of scaffold legs instead of base plates; most come with brakes.
- Do not ride on mobile scaffolding.
- Avoid building scaffolding near power lines.

3.2.2 *Personal Fall-Arrest Systems*

Personal fall-arrest systems catch workers after they have fallen. They are designed and rigged to prevent a worker from free-falling a distance of more than 6 feet and hitting the ground or a lower work area.

Personal fall-arrest systems use specialized equipment that includes the following:

- Body harness (*Figure 23*)
- Lanyard (*Figure 24*)
- Deceleration device (*Figure 25*)
- Lifeline (*Figure 26* and *Figure 27*)
- Anchoring device and equipment connector

NOTE

In the past, body belts were often used instead of a full-body harness as part of a fall-arrest system. As of January 1, 1998, however, they were banned from such use because body belts concentrate all of the arresting force in the abdominal area, causing the worker to hang in an uncomfortable and potentially dangerous position while awaiting rescue.

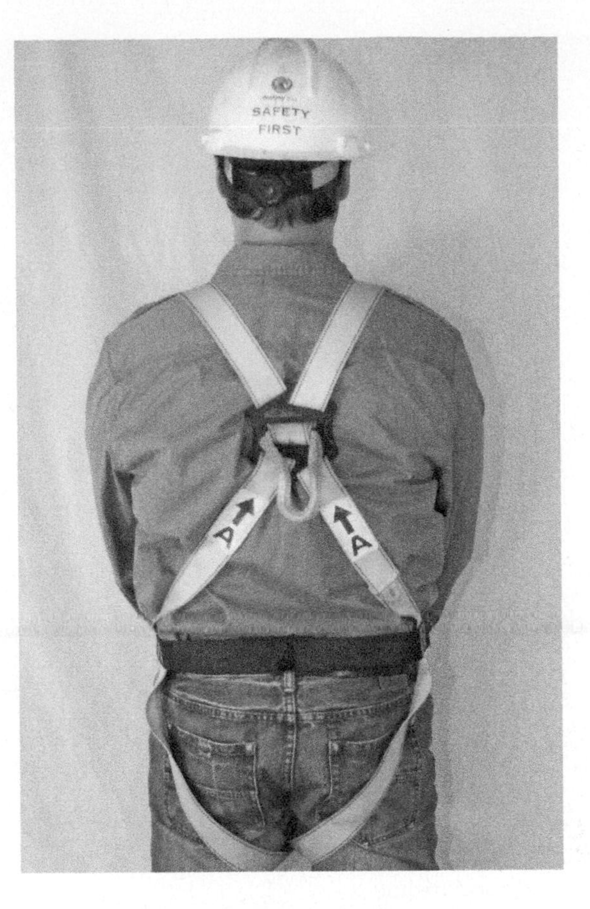

Figure 23 Harness.

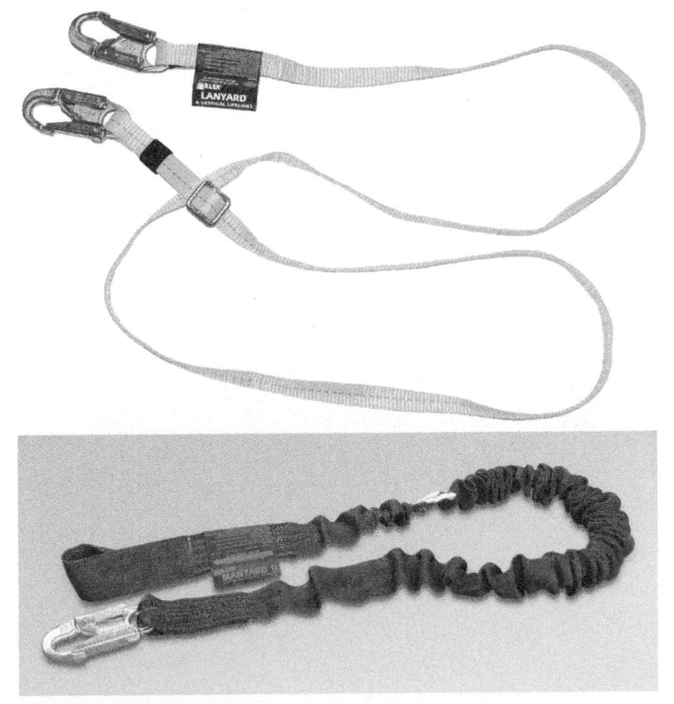

Figure 24 Lanyards.

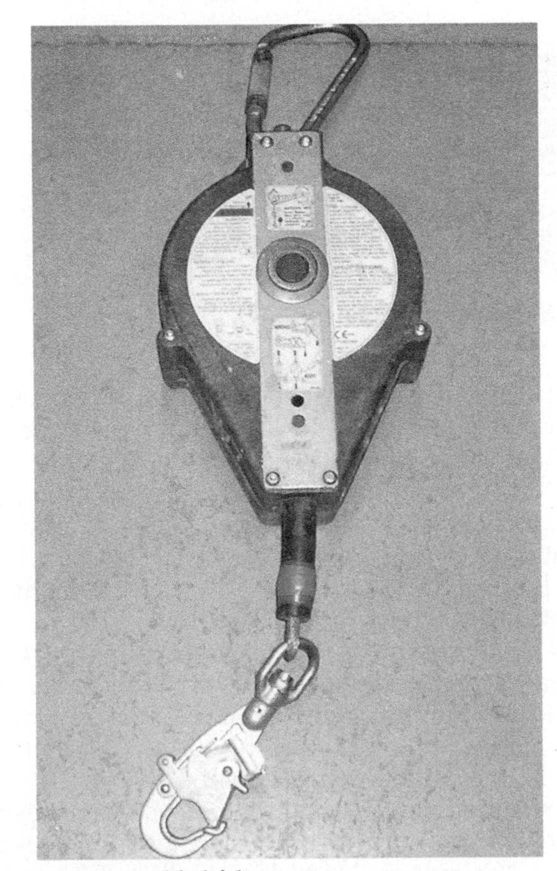

Figure 25 Retractable lifeline.

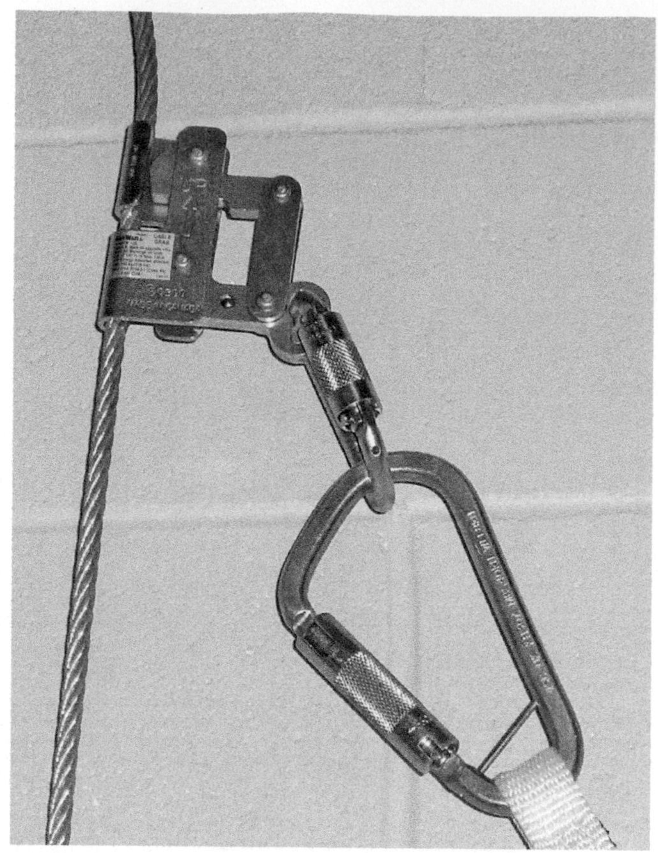

Figure 26 Vertical lifeline.

WARNING!

When activated during the fall-arrest process, a shock-absorbing lanyard stretches in order to reduce the arresting force. This potential increase in length must always be taken into consideration when determining the total free-fall distance from an anchor point.

3.2.3 Rescue After a Fall

Every elevated job site should have an established rescue and retrieval plan. Planning is especially important in remote areas without ready access to a telephone. Before beginning work, make sure that you know what your employer's rescue plan calls for you to do in the event of a fall. Find out what rescue equipment is available and where it is located and learn how to use equipment for self-rescue and the rescue of others.

If a fall occurs, any employee hanging from the fall-arrest system must be rescued safely and quickly. Your employer should have previously determined the method of rescue for fall victims, which may include equipment that lets the victim rescue himself or herself, a system of rescue by co-workers, or a way to alert a trained rescue squad. If a rescue depends on calling for outside

Figure 27 Horizontal lifeline.

help such as the fire department or rescue squad, all the needed phone numbers must be posted in plain view at the work site. In the event a co-worker falls, follow your employer's rescue plan and call any special rescue service needed. Communicate with the victim and monitor him or her constantly during the rescue.

Additional Resource

Personal Fall Protection Systems. The US Department of Labor: Occupational Safety and Health Administration. 2018. Available at: **https://www.osha.gov/pls/oshaweb/owadisp.show_document?p_table=STANDARDS&p_id=1291**.

3.0.0 Section Review

1. You can ride mobile scaffolding ____.
   a. as long as it's not positioned near a power line
   b. safely monitored by employees on the ground
   c. if it's been recently inspected and is in safe working conditions
   d. under no circumstances whatsoever

2. Which of the following is not a type of fall protection?
   a. Guardrail system
   b. Personal fall-arrest system
   c. Safety nets
   d. Toeboard

SUMMARY

Ladders and scaffolding are used in a variety of different construction jobs. The type of ladders or scaffolding will vary with each job, but the dangers won't. There is always a risk of falling and being struck by falling objects, each of which can result in serious injury or death. It is your responsibility to be aware of and follow the safety procedures associated with ladders and scaffolding. If you are involved in assembly of scaffolding, you must make absolutely sure it is properly assembled and leveled, and equipped with guardrails, toeboards, and other safety features.

Review Questions

1. When you are working on ladders or scaffolding, your first responsibility is to _____.
   a. get the job done rapidly and efficiently
   b. work carefully whenever it is possible
   c. keep yourself and your co-workers safe
   d. complete your part of the job correctly

2. The type of ladder that is corrosion-resistant and can be used in situations where it might be exposed to the elements is _____.
   a. aluminum
   b. fiberglass
   c. wooden
   d. steel

3. The type of ladder that is heavier than other ladders and can be used when heavy loads must be moved up and down is _____.
   a. aluminum
   b. fiberglass
   c. wooden
   d. steel

4. The type of ladder that is durable, non-conductive, and useful in situations where some amount of rough treatment is unavoidable is _____.
   a. aluminum
   b. fiberglass
   c. wooden
   d. steel

5. Before climbing a ladder, you must check all of the following *except*:
   a. the ladder's rungs to be sure they are clean and in good repair
   b. that when placed against a window the ladder is securely tied down
   c. the soles of your shoes are free from any slippery substances
   d. that the ladder is placed on stable ground where it will not slip

6. If you lean a straight ladder against the top of a 16-foot wall, the base of the ladder should be _____ feet from the base of the wall.
   a. $1\frac{1}{2}$
   b. 3
   c. 4
   d. 6

7. All of the following are safety precautions when using a stepladder *except*:
   a. Standing on the top of the ladder.
   b. Ensuring all four feet are on a hard, even surface.
   c. Using the stepladder from both signs when it is designed for that type of use.
   d. Locking the spreaders in the fully open position when the ladder is in use.

8. Straight ladders are typically _____ feet in length.
   a. 4 to 8
   b. 8 to 12
   c. 12 to 16
   d. 16 to 12

9. What is the best way to access tools when you are working from a straight ladder?
   a. Carry the tools you need in your hand and be extra careful when climbing the ladder.
   b. Set the tools on a rung before you climb the ladder and move them up as you climb.
   c. Plan to use a rope to lift tools that you will need after you have climbed the ladder.
   d. Using tools on a ladder is not safe. Use scaffolding if tools are needed above ground.

10. At most job sites, scaffold builders _____.
    a. are certified according to company policy
    b. learn to build scaffolds by trial and error
    c. work alone to reduce the odds of a mishap
    d. are used to erect only complex scaffolds

11. Before you erect any scaffolding, you must _____.
    a. examine the area and place the scaffolding so it will not hamper other work
    b. ask all other workers to leave the area to decrease the chances of an accident
    c. reschedule tasks of your co-workers so you can quickly build the scaffolding
    d. ask the local OSHA agent to inspect the area you have selected for the scaffold

12. Fall protection is required on any scaffolding that is _____ feet or more above ground level
    a. 6
    b. 10
    c. 12
    d. 14

13. Because most scaffolding is made of metal, there is a risk of _____ when working around high-voltage wires.
    a. electric shock
    b. burning yourself
    c. falling
    d. getting muscle pain

14. Scaffolds must be capable of supporting at least _____ their maximum intended load weight.
    a. twice
    b. three times
    c. four times
    d. five times

15. When you are assigned to work on scaffolding you need to _____.
    a. take the least amount of equipment onto the scaffold with you including safety equipment
    b. notify your supervisor when you are taking prescription medication that makes you dizzy
    c. ask any non-essential workers to leave the work area to reduce the potential of accidents
    d. climb the scaffolding quickly since that is the time that you are at greatest risk of falling

Trade Terms Quiz

Fill in the blank with the correct term that you learned from your study of this module.

1. Discs or rectangles used to distribute the weight of a scaffold are ________.
2. A(n) ________ is the end section of a scaffold, also known as a buck.
3. Some scaffolds have wheels, or ________, attached to the bottoms of the scaffold legs.
4. To keep tools from falling off the scaffold, attach a(n) ________ around the scaffold floor.
5. The vertical uprights of a scaffold are connected and supported by ________.
6. A temporary built-up framework to support workers, materials, and equipment is called ________.
7. A pin used to hold a base plate or castor wheel onto a leg is called a(n) ________.
8. Part of a personal fall arrest system, the ________ straps around your body and is connected to a lifeline.
9. Ladders are assigned a ________ based on the intended use and maximum load limit of the ladder.
10. Sections of the scaffold are lined up and joined with a(n) ________.
11. A(n) ________ is an elevated work platform.
12. To level a scaffold on an uneven surface, use ________.
13. The top ________ should be 42 inches above the scaffold floor.
14. The platform on top of the scaffolding is the ________.
15. The ________ states that a ladder should be placed so that the distance between the base of the ladder and the supporting wall is one-fourth the working distance of the ladder.

Trade Terms

4-to-1 rule
Base plates
Casters
Coupling pin
Cross braces
Duty rating
Guardrail
Harness
Hinge pin
Leveling jacks
Scaffold
Scaffolding
Scaffold floor
Toeboard
Vertical upright

Trade Terms Introduced in This Module

4-to-1 rule: The safety rule for straight ladders and extension ladders, which states that a ladder should be placed so that the distance between the base of the ladder and the supporting wall is one-fourth the working distance of the ladder.

Base plates: Flat discs or rectangles under scaffold legs that are used to evenly distribute the weight of the scaffold. Base plates come in different sizes for different scaffold heights.

Casters: Wheels that are attached to the bottoms of scaffold legs instead of base plates. Most casters come with brakes.

Coupling pin: A steel pin used to line up and join sections of a scaffold.

Cross braces: Steel pieces used to connect and support the vertical uprights of a scaffold.

Duty rating: American National Standards Institute (ANSI) rating assigned to ladders. It indicates the type of use the ladder is designed for (industrial, commercial, or household) and the maximum working load limit (weight capacity) of the ladder. The working load limit is the maximum combined weight of the user, tools, and any materials bearing down on the rungs of a ladder.

Guardrails: Protective rails attached to a scaffold. The top rail is 42 inches above the scaffold floor, and the middle rail is halfway between the top rail and the toeboard.

Harness: A device that straps securely around the body and is connected to a lifeline. It is part of a personal fall arrest system.

Hinge pin: A small pin inserted through a leg and base plate or caster wheel and used to hold these parts together.

Leveling jack: A threaded, adjustable screw located between the legs and the base plates or caster wheels of a scaffold. It is used to raise or lower parts of a scaffold to level it on uneven surfaces.

Scaffold: An elevated work platform for workers and materials.

Scaffolding: A temporary built-up framework or suspended platform or work area designed to support workers, materials, and equipment at elevated or otherwise inaccessible job sites.

Scaffold floor: A work area platform made of metal or wood.

Toeboard: A 4-inch railing attached around the scaffold floor to prevent tools and materials from falling off the scaffold.

Vertical upright: The end section of a scaffold, also known as a buck, made of welded steel. It supports other vertical uprights or upper end frames and the scaffold floor.

Additional Resource

This module presents thorough resources for task training. The following reference material is recommended for further study.

Fall Protection in Construction. US Department of Labor: Occupational Safety and Health Administration. 2015. Available at: **https://www.osha.gov/Publications/OSHA3146.pdf.**

A Guide to Scaffold Use in the Construction Industry. US Department of Labor: Occupational Safety and Health Administration. 2002. Available at: **https://www.osha.gov/Publications/OSHA3150/osha3150.html.**

Occupational Safety and Health Standards for the Construction Industry, Latest Edition. Occupational Safety and Health Administration. U.S. Department of Labor. Washington, DC: U.S. Government Printing Office.

Personal Fall Protection Systems. The US Department of Labor: Occupational Safety and Health Administration. 2018. Available at: **https://www.osha.gov/pls/oshaweb/owadisp.show_document?p_table=STANDARDS&p_id=1291.**

Figure Credits

DBI/SALA & Protecta, F27

Louisville Ladder Group, F02

Levelok Corporation, F04

Safway Services, Inc., F18, F19

Ridge Tool Co. (RIDGID®), F06, F11

Werner Company, F07

Section Review Answer Key

SECTION 1.0.0

Answer	Section Reference	Objective
1. c	1.0.0	1b
2. a	1.3.0	1a
3. d	1.1.0	1a
4. b	1.2.0	1c
5. c	1.3.0	1c

SECTION 2.0.0

Answer	Section Reference	Objective
1. c	2.4.0	2d
2. a	2.3.0	2c
3. a	2.2.0	2b
4. c	2.1.0	2a

SECTION 3.0.0

Answer	Section Reference	Objective
1. d	3.2.1	3b
2. d	3.1.0	3a

This page is intentionally left blank.

NCCER CURRICULA — USER UPDATE

NCCER makes every effort to keep its textbooks up-to-date and free of technical errors. We appreciate your help in this process. If you find an error, a typographical mistake, or an inaccuracy in NCCER's curricula, please fill out this form (or a photocopy), or complete the online form at **www.nccer.org/olf**. Be sure to include the exact module ID number, page number, a detailed description, and your recommended correction. Your input will be brought to the attention of the Authoring Team. Thank you for your assistance.

Instructors – If you have an idea for improving this textbook, or have found that additional materials were necessary to teach this module effectively, please let us know so that we may present your suggestions to the Authoring Team.

NCCER Product Development and Revision

13614 Progress Blvd., Alachua, FL 32615

Email: curriculum@nccer.org
Online: www.nccer.org/olf

❑ Trainee Guide ❑ Lesson Plans ❑ Exam ❑ PowerPoints Other ____________

Craft / Level: Copyright Date:

Module ID Number / Title:

Section Number(s):

Description:

Recommended Correction:

Your Name:

Address:

Email: Phone:

This page is intentionally left blank.

Motorized Equipment One

Overview

Pipefitters work with various types of motorized equipment to speed up production. From smaller items such as generators and compressors to larger machinery such as forklifts and backhoe loaders, a pipefitter must understand the capacities and limitations of each item in order to work safely and efficiently. This is because increased power means increased hazards. Anyone operating motorized equipment must be properly trained, and in some cases certified, to use it. Always follow the recommended safety precautions, manufacturer's instructions, and maintenance schedule.

Module 08106

Trainees with successful module completions may be eligible for credentialing through the NCCER Registry. To learn more, go to www.nccer.org or contact us at 1.888.622.3720. Our website has information on the latest product releases and training, as well as online versions of our *Cornerstone* magazine and Pearson's product catalog.

Your feedback is welcome. You may email your comments to curriculum@nccer.org, send general comments and inquiries to info@nccer.org, or fill in the User Update form at the back of this module.

This information is general in nature and intended for training purposes only. Actual performance of activities described in this manual requires compliance with all applicable operating, service, maintenance, and safety procedures under the direction of qualified personnel. References in this manual to patented or proprietary devices do not constitute a recommendation of their use.

08106 V4.0

08106
MOTORIZED EQUIPMENT ONE

Objectives

Successful completion of this module prepares trainees to:

1. Describe the types of motorized equipment found in the pipefitting environment and state general safety precautions for their use.
   a. Identify safety precautions common to motorized equipment.
2. Identify and describe how to use portable generators.
   a. Identify typical generator controls.
   b. Explain how to operate a portable generator.
   c. State generator safety guidelines and describe their maintenance needs.
3. Identify and describe how to use portable air compressors.
   a. Identify typical air-compressor controls.
   b. Explain how to operate a portable air compressor.
   c. State air-compressor safety guidelines and describe their maintenance needs.
4. Identify aerial-lift controls and describe how to use aerial lifts.
   a. Identify typical aerial-lift controls.
   b. Explain how to operate aerial lifts.
   c. State aerial-lift safety guidelines and describe their maintenance needs.
5. Identify and describe how to use forklifts.
   a. Identify typical forklift controls.
   b. Explain how to operate forklifts.
   c. State forklift safety guidelines and describe their maintenance needs.
6. Identify and describe how to use trenchers.
   a. Identify typical trencher controls.
   b. Explain how to operate trenchers.
   c. State trencher safety guidelines and describe their maintenance needs.
7. Identify and describe the use of support equipment.
   a. Identify and describe the use of portable welding machines.
   b. Identify and describe the use of portable pumps.
   c. Identify and describe the use of portable compactors.
8. Identify and describe the use of backhoes and mobile cranes.
   a. Identify and describe the use of backhoes.
   b. Identify and describe the use of mobile cranes.

Performance Tasks

Under the supervision of your instructor, demonstrate your abilities with two of the following:

1. Perform all prestart checks for engine-driven generators.
2. Operate engine-driven generators.
3. Set up and operate engine-driven welding machines.
4. Perform all prestart checks for portable air compressors.
5. Operate portable air compressors.
6. Identify forklift trucks and recognize safety hazards involved in working around them.
7. Identify portable pumps to use for specific applications.
8. Identify types of hydraulic cranes and recognize safety hazards involved in working around them.

Trade Terms

Aerial lift
Ampere (amp)
Centrifugal force
Centrifugal pump
Cherry picker
Circuit breaker
Compactor
Compressor
Diaphragm
Diaphragm pump
Forklift
Fuse holder
Generator
Governor
Impeller
Mast
Mobiling
Pneumatic
Powered industrial truck
Proportional control
Straight blade duplex connector
Trencher
Twist-lock connector
Velocity
Volt
Watt

Industry Recognized Credentials

If you are training through an NCCER-accredited sponsor, you may be eligible for credentials from NCCER's Registry. The ID number for this module is 08106. Note that this module may have been used in other NCCER curricula and may apply to other level completions. Contact NCCER's Registry at 888.622.3720 or go to www.nccer.org for more information.

Contents

Contents (continued)

Figures

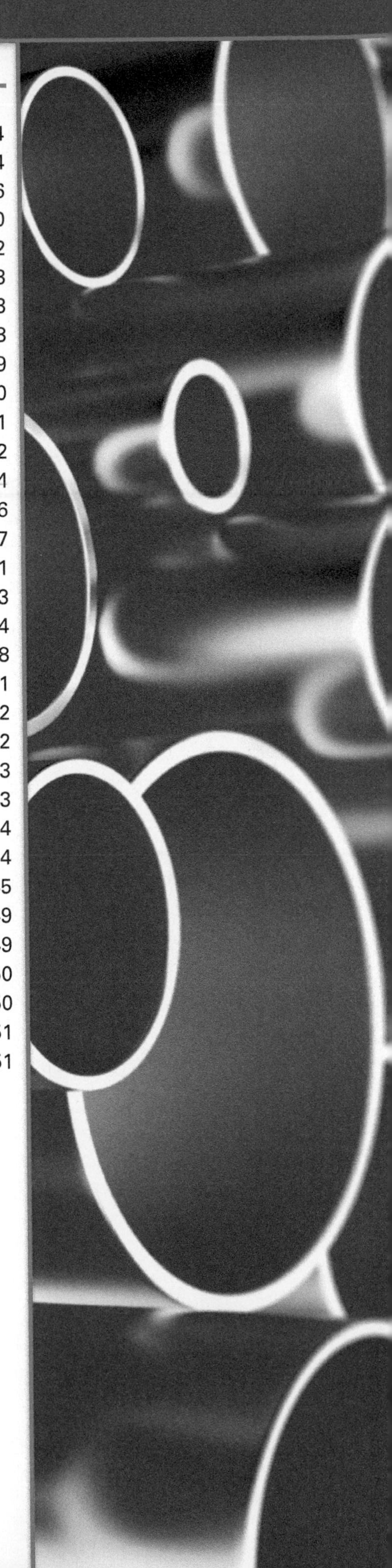

This page is intentionally left blank.

SECTION ONE

1.0.0 MOTORIZED EQUIPMENT IN PIPEFITTING

Objective

Describe the types of motorized equipment found in the pipefitting environment and state general safety precautions for their use.

a. Identify safety precautions common to motorized equipment.

Trade Terms

Compressor: A motor-driven machine used to supply compressed air for pneumatic tools.

Forklift: A machine designed to facilitate the movement of bulk items around the job site.

Generator: A machine used to generate electricity.

Pipefitters work with and around various types of motorized equipment throughout their careers. Motorized equipment includes portable equipment such as generators, compressors, and pumps, as well as larger equipment, such as trucks, mobile cranes, and personnel lifts. A piece of motorized equipment is considered portable if it can be transported from job site to job site or to different areas of the same job site. Portable equipment can either be moved or rolled to a different location by hand or transported on a trailer behind a truck or tractor. A piece of equipment is also considered to be portable if it can provide its own power source in the field. The proper use, care, and preventive maintenance of portable motorized equipment are essential to safe operations; this includes, among other items, forklifts, backhoe loaders, and mobile cranes that pipefitters will be working with and around throughout their careers.

1.1.0 Safety Precautions

Safety is the primary concern no matter what type of equipment is being operated. Appropriate personal protective equipment must be worn while operating or working near any machinery, and all use must be performed under the direct supervision of an instructor or other qualified worker.

While a number of protocols such as these are considered standard, certain pieces come with additional precautions. The manufacturer provides safety instructions which must be followed at all times. Think carefully through the described uses and hazards of each piece of equipment to contribute to a safer work environment. Do not take shortcuts or use workarounds; disregard for safety can cause serious injury or death.

WARNING!

Anyone operating power equipment must be properly trained in its use. Part of this training includes understanding the associated safety precautions. Some localities may require the operator to be trained and qualified in the use of certain equipment. A daily visual inspection (DVI) application (mobile app) may assist with creating a work order when something is wrong.

Remember to:

- Read and fully understand the operator's manual and follow the instructions and safety precautions.
- Know the capacity and operating characteristics of the equipment being operated.
- Inspect the equipment before each use to make sure everything is in proper working order. Have any defects repaired before using the equipment. Never modify or remove any part of the equipment unless authorized to do so by the manufacturer.
- Check for hazards above, below, and all around the job site. Be sure to maintain a safe distance from electrical power lines and other electrical hazards.
- Learn as much about the work area as possible before starting work.
- Fasten the seat belt or operator restraints before starting.
- Set up warning barriers and keep others away from the equipment and job site.
- Know the locations of a first aid kit and fire extinguisher, and how to use both of them.
- Know where to get assistance.

1.1.1 Interlocking Systems

Some equipment comes with safety interlocks to prevent unsafe operation. Keep the following in mind when using this type of equipment:

- Interlock systems must not be modified. Interlock systems help prevent incorrect control operation.
- If an interlock system fails, contact the manufacturer's repair representative immediately.
- Runaway equipment is always possible; learn how to use all the controls and emergency procedures.

1.1.2 Transporting Equipment

When transporting equipment between jobs, remember these safety guidelines:

- Park, unload, and load the equipment from a trailer on level ground.
- To prevent tipping, connect the trailer to the tow vehicle before loading or unloading.
- Follow the manufacturer's requirements for towing and transporting the equipment.
- Inspect the tires and wheels before towing. Do not tow with high or low tire pressures, cuts, excessive wear, bubbles, damaged rims, or missing lug bolts or nuts.
- Do not inflate tires with flammable gases or from systems using an alcohol injector.

WARNING!

Large tires under pressure can explode, causing injury or death. Always maintain the correct tire pressure.

1.1.3 Hydraulic Systems

When operating equipment that uses hydraulics, always remember the following:

- Before disconnecting any hydraulic lines, relieve system pressure by cycling controls.
- Before pressurizing the system, be sure all connections are tight and the lines are undamaged.
- Do not perform any work on the equipment unless you are authorized and qualified to do so.
- Check for leaks using a piece of cardboard or wood; never use bare hands.

Case History

Timing is Everything

A foreman on a construction site noticed that a worker was leaning against the tire of a large scraper while eating lunch. The noonday sun was beating on the tire, which caused a significant increase in air pressure in the tire. The foreman instructed the worker to move away from the tire, which he did. Within minutes, the tire separated from the wheel, resulting in an explosive release of air. Every year, there are numerous reported incidents of death or serious injury from the explosion of truck and heavy equipment tires.

1.1.4 Fueling Safety

Fueling safety precautions are applicable to all light equipment using gasoline. Make sure you follow all the manufacturer's instructions before fueling the equipment. The following are general safety precautions associated with fueling light equipment:

- Never fill the fuel tank with the motor running, while smoking, or near an open flame.
- Be careful not to overfill the tank or spill fuel. If fuel is spilled, clean it up immediately.
- Use clean fuel only.
- Do not operate the equipment if fuel has been spilled inside or near the unit.
- Ground the fuel funnel or nozzle against the filler neck to prevent sparks.
- Replace the fuel tank cap after refueling.

1.1.5 Battery Safety

Battery safety precautions are applicable to all light equipment equipped with a battery for starting or operating electric components.

- Chargers can ignite flammable materials and vapors. They should not be used near fuels, grain dust, solvents, or other flammables.
- To reduce the possibility of electric shock, a charger should only be connected to a properly grounded single-phase outlet. Do not use an extension cord longer than 25 feet.
- If the battery is frozen, do not charge or attempt to jump-start the equipment, as the battery may explode.
- Battery and fuel fumes could ignite and cause explosions and burns. Keep batteries away from flames or sparks. Do not smoke.

Hydraulic Fluid Leaks Can be Dangerous

Hydraulic fluid escaping through pinholes in hoses and hose fittings can be almost invisible. Escaping hydraulic fluid under pressure can penetrate the skin or eyes, causing serious injury. When inspecting light construction equipment, hydraulic hoses, and hose couplings for fluid leaks, always wear the appropriate gloves, protective clothing, and eye protection. Do not use bare hands to trace for leaks along hydraulic hoses and hydraulic system components; instead, use a piece of cardboard or wood. In the event that hydraulic fluid penetrates the skin, seek immediate medical attention from a doctor experienced with treating this type of injury.

Additional Resources

Safe Fueling Procedures. Indiana Constructors. 2018. Available at: **http://indianaconstructors.org/safe-fueling-procedures/?print=pdf**

Hydraulic Systems Safety. Paul D. Ayers. 1992. Colorado State University Cooperative Extension. Available at: **http://nasdonline.org/static_content/documents/1100/d000891.pdf**

1.0.0 Section Review

1. A compressor is a motor-driven machine used to supply compressed air for _____ tools.
   a. nonferrous
   b. hydraulic
   c. battery-operated
   d. pneumatic

2. A trailer should be connected to the tow prior to loading or unloading, in order to _____.
   a. prevent explosions
   b. prevent excessive wear on the brakes
   c. prevent tipping
   d. prevent duplication of efforts

Section Two

2.0.0 Generators

Objective

Identify and describe how to use portable generators.

a. Identify typical generator controls.
b. Explain how to operate a portable generator.
c. State generator safety guidelines and describe their maintenance needs.

Performance Tasks

1. Perform all prestart checks for engine-driven generators.
2. Operate engine-driven generators.
3. Set up and operate engine-driven welding machines.

Trade Terms

Ampere (amp): A unit of electrical current.

Circuit breaker: A device designed to protect circuits from overloads that can be reset after tripping.

Fuse holder: A device used to hold a fuse. It may be part of a circuit or it may simply hold a spare fuse.

Governor: A device used to provide automatic control of speed or power for an internal combustion engine.

Straight blade duplex connector: An electrical connector or style of outlet.

Twist-lock connector: A type of electrical connector.

Watt: A unit of power that is equal to one joule per second.

Generators are used to provide electrical power and/or lighting at the job site. They are available in many different sizes and configurations as shown in *Figure 1* and *Figure 2*, and range from small, portable machines to large, installed backup power systems for emergency power generation. Construction site requirements for power will vary depending on the scale of the work being performed.

TOW-BEHIND GENERATOR

PORTABLE GENERATOR

Figure 1 Generators.

Figure 2 Tow-behind generator for lighting.

2.0.1 Generator Operator Qualifications

Only trained, authorized persons are allowed to operate a generator. Others may do so only during training and under the direct supervision of a qualified trainer.

Portable Welding Machines as Power Sources

Engine-driven portable welding machines are most often a combination welding machine and generator. The machine develops the AC and DC power needed for pipe welding; it also provides auxiliary power that can be used to operate power tools and lighting equipment. This integrated design can help to reduce the amount of equipment on a job site because it eliminates the need for a separate portable generator.

2.1.0 Typical Generator Controls

Tow-behind generators have operator controls for the engine, as well as an electrical control panel to control and monitor the power produced by the generator. *Figure 3 (A)* shows a typical engine control panel used to start and stop the motor and monitor critical motor functions.

Controls and indicators include:

- *Engine over-speed indicator* – This indicator monitors engine speed. If speed exceeds the manufacturer's set limit, the engine will shut down automatically.
- *High engine temperature indicator* – This indicates when the engine coolant temperature has exceeded the manufacturer's set limit. The engine will shut down automatically.
- *Low engine oil pressure indicator* – This indicates when the engine oil pressure has fallen below the manufacturer's preset limit. The engine will shut down automatically.
- *Alternator not charging* – This indicates the engine alternator is not outputting enough energy to charge the unit's battery.
- *Ignition switch* – Place this switch in the ON position to run; place it in the OFF position to stop.
- *Start switch* – Pressing this switch activates the engine starting motor.
- *Safety circuit bypass* – Pressing this switch bypasses automatic shutdowns when starting the engine.
- *Emergency stop* – Pressing this switch causes the engine to shut down with no other operator action required.
- *Engine tachometer gauge* – This gauge indicates engine rpm.
- *Engine oil pressure gauge* – This gauge displays engine oil pressure.
- *Engine coolant temperature* – This gauge indicates the temperature of the engine coolant.

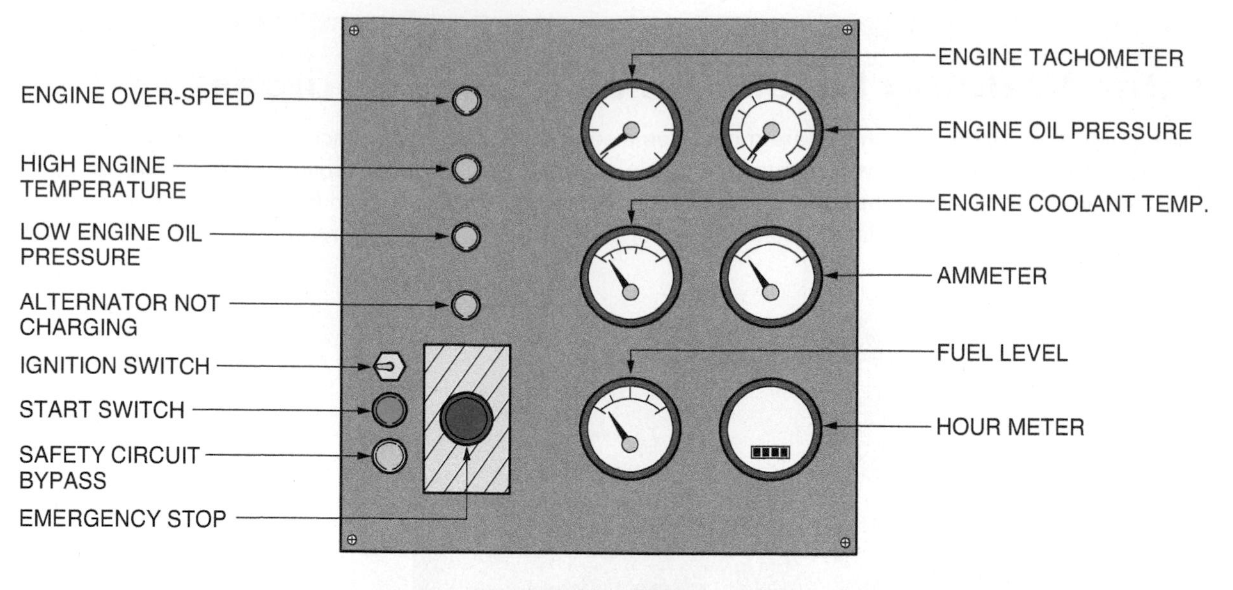

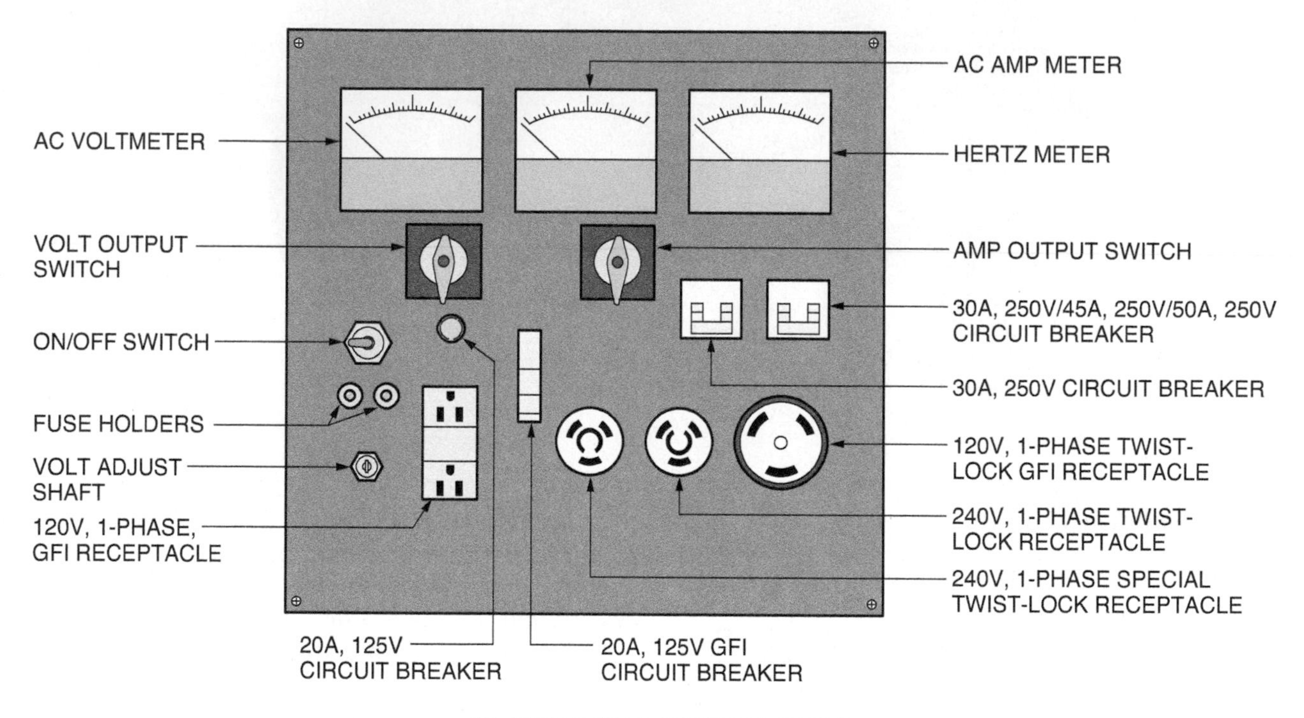

Figure 3 Generator control panels.

- *Ammeter gauge* – This gauge indicates the charge rate of the engine alternator.
- *Fuel level gauge* – This gauge indicates the level of fuel in the fuel tank(s).
- *Hour meter* – This meter records total engine operating hours. It is used in determining the maintenance schedule on the unit.

The second control panel in a tow-behind generator is the generator control panel. This panel contains meters, monitor switches, a voltage regulator, circuit breakers, and receptacles to control and monitor the output of the generator. **Circuit breakers** protect from electrical overload and they may be reset after tripping. *Figure 3 (B)* shows a typical generator control panel.

- *AC voltmeter* – This meter indicates the generator output voltage level.
- *AC amperes meter* – This meter indicates the generator output load in **amperes (amp)**,or, units of electrical current.

- *Hertz meter* – This meter indicates the frequency of the generator output.
- *Voltage output monitor switch* – This switch selects the reference for the generator voltage displayed on the AC voltmeter.
- *Amperage output monitor switch* – This switch selects the line-to-line (phase) amperage displayed on the AC ammeter.
- *Voltage regulator on/off switch* – In the OFF position, this switch removes excitation to the generator field, stopping the generation of power.
- *Voltage regulator fuse holders* – The **fuse holders** house the fuses for the voltage regulator.
- *Voltage regulator voltage adjust shaft* – This shaft is turned to adjust generator output voltage.
- *Circuit breakers* – Various styles of circuit breakers are used, including push-to-reset, ground fault circuit interrupter (GFCI), and flip-to-reset styles.
- *Receptacles* – Various types of receptacles are provided, depending on the size and style of generator used. They will include GFCI **straight blade duplex connectors** and **twist-lock connectors.**

NOTE

Generator selection is based on the total wattage of the tools and devices to be connected to it. The selection must be done by a qualified electrician.

2.2.0 Generator Operation

Before operating the generator, local requirements for grounding must be investigated and followed. The generator set can produce high voltages, which can cause severe injury or death to personnel and damage to equipment. The generator should have proper internal and external grounds when required by the *National Electrical Code®* (*NEC®*). A qualified, licensed electrical contractor, knowledgeable in local codes, should be consulted. Follow all manufacturer's requirements to ensure that the generator is connected properly before operation.

2.2.1 Generator Setup and Preoperational Checks

Many sites have a prestart checklist which must be completed and signed prior to starting or operating an engine-driven generator. Check with your supervisor, and if your site has such a checklist, complete and sign it. A typical procedure for setting up a generator for use at the job site is as follows:

1. Place the unit as level as possible. Follow the manufacturer's directions concerning equipment placement and any special considerations for equipment location.
2. Disconnect the generator from the towing vehicle.
3. Chock the wheels of the generator.
4. Unlock the jack and lower it to the service position.
5. Lock the jack in the service position.
6. Disconnect the safety chains and crank the jack to raise the coupling off the hitch.

Before putting the generator in operation, check for evidence of arcing on or around the control panel. If any arcing is noted, the problem must be located and repaired before beginning operation. Check for loose wiring or loose routing clamps within the housing.

If your site does not have a prestart checklist, do the following before starting the engine:

- Check the oil, using the engine oil dipstick. If the oil is low, add the appropriate grade oil for the time of year.
- If the engine is liquid-cooled, check the coolant level in the radiator. If it's low, add an appropriate amount.

Grounding Portable and Vehicle-Mounted Generators

The grounding requirements for portable and vehicle-mounted generators are listed in *NEC® Section 250.34. NEC® Section 250.20* governs the grounding of portable generators for applications that supply fixed wiring systems.

Each manufacturer provides an operator's manual with detailed instructions for using their equipment. Follow the guidelines in this module, in combination with the specifics of the operator's manual, for safe and efficient use of each machine. Always perform a complete inspection of the generator and correct any deficiencies before using it.

> **CAUTION**
> Do not add plain water to radiators that contain antifreeze. Antifreeze not only protects radiators from freezing in cold weather, it also contains rust inhibitors and additives which aid in cooling. If the antifreeze is diluted, it will not function properly. If the weather turns cold, the system may freeze and cause damage to the radiator, engine block, and water pump.

- Check the fuel. The unit may have a fuel gauge or dipstick. If it's low, add the correct fuel (diesel or gasoline) to the tank. The type of fuel required should be marked on the fuel tank. If it is not marked, contact your supervisor to verify what's needed and have the tank marked.

> **WARNING!**
> Never add fuel to an engine-driven generator while the motor is running. Always make sure that no one is using the generator and shut it down to refuel it.

> **CAUTION**
> Adding gasoline to a diesel engine or diesel to a gasoline engine causes severe engine problems. It can also cause a fire hazard. Always be sure to add the correct fuel to the fuel tank.

- Check the battery water level unless the battery is sealed. Add room-temperature water if the battery water level is low. Never add cold water to a battery.
- If it is a welding machine, check the electrode holder to be sure it is not grounded. If the electrode holder is grounded, it will arc and overheat the welding system when the welding machine is started. This is a fire hazard and can cause damage to the equipment.
- Open the fuel shutoff valve if the equipment has one. The fuel shutoff valve is located in the fuel line between the fuel tank and the carburetor.
- Record the hours from the hour meter if the equipment has one. An hour meter records the total number of hours the engine runs. This information is used to determine when the engine needs to be serviced. The hours are displayed on a gauge similar to an odometer.
- Clean the unit. Use a compressed air hose to blow off the engine and generator or alternator. Use a rag to remove heavier deposits that cannot be removed with the compressed air.

> **WARNING!**
> Whenever using compressed air to clean equipment, you must wear safety glasses with side shields or mono-goggles to protect your eyes from flying debris. You should also warn all others in the immediate area to ensure their safety.

> **NOTE**
> Cleaning may not be required on a daily basis. Clean the unit as required.

2.2.2 Starting the Engine

Read and follow the manufacturer's starting procedures found in the operator's manual. Most engines have an ON/OFF ignition switch and starter. They may be combined into a key switch similar to the ignition on a car. To start the engine, turn on the ignition switch and press the starter. Release the starter when the engine starts. Larger diesel engines have glow plugs that must be warmed up before starting. The engine speed is controlled by the governor. If the governor switch is set for idle, the engine will slow to an idle after a few seconds. Small engine-driven generators may have an ON/OFF switch and a pull cord to start the engine. Engine-driven generators should be started 5 to 10 minutes before they are needed to allow the engine to warm up before a load is placed on it.

> **NOTE**
> Some engines are equipped with a choke, which provides extra combustion air during startup. The choke is closed once the engine is running.

If no power is required for 30 minutes or more, stop the generator by turning off the ignition switch. Never shut down an engine-driven generator before making sure that all workers using the generator have stopped.

2.3.0 Generator Safety and Maintenance

The safe use of a generator is the operator's responsibility. The other safety precautions, already discussed in this module, also apply.

2.3.1 Generator Safety Precautions

WARNING!

High voltages are present when the generator is operating. Exercise caution to avoid electric shock.

When operating a generator:

- Do not operate electrical equipment when standing in water or on wet ground, or with wet hands or shoes.
- Grounding should be performed in compliance with local electrical codes and in accordance with the manufacturer's manual.
- Use a generator as an alternate power supply only after the main feed at the service entrance panel has been opened and locked.
- Do not change voltage selection while the engine is running. Voltage selection, adjustment, and electrical connections may only be performed by qualified personnel.
- Do not exceed the generator power rating during operation. This is measured by **watt**, or a unit of power equal to one joule per second.
- If welding is required on the unit, follow the manufacturer's instructions to prevent damage to the circuitry.
- Never make electrical connections with the unit running.

2.3.2 Generator Operator's Maintenance Responsibility

Figure 4 shows a typical preventive maintenance schedule for a generator. The schedule is laid out with the performance period across the top and the items to be checked listed down the side of the page. For example, under the Daily column, the following items need to be checked:

- Evidence of arcing around electrical terminals
- Loose wire routing clamps
- Engine oil and coolant levels
- Grounding circuit
- Instruments
- Fan belts, hoses, and wiring insulation
- Air vents
- Fuel/water separator
- Service air indicator

Proper maintenance will extend the life and performance of the equipment.

PREVENTIVE MAINTENANCE SCHEDULE						
	DAILY	WEEKLY	MONTHLY/ 150 HRS.	3 MONTHS/ 250 HRS.	6 MONTHS/ 500 HRS.	YEARLY/ 1000 HRS.
Evidence of Arcing around Electrical Terminals	✓					
Loose Wire Routing Clamps	✓					
Engine Oil and Coolant Levels	✓					
Proper Grounding Circuit	✓					
Instruments	✓					
Frayed/Loose Fan Belts, Hoses, and Wiring Insulation	✓					
Obstructions in Air Vents	✓					
Fuel/Water Separator (drain)	✓					
Service Air Indicator	✓					
Precleaner Dumps		✓				
Tires		✓				
Battery Connections		✓				
Engine Radiator (exterior)			✓			
Air Intake Hoses and Flexible Hoses			✓			
Fasteners (tighten)			✓			
Emergency Stop Switch Operation			✓			
Engine Protection Shutdown System			✓			
Diagnostic Lamps			✓			
Voltage Selector and Direct Hook-up Interlock Switches				✓		
Air Cleaner Housing				✓		
Control Compartment (interior)					✓	
Fuel Tank (fill at end of each day)					Drain	
Fuel/Water Separator Element					Replace	
Wheel Bearings and Grease Seals					Repack	
Engine Shutdown System Switches (settings)						✓
Exterior Finish				(as needed)		
Engine			Refer to Engine Operator			
Decals		Replace decals if removed, damaged, or missing.				
✓ = Check or Clean (and Adjust or Replace, if necessary)						

Figure 4 Example of a generator preventive maintenance schedule.

Additional Resource

OSHA Fact Sheet: Using Portable Generators Safely. US Department of Labor: Occupational Safety and Health Administration. 2005. Available at: **https://www.osha.gov/OshDoc/data_Hurricane_Facts/portable_generator_safety.pdf**

2.0.0 Section Review

1. The engine tachometer gauge on a generator control panel ____.
   a. indicates engine failure
   b. indicates excessive speed
   c. indicates engine rpm
   d. indicates engine oil pressure

2. True or false? Generator voltage selection should only occur when the engine is running.
   a. True
   b. False

3. Qualified electricians select generators based on ____.
   a. the total wattage of the tools and devices to be connected to it
   b. the force of the magnetic field surrounding the work area
   c. costs over time according to certified amortization schedules
   d. amperage output monitoring capabilities

4. Generator engine speeds are controlled by the ______.
   a. odometer
   b. governor
   c. regulator
   d. carabiner

5. Grounding circuit inspections, a check of the instruments, and the search for evidence of arcing around electrical terminals should happen ____ when using a generator.
   a. every day
   b. every month
   c. every three months
   d. at least twice a year

SECTION THREE

3.0.0 AIR COMPRESSORS

Objective

Identify and describe how to use portable air compressors.

a. Identify typical air-compressor controls.
b. Explain how to operate a portable air compressor.
c. State air-compressor safety guidelines and describe their maintenance needs.

Performance Tasks

4. Perform all prestart checks for portable air compressors.
5. Operate portable air compressors.

Compressors provide compressed air for pneumatic tools at the job site. A jackhammer is one example of a pneumatic tool. Compressors are available in many different sizes and configurations (*Figure 5*). They range from portable home-use machines to large installed units for industrial applications. Construction site requirements for compressed air will vary depending on the scale of the work being performed.

The primary ratings of air compressor capacity are pounds per square inch (psi) and cubic feet per minute (cfm). The pressure in psi will depend on the pressure ratings of the tools or equipment to be operated by the compressor. Construction compressors typically range from 50 to 125 psi, but can go up to more than 390 psi. Most handheld pneumatic tools run on 90 psi. The term *cfm* refers to the amount of air delivered at the required pressure. The more tools and equipment, the more air delivery is required. A knowledgeable person must determine the size and type of compressor based on the tools and equipment to be used at a site. Only trained, authorized persons are permitted to operate a compressor. Others may do so only during training and under the direct supervision of a qualified trainer.

3.0.1 Compressor Assemblies

Figure 6 shows a tow-behind compressor. Its major components include:

- Towing assembly and frame
- Protective cover and doors
- Engine and compressor assembly
- Operator control panel, located behind the access door

3.1.0 Typical Compressor Controls

Compressors normally have operator controls for the engine and compressor. *Figure 7* shows a typical compressor control panel for a tow-behind compressor. This panel provides the operator with the controls required to start and stop the motor and monitor critical motor and compressor functions. Notice its similarities to a generator control panel: Many of the same functions are controlled in this panel, and they are found in similar locations within the panel.

Controls and indicators include the following:

- *Engine over-speed indicator* – This indicator monitors engine speed. If the speed exceeds the manufacturer's preset limit, the engine will shut down automatically.
- *High engine temperature indicator* – This indicates that the engine coolant temperature has

Figure 5 Types of compressors.

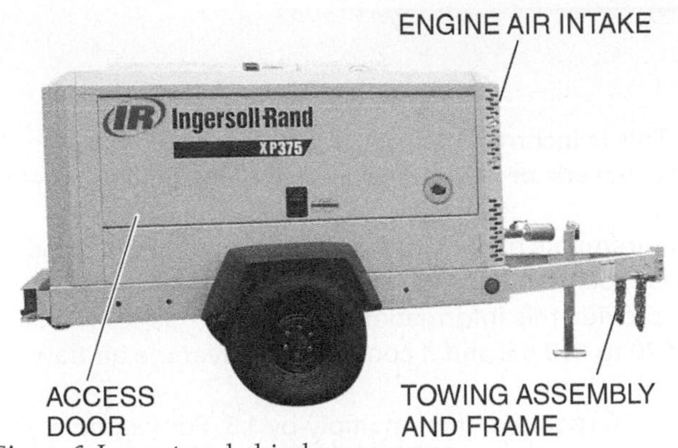

Figure 6 Large tow-behind compressor.

exceeded the manufacturer's preset limit; the engine will shut down automatically.

- *Low engine oil pressure indicator* – This indicates that the engine oil pressure has fallen below the manufacturer's preset limit; the engine will shut down automatically.
- *Alternator not charging* – This indicates that the engine alternator is not outputting enough energy to charge the unit's battery.
- *Ignition switch* – Place this switch in the ON position to run; place to the OFF position to stop.
- *Start switch* – Pressing the start switch activates the engine starting motor.
- *Safety circuit bypass* – Pressing this switch bypasses automatic shutdowns when starting the engine.
- *Emergency stop* – Pressing this switch causes the engine to shut down with no other operator action required.
- *Air pressure gauge* – This gauge indicates the output air pressure level.
- *Engine tachometer gauge* – This gauge indicates engine rpm.
- *Engine oil pressure gauge* – This gauge displays engine oil pressure.
- *Engine coolant temperature* – This gauge indicates the temperature of the engine coolant.
- *Ammeter gauge* – This gauge indicates the charge rate of the engine alternator.
- *Fuel level gauge* – This gauge indicates the level of fuel in the fuel tank(s).
- *Hour meter* – This meter records total engine operating hours. It is used for scheduling maintenance on the unit.
- *Temperature gauge* – This gauge indicates the temperature of the compressed air.

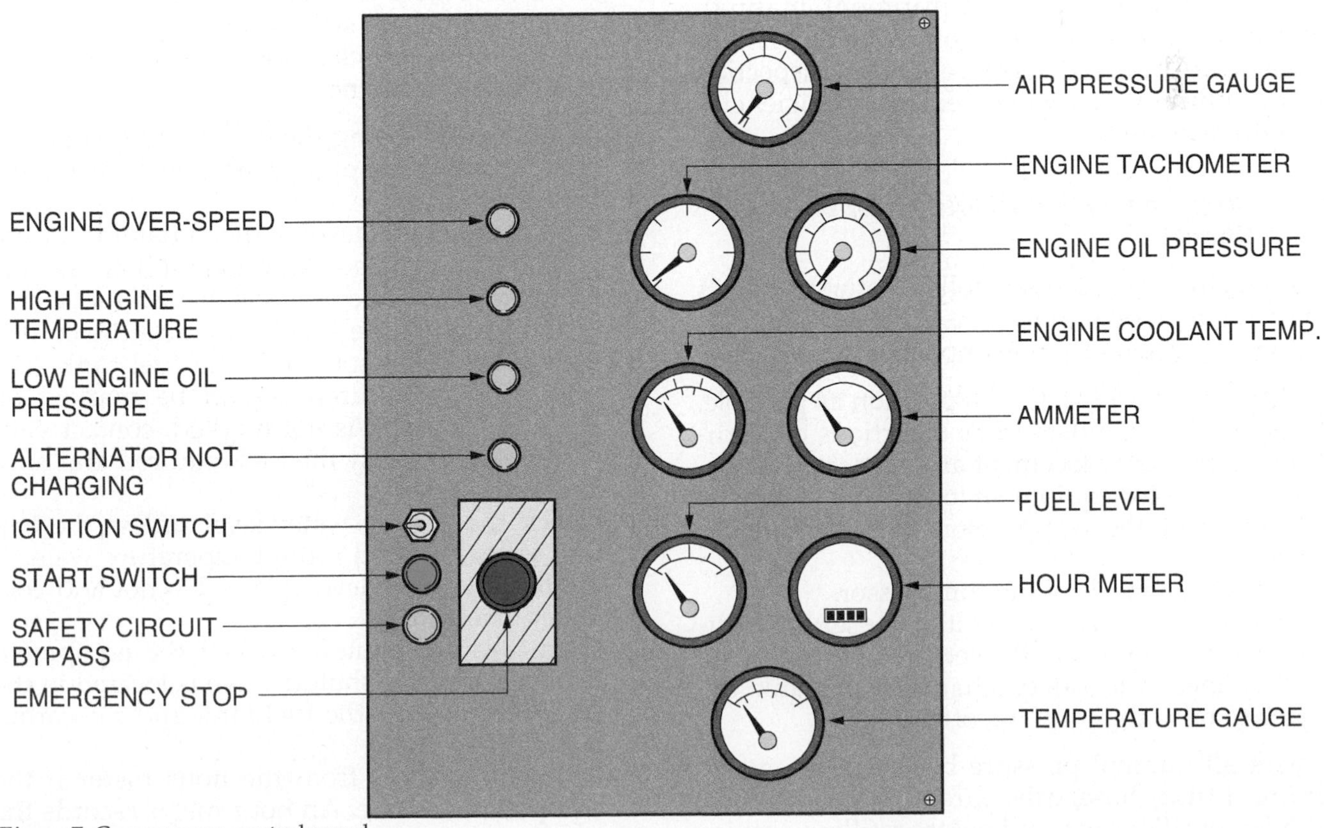

Figure 7 Compressor control panel.

Sizing Compressors

Many people size compressors based on a horsepower rating. This is incorrect. Selection of a larger horsepower compressor motor/engine allows the compressor to run at faster speeds or larger displacement to produce greater air flow at a rated pressure.

The size of a compressor should be based on the total air requirements of the various tools or equipment to be used with the compressor. These are the total air flow in cubic feet per minute (cfm) and the required air pressure in pounds per square inch (psi). Pneumatic tool manufacturers provide this information for each of their tools. For example, a typical framing nailer gun operates at a pressure of 70 to 100 psi and it consumes an average air flow of 9.6 cfm.

A standard way to size a compressor is to determine the total cfm required, then multiply by 1.5. For example, if the total required air flow is 20 cfm, the compressor should be sized to produce 30 cfm (20 x 1.5 = 30). This means that the compressor must produce a minimum of 30 cfm at maximum rated pressure to properly operate the tool.

The compressor rated pressure in psi is determined by the pressure requirements of the tools and/or equipment that the compressor is supplying. It is important that the compressor maintain pressures that are higher than what is required. For example, if the tools/equipment being used require a pressure of 120 psi, then a compressor capable of producing and maintaining a pressure higher than 120 psi is required.

The operating duty cycle of the tools being used with the compressor is another factor to be considered. Tools such as nail guns tend to be operated intermittently so that they only consume small amounts of air over a long period of time. However, tools like sanders and grinders tend to be operated continuously, causing them to consume large amounts of air over a long period of time. If the application involves tools or equipment that will be used continuously, select a compressor capable of producing a higher cfm output.

3.2.0 Compressor Operation

As with generators and other types of equipment, the manufacturer's operating instructions should be followed. General safety guidelines, in concert with specifics from the equipment manual, allow the operator to carry out a job safely and efficiently. Always perform a complete inspection of the compressor and correct any deficiencies before starting it up.

3.2.1 Compressor Setup and Preoperational Checks

Before using a compressor, follow setup protocols involving a preoperational checklist.

As with generators, it is important to:

- Place the unit in as level a position as possible. Follow the manufacturer's directions concerning equipment placement and any special considerations for equipment location.
- Disconnect the compressor from the towing vehicle.
- Chock the wheels of the compressor.
- Unlock the jack and lower it to the service position. Lock the jack in the service position.
- Disconnect the safety chains and crank the jack to raise the coupling off the hitch.

Vent all internal pressure before opening any air line, fitting, hose, valve, drain plug, connector, oil filler, or filler caps, and before refilling anti-ice systems. Be sure all hoses, shutoff valves, flow-limiting valves, and other attachments are connected according to the manufacturer's instructions. If the air compressor is to be used along with other sources of air, be sure there is a check valve installed at the service valve.

The following checks should be performed before starting the engine:

- Check the oil, using the engine oil dipstick. If it's low, add the appropriate grade oil for the time of year.
- Check the coolant level in the radiator if the engine is liquid-cooled. Add coolant if necessary.
- Check the fuel. The unit may have a fuel gauge or a dipstick. If the fuel is low, add the correct fuel (diesel or gasoline) to the fuel tank. The type of fuel required should be marked on the fuel tank. If it is not marked, contact your supervisor to verify the fuel required and have the tank marked.
- Check the battery water level unless the battery is sealed. Add room-temperature water if the battery water level is low. Do not add cold water to a battery.
- Open the fuel shutoff valve if the equipment has one. The fuel shutoff valve is located in the fuel line between the fuel tank and the carburetor.
- Record the hours from the hour meter if the equipment has one. An hour meter records the

total number of hours the engine runs. This information is used to determine when the engine needs to be serviced. The hours are displayed on a gauge similar to an odometer.

- Check the tension and condition of the belts. If the belts are loose, tighten them. If they are worn, replace them.
- Check the safety valve setting to make sure it does not stick and is working properly.
- Make sure that the regulating valve between the compressor and the outlet valve is closed.

3.2.2 Starting the Air Compressor

Most engines have an ON/OFF ignition switch and a starter, which may be combined into a key switch similar to the ignition on a car. To start the engine, turn on the ignition switch and press the starter. Release the starter when the engine starts. On diesel-engine generators, hold the glow plug button to warm up the glow plugs before starting the compressor engine. Small air compressors may have an ON/OFF switch and a pull cord to start the engine. Engine-driven compressors should be allowed to build up to the operating pressure before they are used. While the compressor is building up pressure, connect all outlet lines and tools to the compressor. As soon as the operating pressure is reached, open the regulating valve to the outlet valves.

CAUTION

Before disconnecting any tools from the air lines, close the regulating valve to that particular line and bleed the air from the line. It is then safe to disconnect a tool from that line and connect another tool. Open the regulating valve before trying to use the tool.

When shutting down the air compressor, close the regulating valve at the outlet ports and then bleed the air from the air lines. Turn off the ignition switch and disconnect all tools and air lines from the compressor.

3.3.0 Air Compressor Safety and Maintenance

Air compressors are loud when operated, and they present fire hazards. Because they operate under high pressure, additional precautions must be taken to use them safely.

3.3.1 Compressor Safety Precautions

When working with compressors:

- Make sure everyone in the area is wearing proper hearing protection. Compressed air venting to the atmosphere can cause hearing damage.
- Remember that ether is highly flammable. Do not inject ether into a hot engine or an engine equipped with a glow-plug type of preheater.
- Do not inject ether into the compressor air filter or a common air filter for the engine and compressor.
- Do not attempt to move the compressor or lift the drawbar without adequate personnel or equipment to handle the weight.
- Do not exceed the machine's air pressure rating.
- Do not use compressed air for breathing.
- Never direct compressed air at anyone.
- Before removing the compressor filler cap, make sure there is no pressure in the system.
- Do not use tools that are rated for a pressure lower than that provided by the compressor.
- Be sure all connections are securely made and that hoses under pressure are secured to prevent whipping. Use appropriate safety devices to secure hoses.
- Do not weld or perform any modifications on the air compressor receiver tank.
- Use a safe, nonflammable solvent when cleaning parts.

3.3.2 Air Compressor Operator's Maintenance Responsibilities

The manufacturer's manual provides a schedule and procedures to be followed when performing maintenance on the air compressor. Before performing any maintenance, the operator must be trained, authorized, and have the proper tools to perform any procedures. The following are examples of the types of maintenance that could be performed on an air compressor:

- Changing the engine oil and filter
- Changing the engine and compressor air filters
- Changing the engine coolant
- Performing battery maintenance
- Lubricating parts
- Inspecting all guards and safety devices

Additional Resource

Oregon OSHA Fact Sheet Plus: Compressed Air Piping Systems. Oregon OSHA. 2011. Available at: **https://osha.oregon.gov/OSHAPubs/factsheets/fs44.pdf**

3.0.0 Section Review

1. The term cfm refers to _____.
   a. the amount of hydraulic pressure indicated by the control panel
   b. centrifugal force mechanism, or the power circulating around a spindle
   c. cubic feet per minute, or the amount of air delivered at the required pressure
   d. concentric funnel machine, used to divert excessive air from one source to another

2. True or false? The control panels for compressors are very similar to control panels for generators.
   a. True
   b. False

3. The temperature gauge on a compressor control panel indicates _____.
   a. temperature of the outside air
   b. temperature of the compressed air
   c. internal temperature relative to watts
   d. temperature of the engine coolant

4. True or false? When the battery water level is low, add cold water for a faster start.
   a. True
   b. False

5. All of the following are good safety tips for working with air compressors, EXCEPT:
   a. wearing hearing protection
   b. being careful not to exceed the machine's air pressure rating
   c. using tools that are rated for a lower pressure than what the compressor provides
   d. using safe, nonflammable solvents when cleaning parts

6. True or false? Routine maintenance on air compressors is not necessary, so long as all the gauges read correctly.
   a. True
   b. False

SECTION FOUR

4.0.0 AERIAL LIFTS

Objective

Identify aerial-lift controls and describe how to use aerial lifts.

a. Identify typical aerial-lift controls.
b. Explain how to operate aerial lifts.
c. State aerial-lift safety guidelines and describe their maintenance needs.

Trade Terms

Aerial lift: A mobile work platform designed to transport and raise personnel, tools, and materials to overhead work areas.

Proportional control: A control that increases speed in proportion to the movement of the control.

Aerial lifts are used to raise and lower workers to and from elevated job sites. There are two main types of lifts: boom lifts and scissor lifts. Some are transported on a vehicle to a job site where they are unloaded. Others are trailer-mounted and towed to the job site by a vehicle, while some are permanently mounted on a vehicle. Depending on their design, they can be used for indoor work, outdoor work, or both. *Figure 8* shows two commonly used types of aerial lifts.

Boom lifts are designed for both indoor and outdoor use. They have a single arm that extends a work platform/enclosure capable of holding one or two workers. Some models have a jointed (articulated) arm that allows the work platform to be positioned both vertically and horizontally. Scissor lifts raise a work enclosure vertically by means of crisscrossed supports; these can also be used indoors and outdoors.

Most models of aerial lifts are self-propelled, allowing workers to move the platform as work is performed. The power to move these lifts is provided by several means, including electric motors, gasoline or diesel engines, and hydraulic motors.

4.0.1 Aerial Lift Assemblies

Aerial lifts normally consist of three major assemblies: the platform, a lifting mechanism, and the base. *Figure 9* shows these components for a scissor lift.

The platform of an aerial lift is constructed of a tubular steel frame with a skid-resistant deck surface, railings, toeboard, and midrails. Entry to the platform is normally from the rear. The entry opening is closed either with a chain or a spring-returned gate with a latch. The work platform may also be equipped with a retractable extension platform. The lifting mechanism is raised and lowered either by electric motors and gears or by one or more single-acting hydraulic lift cylinder(s).

A pump, driven by either an AC or DC motor, provides hydraulic power to the cylinder(s). The base provides a housing for the electrical and hydraulic components of the lift. These components are normally mounted in swingout access trays. These allow easy access when performing maintenance or repairs to the unit.

The base also contains the axles and wheels for moving the assembly. In the case of a self-propelled platform, electrical or hydraulic motors will drive two or more of the wheels to allow movement of the lift from one location to another. Brakes will be incorporated on one or more of the wheels to prevent inadvertent movement of the lift.

4.0.2 Aerial Lift Operator Qualifications

Only trained and authorized workers may use an aerial lift. Safe operation requires the operator to understand all limitations and warnings, operating procedures, and operator requirements for maintenance of the lift.

The operator must:

- Understand and be familiar with the associated operator's manual for the lift being used.
- Understand all procedures and warnings within the operator's manual and those posted on decals on the aerial lift.
- Be familiar with employer's work rules and OSHA regulations.
- During training, demonstrate this understanding and operate the associated model in the presence of a qualified trainer.

4.1.0 Aerial Lift Controls

Figure 10 shows a typical electrical panel and its associated controls; each type of lift may vary its configuration. This example contains the following switches and controls:

- *Up/down toggle switch* – By holding this switch in the up position, the platform can be raised to the desired level. The platform will stop

Figure 8 Aerial lifts.

moving when the switch is returned to the center position. By holding this switch in the down position, the platform can be lowered.

- *Buzzer alarm* – An audible alarm sounds when the platform is being lowered. On some models, this alarm may sound when any control function is being performed.
- *Hour meter* – This meter records the number of hours the platform has been operating. The meter will only register when the electric or hydraulic motor associated with operating the aerial lift is running.

Another control that may be available is an emergency battery disconnect switch. When this switch is placed in the off position, it will disconnect power to all control circuits.

The basic controls for an aerial lift with a hydraulic system generally include:

- *Emergency lowering valve* – This valve allows for platform lowering in the event of an electrical/hydraulic system failure.
- *Free wheeling valve* – Opening this valve allows hydraulic fluid to flow through the wheel motors. This allows the aerial lift to be pushed by hand and prevents damage to the motors when the aerial lift is moved between job site locations. There are usually strict limits on how fast the aerial lift may be moved without causing damage to hydraulic system components.

Other types of controls that may be available on aerial lifts include:

- *Parking brake manual release* – This control allows manual release of the parking brake. It should only be used when the aerial lift is located on a level surface.

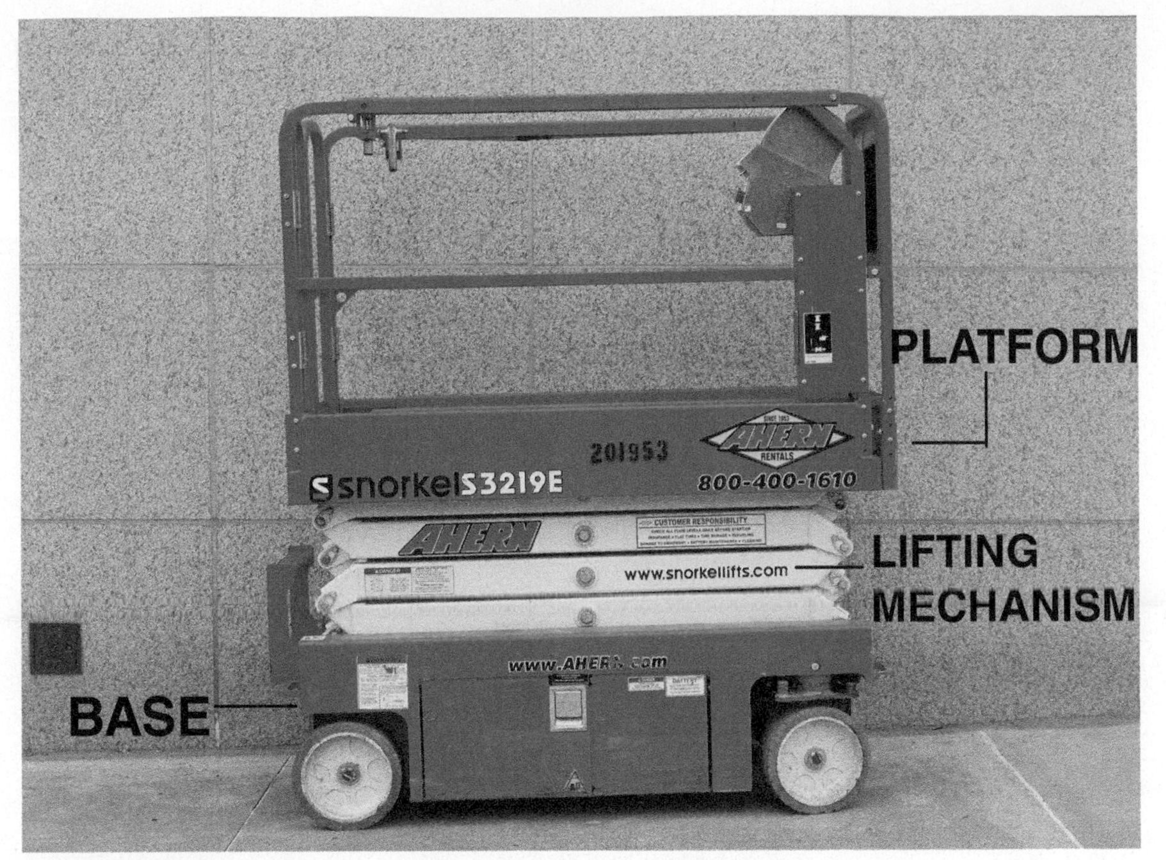

Figure 9 Aerial lift components.

- *Safety bar* – This is used to support the platform lifting hardware into a raised position during maintenance or repair.
- *Up/down selector switch* – This switch is used to raise and lower the platform.
- *Emergency stop button* – This button is used to cut off power to both the platform and base control boxes.

Stabilizer hardware is required on some aerial lifts. At a minimum, this hardware includes a stabilizer leg, stabilizer lock pin and cotter key, and a stabilizer jack at each corner of the aerial lift.

Most aerial lifts are equipped with a base control box. The base control box includes a switch to select whether operation will be controlled from the platform or the base. The base control overrides the platform control.

The worker on the platform designated as the operator uses the platform controls. In some aerial lifts, operation from the platform is limited to raising and lowering the platform. On others, the aerial platform is designed to be driven by the operator using controls provided on the platform. In either case, the operator will have an operator control box located on the platform assembly. *Figure 11* shows a typical platform control box for a self-propelled aerial lift.

Operator controls include:

- *Drive/steer controller* – This is a one-handed toggle (joystick) lever for controlling speed and steering of the aerial platform. This is usually a deadman switch that returns to neutral and locks when released. The handle is moved forward to drive the aerial platform forward. The platform speed is determined by how far forward the handle is moved. The handle is moved backward to drive the aerial platform backward; speed is selected as before. Releasing the stick will stop the motion of the aerial lift. Steering is performed by depressing a rocker switch on the top of the stick in the desired direction of travel, either right or left.
- *Up/down selector switch* – Placing the switch in the up position raises the platform; placing the switch in the down position lowers it. When released, the switch returns to the middle position and stops the movement of the platform.
- *Lift/off/drive selector switch* – The lift position energizes the lift circuit; the off position removes power to the control box; the drive position energizes the drive and steering controls.
- *Emergency stop pushbutton* – When pushed, this button disconnects power to the platform

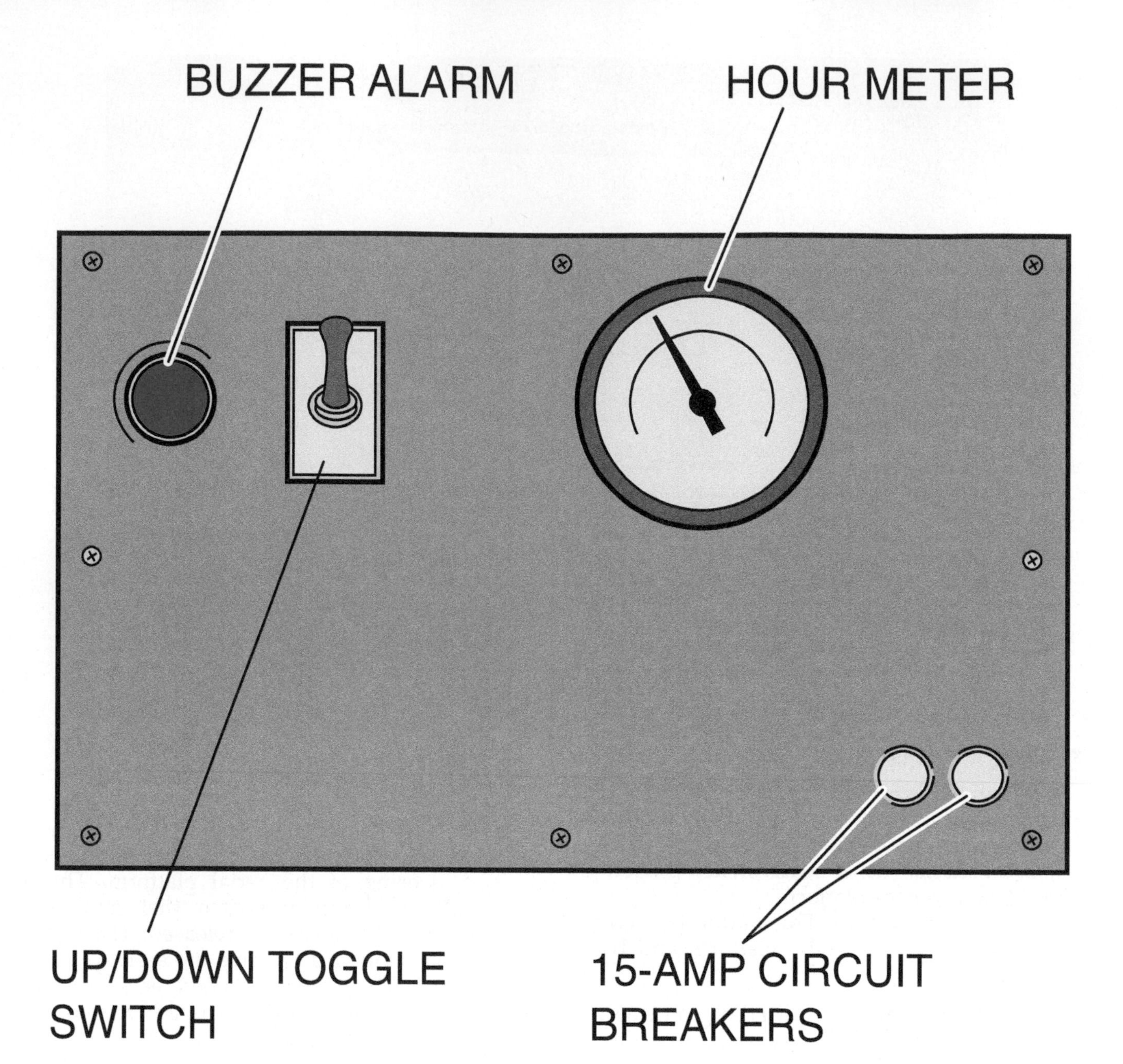

Figure 10 Aerial lift electrical panel.

control circuit. In an emergency, push the button in. To restore power, pull the button out.

- *Lift enable button* – When pushed and held down, this button enables the lift circuit. The button must be held down when raising or lowering the platform. Releasing the button stops the motion of the platform.

4.2.0 Aerial Lift Operation

Scissor-type, self-propelled aerial lifts involve proportional controls. A typical **proportional control** procedure involves the operation of a lever or foot pedal to cause the aerial lift to move. The further the control is moved, the more power is applied to the motor and the faster the aerial lift will move.

The following tasks must be performed before operating the aerial lift:

- Carefully read and fully understand the operating procedures in the operator's manual and all warnings and instruction decals on the work platform.
- Check for obstacles around the work platform and in the path of travel such as holes, dropoffs, debris, ditches, and soft fill. Operate the aerial lift only on firm surfaces.
- Check overhead clearances. Make sure to stay at least 10' away from overhead power lines.

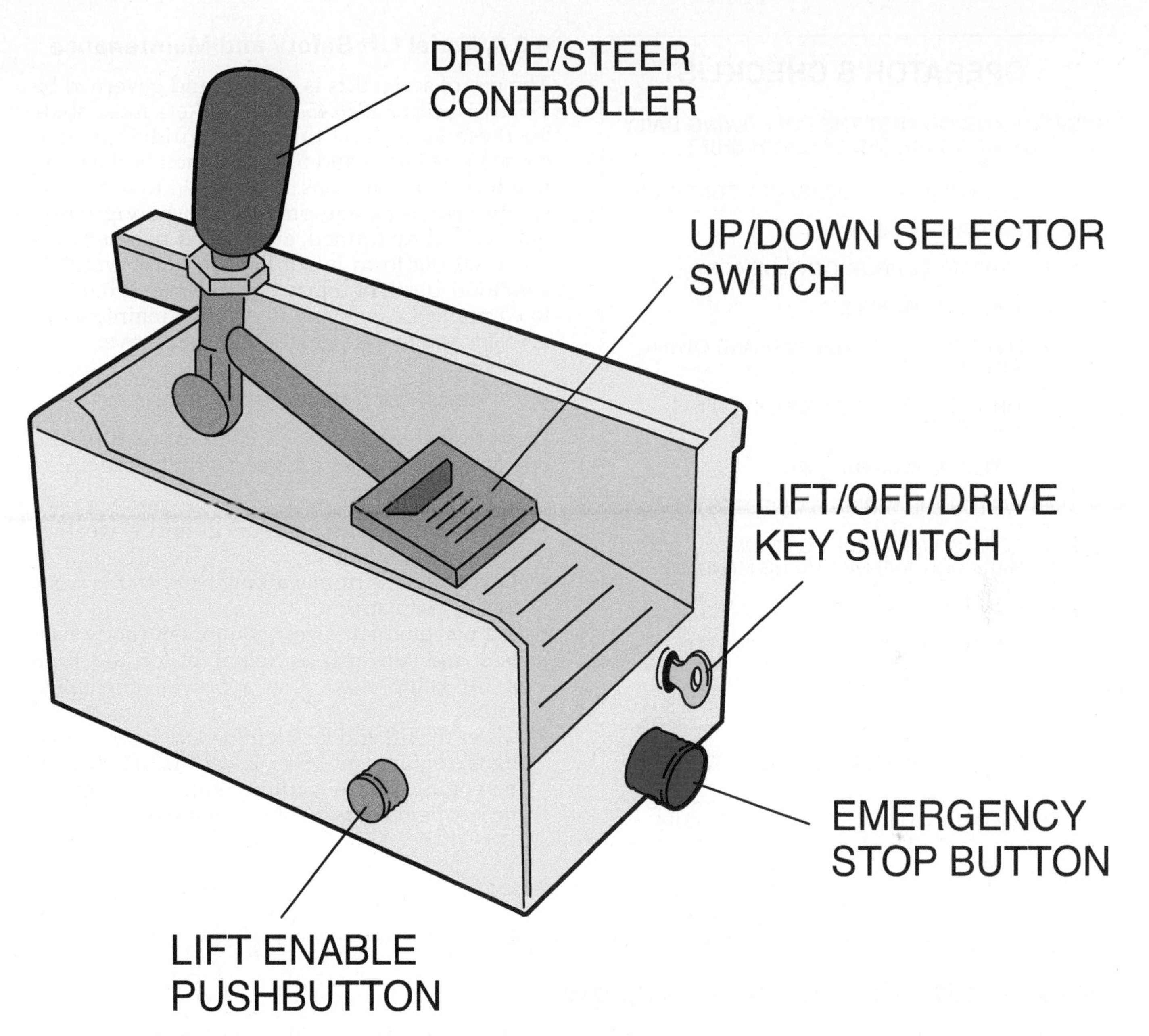

Figure 11 Aerial lift platform control box.

- Make sure batteries are fully charged (if applicable).
- Make sure all guardrails are in place and locked in position.
- Review the operator's checklist to make sure nothing has been overlooked.
- Never make unauthorized modifications to the components of an aerial lift.

Figure 12 shows an example of an operator's checklist.

The following is an example of the operator controls used to drive the aerial lift forward. The operator selects the drive position with the lift drive select switch. The operator then lifts the handle lock ring and moves the drive/steer controller forward. The speed can be adjusted by continuing to move the controller forward until the desired speed is reached. By releasing the controller, the forward motion of the lift is stopped. To drive in reverse, the lever is moved in the opposite direction.

> **NOTE**
> Aerial lifts will be covered in more detail in Level Three.

OPERATOR'S CHECKLIST

INSPECT AND/OR TEST THE FOLLOWING DAILY OR AT BEGINNING OF EACH SHIFT

___ 1. OPERATING AND EMERGENCY CONTROLS

___ 2. SAFETY DEVICES

___ 3. PERSONNEL PROTECTIVE DEVICES

___ 4. TIRES AND WHEELS

___ 5. OUTRIGGERS (IF EQUIPPED) AND OTHER STRUCTURES

___ 6. AIR, HYDRAULIC, AND FUEL SYSTEM(S) FOR LEAKS

___ 7. LOOSE OR MISSING PARTS

___ 8. CABLES AND WIRING HARNESSES

___ 9. DECALS, WARNINGS, CONTROL MARKINGS, AND OPERATING MANUALS

___ 10. GUARDRAIL SYSTEM

___ 11. ENGINE OIL LEVEL (IF SO EQUIPPED)

___ 12. BATTERY FLUID LEVEL

___ 13. HYDRAULIC RESERVOIR LEVEL

___ 14. COOLANT LEVEL (IF SO EQUIPPED)

Figure 12 Example of an aerial lift operator's checklist.

4.3.0 Aerial Lift Safety and Maintenance

The use of aerial lifts is defined and governed by *OSHA Standard 1926.453.* All workers must abide by these as well as the safety guidelines discussed previously and those outlined in the manufacturer's instructions. In addition to following safety protocols, equipment must be vigilantly maintained by trained, authorized personnel. If an aerial platform is not kept in good working condition, death or injury to workers and damage to equipment can result. Preventive maintenance is easier and less expensive than corrective.

4.3.1 Aerial Lift Safety Precautions

Do not overlook the significance of reading the operator's manual for each aerial lift before use.

It is important to:

- Avoid using the lift outdoors in stormy weather or in strong winds.
- Prevent people from walking beneath the work area of the platform.
- Use personal fall arrest equipment (body harness and lanyard) as required for the type of lift being used. Use approved anchorage points.
- Lower the lift and lock it into place before moving the equipment. Also, lower the lift, shut off the engine, set the parking brake, and remove the key before leaving it unattended.

Aerial Lift Operator Controls

Operator controls for aerial lifts will differ by manufacturer and model. Shown here are the ground control station and platform control box for an electric-drive scissor lift produced by a major lift manufacturer.

GROUND CONTROL STATION

PLATFORM CONTROL BOX

- Stand firmly on the floor of the basket or platform. Do not lean over the guardrails of the platform, and never stand on the guardrails. Do not sit or climb on the edge of the basket or use planks, ladders, or other devices to attain additional height.

4.3.2 Aerial Lift Operator's Maintenance Responsibilities

Inspection and maintenance duties should only be performed by authorized personnel. If the operator is not responsible for maintenance of the lift, that person should, at minimum, perform the daily checks. This should happen at the beginning of each shift. The aerial lift must never be used until these checks are completed satisfactorily. Any deficiencies found during the daily check must be corrected before the lift is used. *Figure 13* shows a typical maintenance and inspection schedule for an aerial lift.

The maintenance and inspection schedule shown in *Figure 13* covers four major categories. Each includes the components that should be checked daily, weekly, monthly, every three months, every six months, and once a year. Footnotes (the numbers in parentheses) following each component tell the operator what type of inspection is required. For example, under the Electrical category, battery fluid level should be checked daily or at the beginning of each shift. The (1) refers the operator to the Notes portion of the schedule. Note 1 tells the operator to perform a visual inspection of the battery fluid level.

Each aerial lift will have its own maintenance requirements, depending on the model and manufacturer. Always refer to the operator's manual to determine exactly what checks are required before operation.

Regardless of what type of lift is being maintained, it is important to:

- Disconnect the battery ground negative (–) lead before performing any maintenance.
- Properly position safety devices before performing maintenance with the work platform in the raised position.

Aerial Lift Safety

A study by the Center to Protect Worker's Rights (a research arm of the Building and Construction Trades Department, AFL-CIO) found that aerial lifts used in US construction trades from 1992 to 1999 had been involved in 207 deaths. Thirty-three percent were caused by electrocutions, 31 percent were due to falls, and 22 percent were attributed to lift collapses or tip-overs. The remaining 14 percent were from other causes.

Deaths from electrocutions involved mostly electricians or electrical repair personnel. Of those, about 70 percent involved the use of boom lifts, while 25 percent involved scissor lifts. Falls from boom lifts occurred mainly because the lift was struck by vehicles, cranes, crane loads, falling objects, or because the lift was suddenly jerked. Falls from scissor lifts were mainly due to the lift being struck by an object or because the safety chains or guardrails were removed. Some were caused by the occupant standing on or leaning over the lift railings. The operation of any aerial lift is subject to hazards that can be reduced or removed by being careful, using common sense, and following all safety rules.

	Daily	Weekly	Monthly	3 Months	6 Months	*Annually
Mechanical						
Structural damage/welds (1)	✓					✓
Parking brakes (2)	✓					✓
Tires and wheels (1)(2)(3)	✓					✓
Guides/rollers/slides (1)	✓					✓
Railings/entry chain/gate (2)(3)	✓					✓
Bolts and fasteners (3)	✓					✓
Rust (1)			✓			✓
Wheel bearings (2) King pins (1)(8)	✓					✓
Steer cylinder ends (8)				✓		✓
Electrical						
Battery fluid level (1)	✓					✓
Control switches (1)(2)	✓					✓
Cords and wiring (1)	✓					✓
Battery terminals (1)(3)	✓					✓
Terminals and plugs (3)	✓					✓
Generator and receptacle (2)	✓					✓
Limit switches (2)	✓					✓
Hydraulic						
Hydraulic oil level (1)	✓					✓
Hydraulic leaks (1)	✓					✓
Lift/lowering time (10)				✓		✓
Hydraulic cylinders (1)(2)		✓				✓
Emergency lowering (2)	✓					
Lift capacity (7)			✓			✓
Hydraulic oil/filter (9)					✓	✓
Miscellaneous						
Labels (1)(11) Manual (12)	✓					✓

Notes:

(1) Visually inspect.
(2) Check operation.
(3) Check tightness.
(4) Check oil level.
(5)(6) N/A.
(7) Check relief valve setting. Refer to serial number nameplate.
(8) Lubricate.
(9) Replace.

(10) General specifications.
(11) Replace if missing or illegible.
(12) Proper Operating Manual *must* be in the manual tube.

*Record Inspection Date

Figure 13 Example of an aerial lift maintenance and inspection schedule.

Additional Resource

Aerial Lifts: Protect Yourself. 2018. US Department of Labor: Occupational Safety and Health Administration. Available at: **https://www.osha.gov/Publications/aerial_lifts_safety.html**

4.0.0 Section Review

1. True or false? Aerial lifts are designed for outdoor use only.
   a. True
   b. False
2. The _____ is usually a deadman switch that returns to neutral and locks when released.
   a. drive/steer controller
   b. up/down selector switch
   c. emergency stop pushbutton
   d. lift enable button
3. What type of procedure involves the use of a lever or foot pedal to move an aerial lift?
   a. perpendicular control
   b. parallel control
   c. proportional control
   d. pedestrian control
4. True or false? Because aerial lifts cannot go as high as mobile cranes, their platforms are designed for use with a ladder when additional height is needed.
   a. True
   b. False

SECTION FIVE

5.0.0 FORKLIFTS

Objective

Identify and describe how to use forklifts.

a. Identify typical forklift controls.
b. Explain how to operate forklifts.
c. State forklift safety guidelines and describe their maintenance needs.

Performance Task

6. Identify forklift trucks and recognize safety hazards involved in working around them.

Trade Terms

Powered industrial truck: Another term for forklift.

Mast: The upright member along which forks on a forklift travel.

Forklifts, which are sometimes called **powered industrial trucks**, are used to move, unload, and place material at specific locations on the job site. Forklifts are often misused, but a firm foundation in forklift safety allows for greater productivity and a safer working environment.

5.0.1 Forklift Assemblies

Many different types of forklifts are available to meet different needs. Most forklifts used in construction are diesel-powered and fall into one of two broad categories: fixed mast and telescoping boom. Within each of these categories, forklifts are further differentiated by their drive train, steering, and capacity.

There are two types of fixed-mast forklifts: rough terrain and warehouse. Rough terrain forklifts are by far the most common type of fixed-mast forklift used in construction (*Figure 14*). These are made for outside use. They have higher ground clearances than warehouse forklifts, as well as larger tires and leveling devices. When the term *fixed-mast forklift* is used in this module, it refers to the rough terrain forklift.

Warehouse forklifts are designed for inside use. They have hard rubber tires that are usually the same size.

Figure 14 Fixed-mast rough terrain forklifts.

5.0.2 Forklift Operator Qualifications

Only trained, authorized persons are permitted to operate a powered forklift. Forklift operators must have the visual, auditory, physical, and mental abilities to run the equipment safely. Those who are learning to use a forklift must only operate them in training conditions, and under the direct supervision of a qualified trainer.

5.1.0 Typical Forklift Controls

Forklifts have either a two-wheel drive or four-wheel drive. The distinction is the same as that of other vehicles.

A two-wheel-drive forklift has a drive train that transmits power to the front wheels. Many times, the front wheels are larger than the rear wheels and do not steer the forklift (the rear wheels move when the steering wheel is turned). This standard feature provides increased maneuverability.

Two-wheel-drive forklifts are good for general use, but they may lack power in rough terrain.

A four-wheel-drive forklift has a drive train that transmits power to all four wheels. These are well suited for rough terrain and conditions that require additional traction. When four-wheel steering is engaged, all the wheels steer. Some forklifts of this type offer the following three options:

- The rear wheels may be locked to allow the front wheels to steer.
- All wheels may move in the same direction. This is called crab steering or oblique steering.
- The front and rear wheels may move in opposite directions. This is called articulated steering or four-wheel steering.

5.1.1 Fixed-Mast Forklifts

The upright member along which the forks travel is called the **mast**. A typical fixed-mast forklift has a mast that may tilt as much as 19 degrees forward and 10 degrees rearward. These types of forklifts are suitable for traveling with loads and placing them vertically, but their horizontal reach is limited to how close the machine can be driven to the pick-up or landing point. For example, a fixed-mast forklift cannot place a load of pipe beyond the edge of a trailer because of its limited reach.

5.1.2 Telescoping-Boom Forklifts

Telescoping-boom forklifts (*Figure 15*) provide more versatility in horizontal and vertical placement than a fixed-mast forklift. A telescoping-boom forklift is really a combination of a telescoping-boom crane and a forklift.

The mast can be either two-stage or three-stage. This designation refers to the number of telescoping channels built into the mast. A two-stage mast has one telescoping channel; a three-stage mast has two telescoping channels. These provide greater lift height. Some models of telescoping-boom forklifts have a level-reach fork carriage, which allows the fork carriage to be moved horizontally while the boom remains in a stationary position.

5.2.0 Forklift Operation

The most important factor to consider when using any forklift is its intended capacity, which must

Figure 15 Telescoping-boom forklift.

never be exceeded. The capacity for one forklift will be different from another; exceeding its limits may harm the equipment and jeopardize the safety of everyone around it. Each manufacturer supplies a capacity chart for each forklift model-be sure to read and follow it.

5.2.1 Picking Up a Load

Some forklifts are equipped with a sideshift device that allows the operator to horizontally shift the load several inches in either direction with respect to the mast. This device enables more precise placing of loads, but it changes the center of gravity and must be used with caution.

CAUTION

If a forklift has a sideshift device, be sure to return the fork carriage to the center position before attempting to pick up a load.

To pick up a load using a forklift:

Step 1 Check the position of the forks with respect to each other. They should be centered on the carriage. If the forks have to be moved, check the operator's manual for the proper procedure. Usually, there is a pin at the top of each fork that, when lifted, allows each fork to be slid along the upper backing plate until the fork centers over the desired notch.

Step 2 Before picking up a load with a forklift, make sure the load is stable. If it looks like the load might shift when picked up, secure the load. Knowing the center of gravity is crucial, especially when picking up tapered sections. Make a trial lift, if necessary, to determine and adjust the center of gravity.

Step 3 Approach the load so that the forks straddle the load evenly. It is important that the weight of all loads be distributed evenly on the forks. Overloading one fork at the expense of the other can damage the forks. In some cases, it may be advisable to measure the load and mark its center of gravity.

Step 4 Drive up to the load with the forks straight and level. If the load being picked up is on a pallet, be sure the forks are low enough to clear the pallet boards.

Step 5 Move forward until the leading edge of the load rests squarely against the back of both forks. If you cannot see the forks engage the load, ask someone to signal for you. This prevents expensive damage and injury.

Step 6 Raise the carriage, then tilt the mast rearward until the forks contact the load. Raise the carriage until the load safely clears the ground. Then tilt the mast fully rearward to cradle the load. This minimizes the chance that the load may slip during travel.

NOTE

If there is a load on the forks, the operator may not leave the seat.

5.2.2 Traveling with a Load

Never travel with forks in the air. Always travel with a load at a safe rate of speed. Never travel with a raised load. Keep the load as low as possible and be sure the mast is tilted rearward to cradle the load.

As you travel, keep your eyes open and stay alert. Watch the load and the conditions ahead of you, and alert others of your presence. Avoid sudden stops and abrupt changes in direction. Be careful when downshifting, because sudden deceleration can cause the load to shift or topple. Be aware of front and rear swing when turning.

If you are traveling with a telescoping-boom forklift, be sure the boom is fully retracted.

If you have to drive on a slope, keep the load as low as possible and back down the slope. Do not drive across steep slopes. If you have to turn on an incline, make the turn wide and slow. If the load is large enough to obstruct your view, operate in reverse.

5.2.3 Placing a Load

Position the forklift at the landing point so that the load can be placed where you want it. Be sure everyone is out of the way and away from the load.

The area under the load must be clear of obstructions and must be able to support the weight of the load. If you cannot see the placement, use a signaler to guide you.

With the forklift in the unloading position, lower the load and tilt the forks to the horizontal position. When the load has been placed and the forks are clear from the underside of the load, back away carefully to disengage the forks or retract the boom on variable-reach units.

5.2.4 Placing Elevated Loads

Special care needs to be taken when placing elevated loads. Some forklifts are equipped with a leveling device that allows the operator to rotate the fork carriage to keep the load level during travel. When placing elevated loads, it is extremely important to level the machine before lifting the load.

One of the biggest potential safety hazards during elevated load placement is poor visibility. There may be workers in the immediate area who cannot be seen. The landing point itself may not be visible. Your depth perception decreases as the height of the lift increases. To be safe, use a signaler to help you spot the load.

Use tag lines to tie off long loads.

Drive the forklift as close as possible to the landing point with the load kept low. Set the parking brake. Raise the load slowly and carefully while maintaining a slight rearward tilt to keep the load cradled. Under no circumstances should the load be tilted forward until the load is over the landing point and ready to be set down.

If the forklift's rear wheels start to lift off the ground, stop immediately, but not abruptly. Lower the load slowly and reposition it or break it down into smaller components if necessary. If surface conditions are bad at the unloading site, it may be necessary to reinforce the surface conditions to provide more stability.

5.2.5 Traveling with Long Loads

Traveling with long loads presents special problems, particularly if the load is flexible and subject to damage. Traveling multiplies the effect of bumps over the length of the load. A stiffener may be added to the load to give it extra rigidity.

To prevent slippage, secure long loads to the forks. This may be done in one of several ways. A field-fabricated cradle may be used to support the load. While this is an effective method, it requires that the load be jacked up.

In some cases, long loads may be snaked through openings that are narrower than the load itself. This is done by approaching the opening at an angle and carefully maneuvering one end of the load through the opening first. Avoid making quick turns because abrupt maneuvers cause the load or its center of gravity to shift.

5.2.6 Using the Forklift to Rig Loads

The forklift may be used to carry pieces of rigging equipment. This method requires slings and a spreader bar. The forklift can be a very useful piece of rigging equipment if it is properly and safely used. Loads can be suspended from the forks with slings, moved around the job site, and placed.

All the rules of careful and safe rigging apply when using a forklift to rig loads. Be sure not to drag the load or let it swing freely. Use tag lines to control the load.

Never attempt to rig an unstable load with a forklift. Be especially mindful of the load's center of gravity when rigging loads with a forklift.

When carrying cylindrical objects, such as oil drums, keep the mast tilted rearward to cradle the load. If necessary, secure the load to keep it from rolling off the forks.

5.3.0 Forklift Safety and Maintenance

The safe operation and preventive maintenance of a forklift is the responsibility of the operator. Always be aware of unsafe conditions to protect the load and the forklift from damage. Work proactively to keep the equipment clean and running efficiently. Take the initiative to identify any areas of concern that could affect the safety of the forklift's operation and report any mechanical issues that need to be addressed.

5.3.1 Forklift Safety Precautions

Be familiar with the operation and function of all controls and instruments before operating a forklift. This means reading and fully understanding the operator's manual, as well as the layout of the work area and the company's rules regarding forklift use.

When using a forklift:

- Never put any part of your body into the mast structure or between the mast and the forklift.
- Never put any part of your body within the reach mechanism.
- Understand the limitations of the forklift (know the capacity it was designed to handle).
- Do not allow passengers to ride unless a safe place to ride has been provided by the manufacturer.
- Never leave the forklift running unattended.

For the safety of pedestrians:

- Always look where you're going.
- Do not drive the forklift up to anyone standing in front of an object or load.
- Make sure that everyone stands clear of the rear swing area before turning.
- Be extra careful to watch out for blind spots, cross aisles, doorways, and other locations where pedestrians may step into the travel path.
- The use of a spotter is recommended when landing an elevated load with a telescoping-boom forklift.

Case History

Forklift Accident

A hardware store employee was killed after being pinned between the forklift he was operating and a tractor trailer. He was using the forklift to move tubing in the shipping area of a store and jumped off the forklift to retrieve some tubing that had fallen. The forklift continued to move and pinned the man to the side of the tractor trailer.

The Bottom Line: Leaving an unattended forklift (or any machine) running can result in a disabling or fatal injury.

5.3.2 Forklift Operator's Maintenance Responsibility

Forklift operators are responsible for preventive maintenance procedures of the machines they are using. Most companies have their own inspection checklist like the one below; pipefitters should follow company policies in addition to the manufacturer's recommendations.

Parking the Forklift

At the end of the day, park the forklift on level ground. Lower the forks to the ground and set the parking brake. If you have to park on a slope, position the forklift at a right angle to the slope and block the wheels.

Be sure to:

- Check the operator's manual for lubrication points and suggested lubrication periods. These are given in terms of the number of hours of operation.
- Check all fluid levels and use the recommended fluids when refilling.
- Keep the moving parts of the machine well-greased since forklifts are often used in dusty and dirty environments. Follow the manufacturer's recommendations for lubricants. This reduces wear, prolongs the life of the machine, and ensures its safe operation.

Forklifts are also provided with a daily checklist from the manufacturer, as shown in *Figure 16*. Checklists like these may differ from unit to unit.

OPERATORS' DAILY CHECKLIST

Check Each Item Before Start Of Each Shift Date:____________

Check One: ☐ Gas/LGP/Diesel Truck ☐ Electric Sit-down ☐ Electric Stand-up ☐ Electric Pallet

Truck Serial Number:________________ Operator:________________ Supervisor's OK: ________________

Hour Meter Reading: ________________

Check each of the following items before the start of each shift. Let your supervisor and/or maintenance department know of any problem. DO NOT OPERATE A FAULTY TRUCK. Your safety is at risk.

After checking, mark each item accordingly. Explain below as necessary.

Check boxes as follows: ☐ OK ☐ NG, needs attention, or repair. Circle problem and explain below.

OK	NG	Visual Checks
		Tires/Wheels: wear, damage, nuts tight
		Head/Tail/Working Lights: damage, mounting, operation
		Gauges/Instruments: damage, operation
		Operator Restraint: damage, mounting, operation, oily, dirty
		Warning Decals/Operators' Manual: missing, not readable
		Data Plate: not readable, missing adjustment
		Overhead Guard: bent, cracked, loose, missing
		Load Back Rest: bent, cracked, loose, missing
		Forks: bent, worn, stops OK
		Engine Oil: level, dirty, leaks
		Hydraulic Oil: level, dirty, leaks
		Radiator: level, dirty, leaks
		Fuel: level, leaks
		Battery: connections loose, charge, electrolyte low
		Covers/Sheet Metal: damaged, missing
		Brakes: linkage, reservoir fluid level, leaks
		Engine: runs rough, noisy, leaks

OK	NG	Visual Checks
		Steering: loose/binding, leaks, operation
		Service Brake: linkage loose/binding, stops OK, grab
		Parking Brake: loose/binding, operational, adjustment
		Seat Brake (if equipped): loose/binding, operational, adjustment
		Horn: operation
		Backup Alarm (if equipped): mounting, operation
		Warning Lights (if equipped): mounting, operation
		Lift/Lower: loose/binding, excessive drift, leaks
		Tilt: loose/binding, excessive drift, "chatters," leaks
		Attachments: mounting, damaged, operation, leaks
		Battery Test (electric trucks only): indicator in green
		Battery: connections loose, charge, electrolyte low while holding full forward tilt
		Control Levers: loose/binding, freely return to neutral
		Directional Control: loose/binding, find neutral OK

Explanation of problems marked above: ____

____

____

____

____

____

Figure 16 Example of a forklift operator's daily checklist.

5.0.0 Section Review

1. The most common type of fixed-mast forklift is the _____ forklift.
   a. all-wheel drive
   b. telescoping
   c. rough terrain
   d. warehouse

2. True or false? When it's necessary to leave the forklift for just a few minutes, it's best to leave it running so that excess fuel is not consumed in the restart.
   a. True
   b. False

3. When traveling on a forklift with a load, make sure the mast is tilted _____.
   a. frontward
   b. eastward
   c. rearward
   d. westward

Section Six

6.0.0 Trenchers

Objective

Identify and describe how to use trenchers.

a. Identify typical trencher controls.
b. Explain how to operate trenchers.
c. State trencher safety guidelines and describe their maintenance needs.

Trade Terms

Mobiling: Term used for driving a trencher around the job site when not digging.

Trencher: A small machine used for digging trenches.

Trenchers are used to dig trenches for the installation of underground services. There are different styles of trenchers available depending upon the scale of the work to be performed. The two styles of trenchers are a pedestrian (walk-behind) and compact (ride-on) (*Figure 17*). Trenchers use either a gasoline or diesel engine to drive a hydraulic system that provides power to the drive wheels and the digging boom.

Figure 17 Trenchers.

6.0.1 Trencher Assemblies

Each type of trencher has a boom. The boom provides the path that the digging chain follows when in operation. The boom is raised, lowered, and moved from side to side using hydraulic cylinders. The size of the boom determines the maximum depth of the trench. Digging chains have replaceable digging teeth that remove material from the ditch.

Both types of trenchers use an engine to power hydraulic system components and move the trencher. The pedestrian machine uses a gasoline engine, and the compact trencher has either a gasoline or diesel engine. The engine provides power to move the machine around the work area and to move the trencher forward as materials are removed from the trench. The control areas are different for each type of machine, because of the way each is operated.

The pedestrian machine is used by the operator walking along behind the trencher. The compact trencher is a ride-on machine with associated controls for driving the trencher, and it requires a certified operator. It is also equipped with a blade for backfilling the trench when work is complete.

6.0.2 Trencher Operator Qualifications

Do not use a trencher unless you have been trained and authorized to do so, or if you are in training under the direct supervision of a qualified trainer. Trench operators must have the visual, auditory, physical, and mental abilities to operate the equipment safely, in addition to understanding the proper functioning and safety requirements of the trencher they're operating.

6.1.0 Typical Trencher Controls

Both pedestrian-style and compact trenchers have sets of controls that may include features not found on others. The operator must read and fully understand the operator's manual provided by the equipment manufacturer.

Figure 18 shows a typical pedestrian trencher control panel. The panel contains the controls and indicators the operator uses when moving or digging with the trencher.

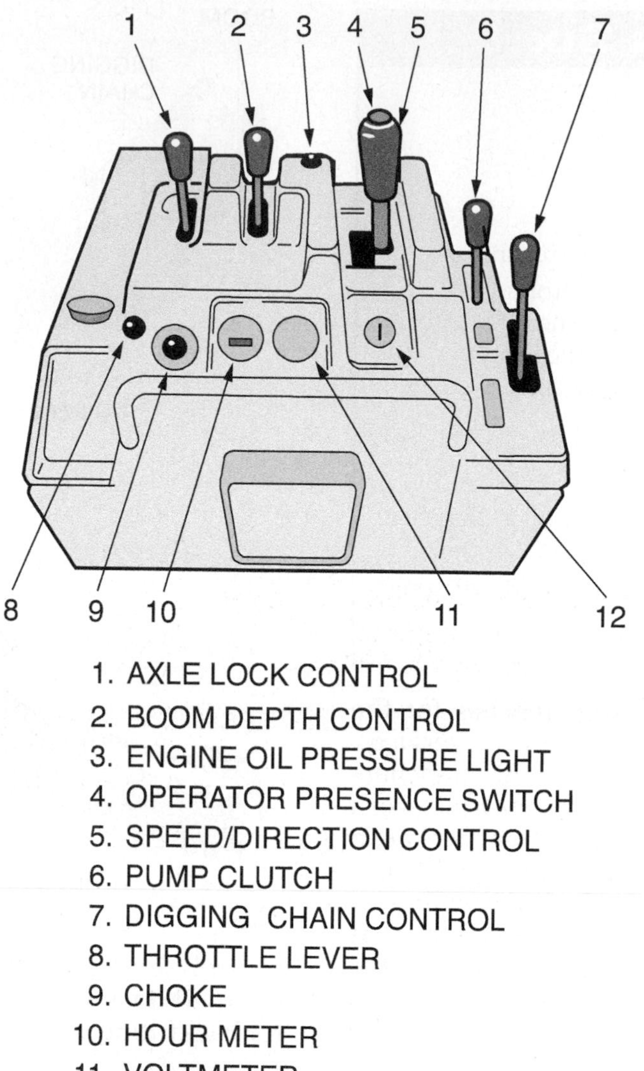

Figure 18 Pedestrian trencher control panel.

Trench controls and indicators are designed for very specific purposes:

- *Axle lock control* – This control shifts the trencher in and out of two-wheel drive. Two-wheel drive is used for digging, driving over rough ground, loading and unloading the trencher from a trailer, and for parking the trencher or leaving it unattended. Trencher maneuverability is reduced in two-wheel-drive mode.
- *Boom depth control* – This control raises and lowers the digging boom.
- *Engine oil pressure light* – This light indicates when engine oil pressure has dropped below an acceptable level. If the indicator lights during operation, the engine should be turned off and the oil level checked.
- *Operator presence (deadman) switch* – This switch must be depressed whenever the speed and direction control is moved from the neutral position or the digging chain control is moved from the stop position. Releasing the switch causes the engine to shut off.
- *Speed/direction control* – This control selects the speed and direction of machine travel and provides steering control. The speed/direction control must be in neutral when starting the trencher.
- *Pump clutch* – The pump clutch controls the application of engine power to the hydraulic pump and should be disengaged when starting the engine. The pump clutch must be engaged before using the speed/direction control, boom depth control, or digging chain control.
- *Digging chain control* – This control starts and stops rotation of the digging chain. It must be in the stop position before starting the engine.

If it's not in the stop position, the engine will not start.

- *Throttle lever* – The throttle lever regulates engine speed.
- *Choke* – The choke helps to start a cold engine.
- *Hour meter* – The hour meter registers the engine operating time. It is used in the scheduling of lubrication and maintenance.
- *Voltmeter* – The voltmeter shows battery charge. With the ignition switch in the ON position, but with the engine not running, it should read at or about 12 volts.
- *Ignition switch* – This is generally a three-position (stop/on/start) switch with a removable key.

6.2.0 Pedestrian Trencher Operation

A careful inspection of the trencher must be performed before beginning work. Each manufacturer provides specific inspection procedures in the operator's manual.

The checklist will normally include recommendations to:

- Check the engine oil level, hydraulic oil level, fuel level, fuel filter, hoses and clamps, air filter system, and all of the lubrication points.
- Inspect tires for proper pressure and general condition.
- Check all control function positions. Verify that all guards and shields are in place and secure. Inspect the condition of drive belts. Inspect the digging chain for tension and condition.
- Inspect the condition of the battery and cables.
- Verify that all safety warning signs are in place and readable.

6.2.1 Starting the Trencher

A typical starting procedure for pedestrian trenchers is:

Step 1 Begin by disengaging the pump clutch. Place the speed/directional control in neutral and move the digging chain

Protection of Underground Utilities

Make the appropriate contacts for a utility stake-out prior to any digging. This helps prevent damage to and interruptions of underground utilities. Most states have a One-Call Notification System center that makes the process easy. Sometimes this can be done online. Typically, the contact must be made at least two or three working days before digging begins. **The number to call is 811.** This could save your life.

A Dig Safely card and other materials are readily available from the various state Dig Safely notification centers to remind contractors and excavators of this requirement. Shown here is a Dig Safely card available from the Dig Safely New York notification center. In addition to giving the procedure for contacting the center, it shows the American Public Works Association (APWA) universal color codes used for the temporary marking of underground facilities. All states have cards similar to the one shown for New York.

FRONT BACK

control to stop. If the engine is cold, use the choke.

Step 2 Insert the key in the ignition switch and turn it to the on position. Turn the key to start; when the engine starts, release the key. Idle the engine five to 10 minutes before moving to the job site.

Step 3 Engage the pump clutch slowly while warming up the engine to warm the hydraulic oil.

Step 4 While the engine is idling, check the operation of the speed/directional control, boom depth control, and digging chain control.

6.2.2 General Operation

The operator's manual usually contains procedures for driving a trencher around the job site when not digging. This is called **mobiling** and it normally proceeds as follows:

Step 1 Using the boom depth control, raise the boom to the transport position. On rough terrain, keep the boom as low as possible to help with stability.

Step 2 Engage or disengage the axle lock to match ground conditions. On level ground, the axle lock should be disengaged to increase maneuverability. On rough ground, the axle lock should be engaged to improve traction. The axle lock should always be in the engaged position when the machine is left unattended.

Step 3 Press the operator presence switch down. Use the speed/directional control to move the machine in the desired direction.

Step 4 Steer the machine with the speed/directional control.

The stop mobiling procedure is also contained in the operator's manual. Typically, the operator would:

Step 1 Move the speed/direction control to the neutral position.

Step 2 Release the operator presence switch and place the throttle in the idle position.

Step 3 Using the boom depth control, lower the digging boom to the ground. Engage the axle lock and disengage the pump clutch.

Step 4 Turn the ignition key to off.

When leaving the machine, remove the key from the ignition switch.

6.2.3 Trench Digging

The trenching procedure is contained in the operator's manual. A typical procedure is as follows:

Step 1 Begin digging the trench by slowly engaging the pump clutch. Drive the trencher to a point in line with the intended trench. Digging is done with the trencher moving away from the digging boom. Align the trencher with the boom pointing at the starting end of the trench.

Step 2 Place the speed/directional control in neutral. Engage the axle lock. Lower the digging chain to within 1 inch of the ground using the boom depth control.

Step 3 Place the digging chain control in the rotate position. The digging chain will begin to move. Slowly lower the digging boom to the desired digging depth using the boom depth control.

Step 4 When the digging boom reaches the desired depth, adjust the engine throttle for optimum digging speed. Do not overload the engine at low rpm. Use the speed/directional control to begin trenching and to regulate the ground travel speed. Find the best ground travel speed for digging without causing the engine to lug down. Minor, slow directional changes can be made using the speed and direction control.

The stop trenching procedure is also contained in the operator's manual. A typical procedure is as follows:

Step 1 To stop digging, place the speed/direction control in neutral. Using the boom depth control, raise the boom to ground level.

Step 2 Place the digging chain control in the stop position. Raise the boom all the way up using the boom depth control.

Step 3 Use the speed/direction control to drive away from the trench. On level ground, disengage the axle lock and shut down the machine.

Case History

Trencher Safety

While following a trencher and shoveling dirt back into the partially dug trench, an apprentice fell into the trench and was killed. Witnesses standing at the opposite end of the trench line noticed a rapid movement and something bright thrown into the air (the apprentice's hardhat) near the trencher. They immediately alerted the trencher operator who stopped the trencher and moved it away from the trench. Co-workers went into the trench to aid the victim as one of the witnesses called for emergency assistance. Unfortunately, the victim was pronounced dead at the scene.

The Bottom Line: Since no one actually saw what caused this accident, it was theorized that the apprentice either lost his balance and fell toward the machine, or that his clothing got caught in the digging chain, causing him to be pulled into the moving chain and the trench. Had the following precautions been taken, this accident could have been avoided:

The apprentice should have maintained a minimum 10-foot safety zone away from the moving digging chain.

Before starting work, a job safety analysis for all the tasks involved with digging the trench should have been performed.

Note: Safety decals that alert the trencher operator and others to the dangers associated with trencher machine operation should be installed on all older trenching machines that don't have them.

6.3.0 Trencher Safety and Maintenance

Before starting any trenching, digging, or excavation work, always call 811. It is critical to know whether the area to be trenched has underground cables, wires, pipes, or other types of utility equipment buried within it. Call at least 2 days ahead of the scheduled dig to allow time for the utility company to inspect and mark up any areas that must be avoided.

As with any type of equipment, make sure it is good working order before staring, and use each tool or piece of machinery as recommended. Wear appropriate protective gear and follow all company and governmental rules and regulations.

6.3.1 Trencher Safety Precautions

There are no appropriate shortcuts when it comes to safety. Whether operating a trencher or working near one, always act with good judgment, common sense, and an understanding of appropriate behavior for the environment. Show respect for the equipment being used.

Remember:

- Incorrect procedures could result in death, injury, or property damage. Learn how to use the equipment correctly.
- Moving digging teeth can cut off an arm or leg-and also kill. Stay away from the boom when it is operating.
- The trencher may move when the chain starts to dig. Allow 3 feet between the end of the chain and any obstacles.
- Keep the digging boom low when operating on a slope.
- If there is any doubt about the classification of a job site or whether it contains electrical utilities, treat it as electrical.

The following are general work site precautions to be followed by the operator. When practical, stay upwind of the trencher. This reduces the amount of dust and dirt blown toward you. It may also reduce the amount of gas around you if you hit a gas line. If a gas line is hit, do the following to minimize any danger to personnel:

- Quickly turn off the ignition switch key.
- Leave the machine as quickly as possible.
- Contact the gas company to shut off the gas.
- Do not return to the area until given permission by the gas company.

In the event an underground electrical cable is hit, do the following to minimize any danger to personnel:

- If on the ground, stay where you are and do not touch any equipment.
- Warn other personnel that a strike has occurred and that they should stay away from any equipment and the immediate area.
- Contact the utility company to shut off the power.
- Do not return to the area until given permission by the utility company.

6.3.2 Trencher Operator's Maintenance Responsibility

Maintaining a trencher requires changing the engine oil, lubricating the digging chain clutch, and changing the hydraulic oil filter, among other things. Proper maintenance will extend the life and performance of the machine. *Figure 19* shows a typical maintenance schedule for a trencher. The interval hours column shows the number of operating hours between each required maintenance function. The reference number column provides a general location on the machine where the maintenance is to be performed. The description column provides the operator with the maintenance task that needs to be performed. Some charts have a lube column that tells the operator the type of lubricant to use and a point column that provides the total number of points at which the lubricant is applied.

Use Your Senses

During normal operation of the trencher, listen for thumps, bumps, rattles, squeaks, squeals, or other unusual sounds. Smell for odors such as burning insulation, hot metal, burning rubber, or hot oil. Feel for any changes in the way the trencher is operating. Look for problems with wiring and cables, hydraulic connections, or other equipment. Correct anything you hear, smell, feel, or see that is different from what is expected, or that seems to be unsafe.

REF.	HOURS	DESCRIPTION
1	10	CHECK TIRE PRESSURE
2	10	ADJUST DIGGING CHAIN TENSION
3	25	WASH & OIL AIR FILTER PRECLEANER
4	50	CHECK BATTERY
5	50	CHECK BELT TENSION
6	100	REPLACE AIR FILTER PAPER ELEMENT & PRECLEANER
7	100	CHECK DRIVE WHEEL END PLAY AND LUG NUTS
8	100	CHECK FUEL FILTER
9	100	CLEAN COOLING FINS ON CYLINDER HEAD AND BARREL
10	200	REPLACE FUEL FILTER
11	AS NEEDED	RE-TORQUE HEADSHAFT SPROCKET BOLTS
12	AS NEEDED	CHECK LEVEL IN FUEL TANK

Figure 19 Example of trencher maintenance schedule.

Additional Resources

Excavations and Trenching. 2018. Princeton University: Environmental Health and Safety. Available at: **https://ehs.princeton.edu/workplace-construction/construction-safety/excavations-and-trenching**

Trenching and Excavation Safety. The INGAA Foundation, Inc. 2012. Available at: **http://www.ingaa.org/file.aspx?id=18983**

6.0.0 Section Review

1. What part of the trencher creates the path that is followed by the digging chain?
   a. Motor
   b. Controls
   c. Boom
   d. Lift

2. The ____ shifts the trencher in and out of two-wheel drive.
   a. deadman switch
   b. digging chain control
   c. boom depth control
   d. axle lock control

3. When operating a pedestrian trencher, before turning inserting the key in the ignition, it is necessary to ____.
   a. disengage the pump clutch
   b. slowly engage the pump clutch
   c. lower the boom
   d. chock the wheels

4. If an underground electrical cable is hit while trenching, the first concern is to _____.
   a. write up a report and turn it in at the end of your shift
   b. collect all construction debris from the area and load up the equipment.
   c. stay where you are and not touch any equipment, if you're on the ground.
   d. leave to get the foreman, then return with that person to show exactly where the line was hit.

SECTION SEVEN

7.0.0 PORTABLE WELDING MACHINES, PUMPS, AND COMPACTORS

Objective

Identify and describe the use of support equipment.

a. Identify and describe the use of portable welding machines.
b. Identify and describe the use of portable pumps.
c. Identify and describe the use of portable compactors.

Performance Task

7. Identify portable pumps to use for specific applications.

Trade Terms

Centrifugal force: The force that moves a substance outward from a center of rotation.

Centrifugal pump: A machine that moves a liquid by accelerating the liquid radially from the center of an impeller in motion.

Compactor: A machine used to compact soil to prevent the settling of soil after construction.

Diaphragm: A thin, flexible separating wall that flexes back and forth to move a fluid.

Diaphragm pump: A pump that uses a diaphragm to isolate the liquid being pumped from the operating parts in a mechanically driven pump, or from the hydraulic fluid in a hydraulically actuated pump.

Impeller: A rotating part inside a pump that increases the speed of the fluid moving through the pump.

Pneumatic: Run by or using compressed air.

Velocity: The speed at which a fluid is ejected from a pump.

Volt: A unit of electrical potential and electromotive force.

Welding machines, portable pumps, and compactors are essential pieces of support equipment in pipefitting and other fields. While not every type of equipment is necessary for all jobs, it is important to select exactly the right one for the task at hand. Electrical and other hazards are introduced with certain types of machinery, so understanding the power and control mechanisms of each piece is essential. Regular maintenance will extend the life of supportive equipment and contribute to safe operations.

7.1.0 Welding Machines

To produce welding current, the engine must turn the generator at a required number of revolutions per minute (rpm). The engines used to power generators have governors that control the engine speed. Most governors have a welding speed switch that can be set to idle the engine speed when no welding is taking place. When the welding electrode is touched to the base metal, the governor automatically increases the speed of the engine to the required rpms for welding. After about 15 seconds of no welding, the engine automatically returns to an idle.

Engine-driven welding machines often have an auxiliary power unit to produce 110-volt alternating current for lighting, power tools, and other electrical equipment. All power tools powered by a welding machine must have AC and DC capabilities. When 110-volt power is required, the engine-driven generator must run continuously at the welding speed.

Engine-driven welding machines have both engine controls and welding machine controls. The engine controls vary with the size and type of machine but are essentially the same as the generator engine controls previously discussed. The welding machine controls usually include an amperage control switch and gauge, a current-control switch and gauge, and a polarity switch. If the welding machine is not equipped with a polarity switch, manually change the lead cables at the welding current terminals to change the polarity. *Figure 20* shows a typical welding machine control panel.

Extension Cords

When an extension cord is used to supply power from a portable generator to electric power tools or other devices, it is important that it has an adequate current-carrying capacity for the job. If an undersized extension cords is used, excessive voltage drops will result. This causes excess heating of the extension cords and portable tools, as well as additional generator loading.

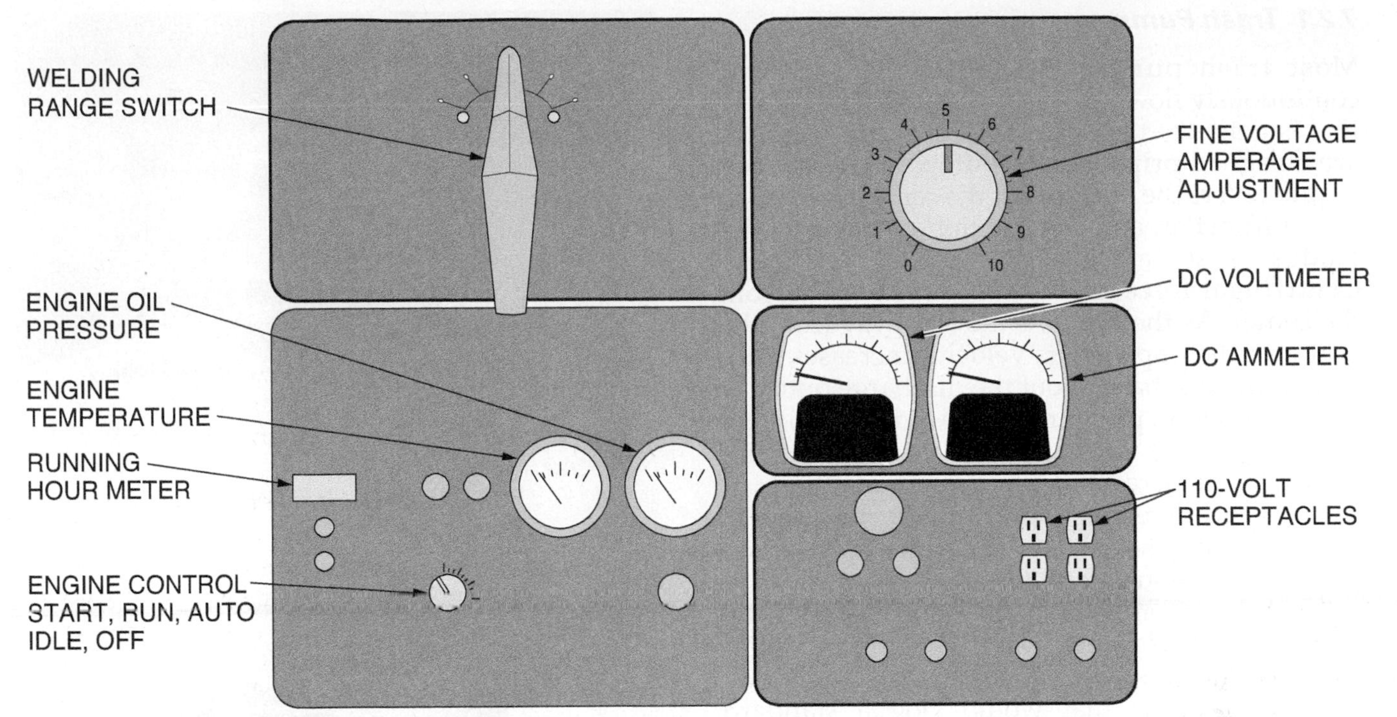

Figure 20 Typical welding machine control panel.

The size of a welding machine is determined by the amperage output of the machine at a given duty cycle. The duty cycle of a welding machine is based on a 10-minute period, which is the percentage of 10 minutes that the machine can continuously produce its rated amperage without overheating. For example, a machine with a rated output of 300 amps at 60 percent duty cycle can deliver 300 amps of welding current for 6 minutes out of every 10 without overheating. The duty cycle of a welding machine will be 10, 20, 30, 40, 60, or 100 percent. A welding machine having a duty cycle of 10 to 40 percent is considered a light- to medium-duty machine. Most industrial, heavy-duty machines for manual welding are 60 percent duty cycle. With the exception of 100 percent duty cycle machines, the maximum amperage that a welding machine produces is always higher than its rated capacity. A welding machine rated 300 amps at 60 percent duty cycle generally puts out a maximum of 375 to 400 amps.

The advantage of using engine-driven welding machines is that they are portable and can be used in the field where electricity is not available for other types of welding machines. The disadvantage of engine-driven welding machines is that they are expensive to purchase, operate, and maintain.

To safely operate a welding machine:

- Operate arc welding machines and equipment only in clean, dry locations.
- Make sure that exhausts of gasoline- or diesel-powered welding generators are sufficiently ventilated when used indoors.
- Ensure that the arc welding machine is properly grounded to either your workpiece or to a ground rod driven into the ground, according to company policies.
- Check all connections to the machine to make sure they are secure before beginning to weld.
- Do not attempt to install or repair welding equipment. Call a qualified electrician.
- When working in the field, protect arc welding machines from weather conditions as much as possible.
- Make sure all welding machines are equipped with a ground fault circuit interrupter (GFCI). This serves as an automatic shutoff to protect the user from injury due to a short circuit.
- In the event of a plant or job site evacuation, shut down welding machines and all motorized equipment.

7.2.0 Portable Pumps

Portable pumps are widely used by pipefitters for various applications. The three most common types of portable pumps are trash pumps, mud pumps, and submersible pumps. These can be pneumatic, electric, or gasoline engine-driven pumps. **Pneumatic** pumps are those that are using or are run by compressed air.

7.2.1 Trash Pumps

Most trash pumps are **centrifugal pumps**, continuously flowing by way of **centrifugal force** that is generated by rotation. Centrifugal pumps are designed primarily for pumping clear water. Liquid enters the **impeller** at the center, or the eye, of the impeller, and the rotation of the impeller blades causes a rotary motion of the liquid. Centrifugal force moves the liquid away from the center. As the liquid is moved away from the center of the impeller, its **velocity** increases until it is finally discharged out the discharge outlet.

In addition to pumping clear water, centrifugal pumps can be used where small quantities of up to 10 percent volume of mud, sand, or silt are encountered. They can also be used to control heavier seepage in excavations, trenches, pipelines, manholes, and foundations.

Trash pumps are centrifugal pumps specifically designed for use with water containing solids, such as sticks, stones, sand, gravel, and other foreign materials that would clog a standard centrifugal pump. The trash pump has a compartment that collects solids to prevent them from damaging the pump impeller. Trash pumps are adequate for pumping water from a maximum depth of 20 feet. To remove water from a depth greater than 20 feet, a submersible pump should be used. Any time pumps are to be left outside during cold weather, remove all water from them to keep the pump housing from freezing and cracking. *Figure 21* shows portable trash pumps.

PORTABLE TRASH PUMP IN USE

CENTRIFUGAL TRASH PUMP

Figure 21 Portable trash pumps.

7.2.2 Mud Pumps

Most mud pumps are **diaphragm pumps**. They are designed for use where low, continuous flow is required for highly viscous fluids, such as thick mud, air mixed with water, and water containing a considerable amount of solids or abrasive materials. *Figure 22* shows a mud pump.

Most diaphragm pumps use hydraulic pressure delivered by a piston to actuate the flexible diaphragm. The side of the **diaphragm** that produces the pumping action is called the power side, and the side doing the pumping is called the fluid side. A suction check valve and a discharge check valve open and close to move fluids through the pump. *Figure 23* shows the operation of a diaphragm pump.

Figure 22 Mud pump.

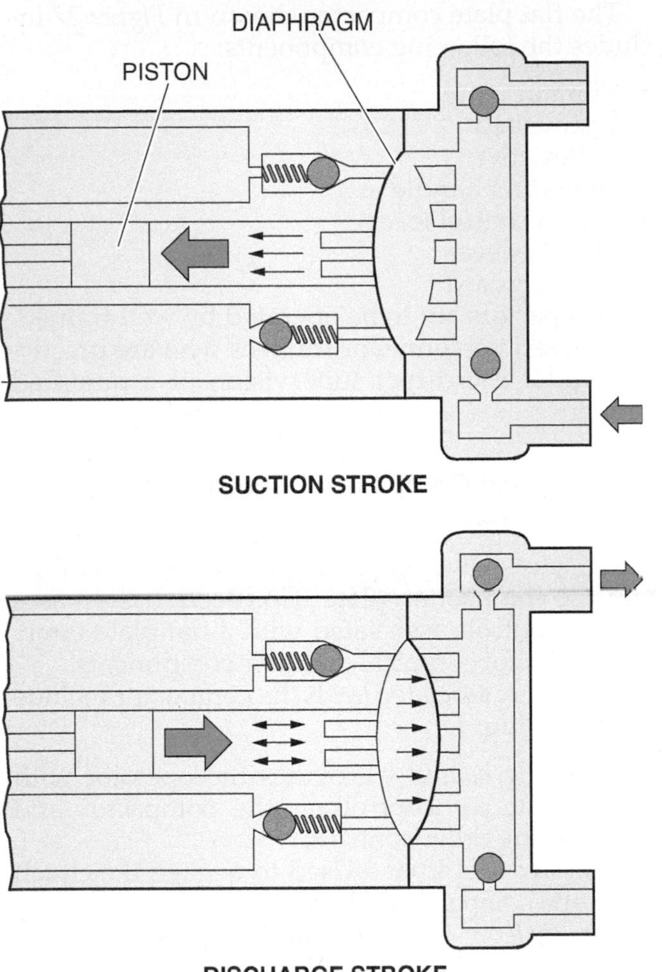

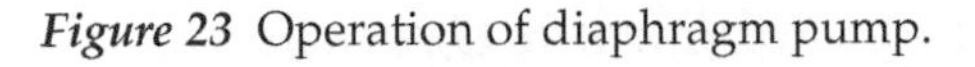

Figure 23 Operation of diaphragm pump.

7.2.3 Submersible Pumps

A submersible pump (*Figure 24*) is encased with its motor in a protective housing that allows the entire unit to operate under water. Submersible pumps are used in many residential, commercial, and industrial wells. The operation of these pumps can be controlled by a float switch located inside the well that contains the pump or by manual or automatic control switches located outside the well. The float switch is a mercury-to-mercury contact switch that is controlled by the position of an attached float. As long as the water level is at an adequate height to remove water from the well, the pump is on. When the water level falls to a level too low for pumping, the float also drops and turns the pump off. As the water level builds back up, the float rises and turns the pump on again. *Figure 25* shows a submersible pump controlled by a float switch.

Figure 24 Submersible pump.

7.3.0 Compaction Equipment

Compactors are used to eliminate soil settlement. If the soil under a structure is not compacted, it will eventually settle and cause damage to the structure. This settling will cause the structure to deform. If the settling occurs more at one side or corner, cracks or structural failures can occur.

Compactors come in a wide variety of different styles and sizes. Each type of compactor is recommended for differing soil conditions. Compactors can also be installed on other equipment as an attachment. *Figure 26* shows two common types of compactors.

The various soil conditions that can be handled by common types of compactors includes:

- *Cohesive soils (silts and clay)* – Use an upright tamp or sheepsfoot vibrator roller.
- *Granular soils (sand or gravel)* – Use a flat plate tamp or smooth drum vibrator roller, or an upright tamp.
- *Mixed soil* – Use an upright tamp, sheepsfoot vibrator roller, smooth drum vibrator roller, or plate tamp.

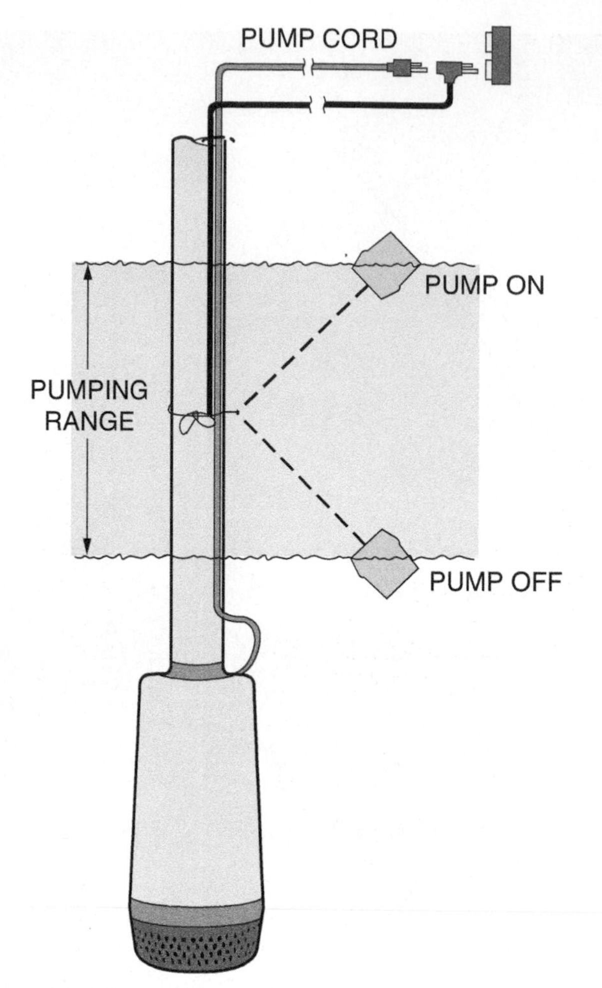

Figure 25 Submersible pump controlled by float switch.

The flat plate compactor shown in *Figure 27* includes the following components:

- Vibrator plate
- Engine bed
- Pulley cover
- Operating handle
- Clutch control lever
- Throttle lever
- Engine guard

Compactors are to be operated by well-trained, authorized personnel or trainees who are practicing under the direct supervision of a qualified trainer.

7.3.1 Typical Compaction Equipment Controls

Compactors have operator controls for the engine and the compaction plate. This section discusses typical controls associated with a flat plate tamp. *Figure 27* shows typical operator components.

Controls associated with the compactor include the following:

- *Operating handle* – Provides the operator with a handle for controlling the compactor and mounting other controls.
- *Clutch control lever* – Used to engage the clutch for the compacting plate.
- *Throttle lever* – Controls engine speed.

7.3.2 Compaction Equipment Operation

Before working with any type of compactor, familiarize yourself with safety recommendations and operational guidelines found in the opera-

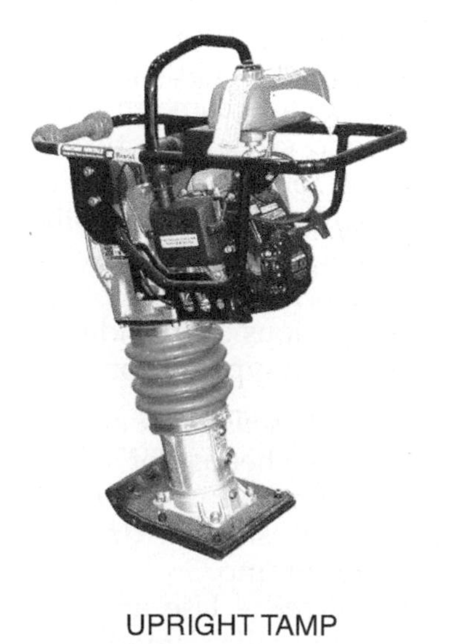

Figure 26 Compactors.

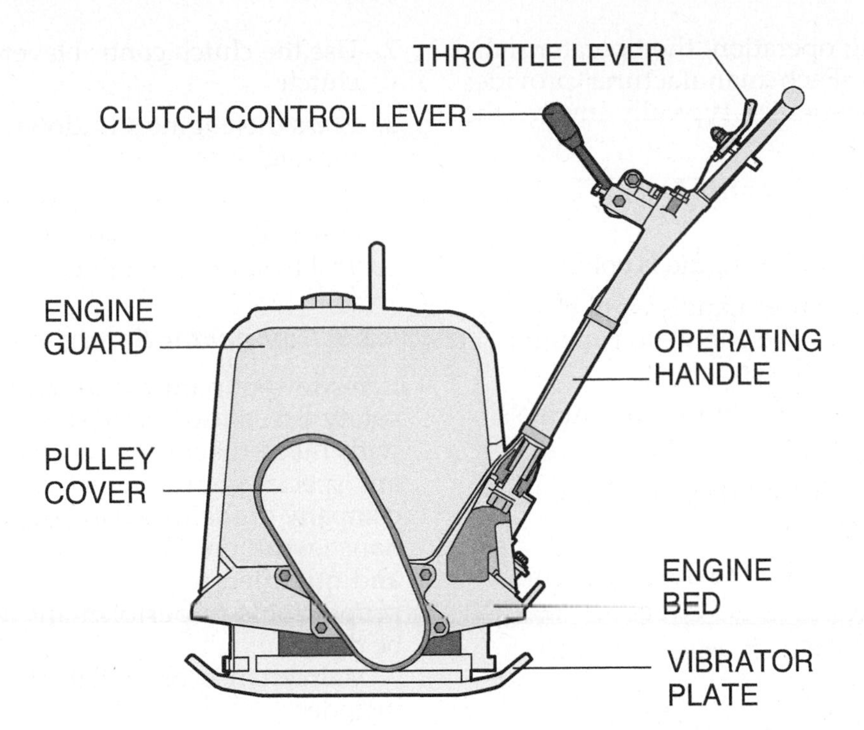

Figure 27 Flat plate compactor components.

tor's manual. Check the machine prior to turning it on; suggestions that may be found in the operator's manual include:

- Inspecting the machine to verify that all guards and safety items are properly installed.
- Checking the fuel level and adding fuel if required.
- Checking the engine oil level and adding oil if required.
- Checking the engine air filter and cleaning or replacing it in accordance with the manufacturer's procedures.

Safety Manuals

The Association of Equipment Manufacturers (AEM) (formerly the Equipment Manufacturers Institute [EMI]) offers a series of safety manuals that are easy to read and use. The manuals were produced by AEM to provide equipment owners, operators, service personnel, and mechanics with information on basic safety procedures and precautions that apply to day-to-day operations and the maintenance of specific equipment. The manuals include information on topics that are common to many types of equipment and are to be used in conjunction with manufacturers' operator manuals for specific pieces of equipment.

Soil Compaction Specifications

Earthwork specifications will determine the depth of fill that can be placed and compacted in each layer. This helps prevent excessive settling. The job specifications will also dictate the frequency of soil compaction testing.

Soil Compaction

Some form of soil compaction is needed before pouring foundations, slabs, or floors. Improperly prepared soil can settle, causing foundations or slabs to crack, and even causing the structure to lean. The most famous example of settling is the Leaning Tower of Pisa. It was built plumb, but now leans at an angle of about 12 degrees.

Before beginning operation, the operator must perform a startup. Each manufacturer provides detailed instructions which typically involve the following steps:

1. Verify that the clutch control lever is in the disengaged position.
2. Engage the choke if the engine is cold.
3. Pull on the recoil starter handle until the engine starts. Allow the engine to run until warm, then release the choke.
4. Use the throttle lever to control the engine's rpm during operation.

Normal operating instructions are likely to include the following steps:

1. Adjust the throttle until the desired engine speed is reached.
2. Slowly engage the clutch using the clutch control lever.
3. Use the operating handle to control the direction and movement of the compactor.

For shutting down the equipment it is likely that you will need to:

1. Use the throttle lever to decrease the engine's speed.
2. Use the clutch control lever to disengage the clutch.
3. Press the engine shutdown control to turn off the engine.

Emergency shutdown of the compactor is accomplished by pressing the engine shutdown control to stop the engine.

7.3.3 Compactor Safety and Maintenance

Refer to the manufacturer's manual for specific safety precautions and use them in conjunction with those discussed in this manual, as well as any special instructions from your supervisor or company guidelines. Do not perform any maintenance work on the compactor unless authorized and qualified to do so. Make sure you have the proper tools to perform any work that needs to be done.

Maintenance on a flat plate tamp compactor includes:

- Changing the engine oil
- Changing the engine air filters
- Lubricating parts
- Inspecting all guards and safety devices

After performing maintenance, make sure all the guards are correctly installed and that all safety devices are functioning.

Additional Resource

Soil Compaction Handbook. 2011. Multiquip Inc.® Available at: **http://www.multiquip.com/multiquip/pdfs/Soil_Compaction_Handbook_low_res_0212_DataId_59525_Version_1.pdf**

7.0.0 Section Review

1. What term is used to describe the part inside of a pump that increases the speed of the fluid moving through it?
   a. Compactor
   b. Diaphragm
   c. Impeller
   d. Floe valve

2. Engine-driven welding machines are attractive because ____.
   a. they are inexpensive to purchase and require little maintenance
   b. they introduce few, if any, safety concerns
   c. they are portable and can be used in places where electricity is not available
   d. entry-level workers can use them with little to no training

3. What type of pump is specifically designed to handle sticks, stones, and gravel?
   a. Standard centrifugal pump
   b. Pneumatic pump
   c. Trash pump
   d. Mud pump

4. True or false? Compacting the soil is recommended, but not essential for enhancing the structural integrity of a building.
   a. True
   b. False

SECTION EIGHT

8.0.0 BACKHOES AND MOBILE CRANES

Objective

Identify and describe the use of backhoes and mobile cranes.

a. Identify and describe the use of backhoes.
b. Identify and describe the use of mobile cranes.

Performance Task

8. Identify types of hydraulic cranes and recognize safety hazards involved in working around them.

Trade Terms

Cherry picker: A truck-mounted or trailer-mounted lift designed for both indoor and outdoor use.

Working with backhoes and mobile cranes is an everyday part of construction work. Each type of equipment is designed for specific purposes, and pipefitters must be knowledgeable about those available. Selecting the right tool or machinery for the job is step one; step two is knowing how to maintain it and use it safely. Prior to using a backhoe or any type of digging equipment, call 811 to make sure no electrical, cable, gas, or other lines will be compromised.

8.1.0 Backhoe/Loader

A backhoe/loader (*Figure 28*) is a dual-purpose, highly maneuverable machine used to dig trenches, foundations, and similar excavations. It is also used to move dirt, crushed stone, gravel, and other materials around the job site.

Backhoes/loaders are equipped with either a gasoline or diesel engine. The backhoe bucket and loader bucket attachments are hydraulically operated under the control of a single operator who performs all backhoe and loader operations.

8.1.1 Backhoe/Loader Operator Qualifications

Before operating a backhoe/loader, the following requirements must be met:

- The operator must successfully complete a training program that includes actual operation of the backhoe/loader.
- The operator is responsible for reading and understanding the safety manual and operator's manual provided with the equipment.
- The operator must know the safety rules and regulations for the job site.

8.1.2 Backhoe/Loader Safety Precautions

The following is a list of general safety rules specific to the operation of a backhoe/loader. Remember that the other safety precautions, already discussed in this module, also apply.

- Mount the backhoe/loader only at points that are equipped with steps and/or handholds.
- Face the backhoe/loader while entering and leaving the operator's compartment.
- Do not mount a backhoe/loader while carrying tools or supplies.
- Do not mount a moving backhoe/loader.
- Do not use controls as handholds when entering the operator's compartment.
- Do not obstruct your vision when traveling or working.
- Never lift, move, or swing a load over a truck cab or over workers.
- When traveling, operate at speeds slow enough so that you have complete control of the backhoe/loader at all times. This is especially true when traveling over rough or slippery ground and when on hillsides. Never place the transmission in neutral to allow the backhoe/loader to coast.
- Never approach overhead power lines with any part of the backhoe/loader unless you are in strict compliance with all local, state, and federal safety regulations.
- Make sure that you know the location of all underground gas and water pipelines and all electrical or fiber-optic cables. Call 811 a few days ahead of the dig to ensure the area is safe for excavation.

When operating the backhoe the following safety precautions apply:

- Never enter or allow anyone to enter the backhoe swing pivot area.

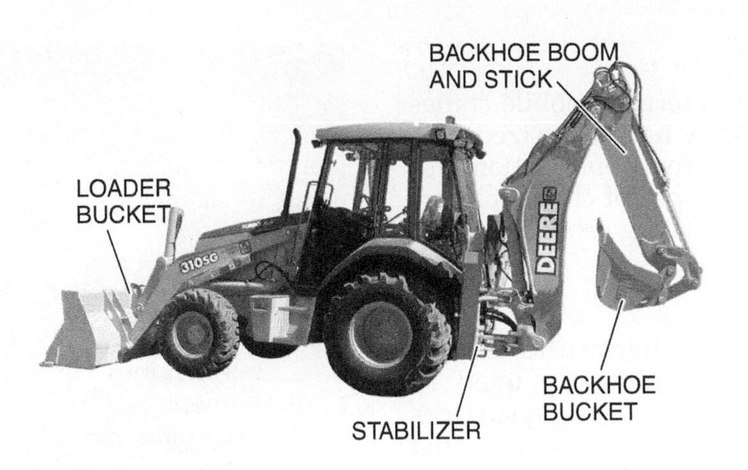

Figure 28 Backhoe/loader.

- Operate the backhoe from the correct backhoe operating position. Never operate the backhoe controls from the ground. Never allow riders in or on the backhoe.
- Do not dig under the backhoe or its stabilizers. This is prohibited in order to prevent cave-ins and the chance of the backhoe falling into the excavation.
- When operating the backhoe on a slope, swing to the uphill side to dump the load, if possible. If required to dump downhill, swing only as far as required to dump the bucket.
- Always dump the soil far enough away from the trench to prevent cave-ins.

When operating the loader the following safety precautions apply:

- Carry the bucket low for maximum stability and visibility.
- Stay in gear when traveling downhill. Use the same gear range as you would for traveling up a grade.
- When on a steep slope, drive up or down the slope; do not drive across the slope. If the bucket is loaded, drive with the bucket facing uphill. If empty, drive with the bucket pointed downhill.
- When operating the loader, make sure that the backhoe is in the transport lock position to prevent backhoe movement.
- When working at the base of a bank or overhang, never undercut a high bank and/or operate the loader close to the edge of an overhang or ditch.

Backhoe Damage

When digging with the backhoe, do not swing the bucket against the sides of the excavation and/or dump pile. Repeatedly doing so can cause structural damage to the boom. It might also result in premature wear of the boom pin and bushings.

8.2.0 Mobile Cranes

The primary function of a mobile crane is to lift and swing loads. A mobile crane consists of a rotating superstructure, operating machinery, and a boom mounted on a vehicle such as a truck, crawler, or railcar. Since their introduction in the 1800s, mobile cranes have gone through constant change, from steam powered to air-cooled diesel engines. Manufacturers continually strive to meet the needs of an ever-changing industry. Because of the wide variety of uses for cranes, many different types have been developed.

8.2.1 Boom Trucks

Boom trucks are unlike any of the other mobile cranes in that they are mounted on carriers that are not solely designed for crane service. These cranes are mounted onto a commercial truck chassis that has been strengthened to carry the crane. The boom truck can also be used to haul loads and it can be driven on the highway system when the boom is in the transport position. Boom trucks typically range in capacity from 6 to 22 tons. *Figure 29* shows a typical boom truck.

Figure 29 Boom truck.

8.2.2 Cherry Pickers

Cherry pickers are rough-terrain mobile cranes as shown in *Figure 30*. They have oversized tires that allow them to move across construction sites and broken ground. Two types of cherry pickers are the fixed cab and the rotating cab.

8.2.3 Carry Decks

A carry deck is a small hydraulic mobile crane. Carry decks are primarily used in industrial applications where working surfaces are significantly better than those of most construction sites. The compact size of the carry deck (15 feet long by 7 feet wide) allows it to be easily maneuvered around the job site. Most offer up to $8\frac{1}{2}$ tons of lifting capacity with the outriggers in place and up to $7\frac{1}{2}$ tons for lifting and transporting work. The carry deck also has a transport area that is designed to lock down and carry up to 15,000 pounds of materials, as shown in *Figure 31*. Most have water-cooled diesel engines, but some use gasoline-powered or propane (LP) engines.

Figure 31 Carry deck.

Figure 30 Cherry pickers.

8.2.4 Lattice Boom Cranes

Lattice boom cranes are capable of reaching much higher boom lengths than hydraulic boom cranes. Site preparation and erection of lattice boom cranes may take anywhere from a day for a smaller crane to weeks for larger cranes. The length of setup time depends on the height of the boom and the lift attachments to be used. Modern lattice boom cranes are powered by computer-controlled hydraulic systems. They have redundant fail-safe devices for operational safety. These consist of, but are not limited to automatic braking systems, load moment measuring devices, mechanical system monitoring devices, and function lockout systems that stop operation when a system fault occurs. These features allow for better control and accuracy than older friction-operated devices.

Lattice boom crawler cranes are specifically designed for the extreme duty associated with the use of a crawler crane. Lattice boom crawler cranes are used to pick up a heavy load and track with it to a nearby location. The reliability and versatility of the crawler-mounted lattice boom crane makes it the most widely applied crane design in use today. Besides lifting heavy loads to great heights, these cranes are ideal for applications with a high duty cycle. *Figure 32* shows a lattice boom crawler crane.

A lattice boom truck crane, shown in *Figure 33*, provides the mobility of a truck crane with the extreme lifting capacity of a lattice boom crane. Depending on the size of the crane and its gross vehicle weight, some components, such as counterweights and outriggers, may have to be removed before highway travel to meet local, state, and federal weight restrictions.

Figure 32 Lattice boom crawler crane.

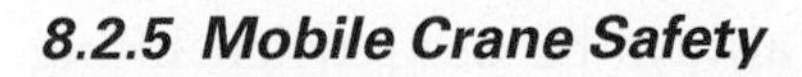

8.2.5 Mobile Crane Safety

Safely lifting and moving machinery and equipment involves decisions regarding the objects being moved and the appropriate lifting and rigging equipment for the job. Everyone must be alert, responsible, and aware of all the factors involved in the safe use of mobile cranes.

When working with or around mobile cranes:

- Operators must be physically fit, properly trained, competent and alert, and free from the influence of alcohol, drugs, and medications. If required, they must be licensed to operate mobile cranes. Good vision, sound judgment, physical coordination, and mental ability are also required. Operators who do not possess all these qualities must not be allowed to operate the equipment.
- Signalmen must also have good vision and sound judgment. They must know the standard crane signals and be able to give signals clearly. They must also have enough experience to be able to recognize hazards and signal the operator to avoid them. An operator must understand the hand signals before lifting a single load with a crane.

Figure 33 Lattice boom truck crane.

- Outriggers must always be used when lifting loads, and good judgment must be used for the placement of outriggers.
- Riggers must be trained to determine weights and distances and to properly select and use lifting tackle.
- Crew members must understand their specific safety responsibilities and report any unsafe conditions or practices to the proper personnel immediately.
- All crew members working around mobile cranes must obey all warning signs and watch out for their own safety and the safety of others.
- Crew members setting up machines or handling loads must be properly trained and aware of proper machine erection and rigging procedures.
- All employees must watch for hazards during operations, and alert the operator and signalman of dangers, such as power lines, the presence of people, other equipment, or unstable ground conditions.

Additional Resource

Preventing Injuries when Working with Hydraulic Excavators and Backhoe Loaders. The Centers for National Institute for Occupational Safety and Health. 2003. Available at: **https://www.cdc.gov/niosh/docs/wp-solutions/2004-107/pdfs/2004-107.pdf**

8.0.0 Section Review

1. When operating the loader of a backhoe, it's best to _____.
   a. carry the bucket as high to prevent interruption with work on the ground
   b. carry the bucket low for maximum stability and visibility
   c. drive across steep slopes instead of up and down
   d. use a different gear for traveling uphill than traveling downhill

2. Which of the following can be used as a mobile crane, and also used to haul loads across highway systems when its components are properly positioned?
   a. Cherry picker
   b. Carry deck
   c. Boom truck
   d. Lattice boom crane

Summary

Selecting the right type of motorized equipment is essential to the overall success of pipefitting operations. Each tool, machine, and piece of equipment introduces its own safety considerations with respect to the item itself and the environment in which it is used. When working with any type of excavation equipment, and prior to any amount of digging at any depth, call 811 a few days before to schedule an inspection. Underground cables, wires, and lines introduce as many hazards as equipment itself. Proper maintenance of tools and machinery, effective training and respect for regulations, and attention to one's surroundings will help with creating a safe, efficient work site for all employees.

Review Questions

1. The operator should check for leaks in a pressurized hydraulic system by _____.
   a. checking the hydraulic oil level in the supply tank
   b. feeling all the hoses and connectors
   c. using a piece of wood or cardboard
   d. using a flashlight and an inspection mirror

2. Before disconnecting any hydraulic system hoses, the operator should shut down the engine and _____.
   a. check the hydraulic fluid level
   b. cycle the hydraulic controls
   c. set the parking brake
   d. chock the wheels

3. When refueling a piece of equipment, keep the nozzle or fuel funnel in contact with the filler neck to _____.
   a. prevent the possibility of sparks
   b. lower the possibility of a fuel spill
   c. protect you from breathing the fuel fumes
   d. make it easier to see when the fuel tank is full

4. When a battery is frozen, _____.
   a. disconnect the negative lead before attaching the charger
   b. do not charge it or attempt to jump-start it
   c. jump-start the equipment following the manufacturer's instructions
   d. replace the battery and charge the old battery

5. A _____ should be consulted about the grounding requirements of a generator.
   a. fellow worker
   b. qualified operator
   c. licensed electrical contractor
   d. representative from OSHA

6. If you find evidence of arcing on the control panel of a tow-behind generator, _____.
   a. continue with operation of the generator
   b. replace the control panel
   c. locate and repair the problem before operating
   d. call the manufacturer for additional information

7. The typical range for construction compressors is _____.
   a. 10 to 50 psi
   b. 50 to 125 psi
   c. 125 to 150 psi
   d. 150 to 200 psi

8. The jackhammer is an example of a _____ tool.
   a. pneumatic
   b. hydraulic
   c. hydrostatic
   d. shock-absorbing

9. On a compressor control panel, the gauge which indicates the charge rate of the engine alternator is the _____.
   a. fuel level gauge
   b. hour meter
   c. ammeter gauge
   d. temperature gauge

10. When a portable air compressor is used along with another compressed air source, make sure _____.
    a. there is a check valve installed at the service valve
    b. all air vents are clear
    c. the engine and air compressor air filters are clean
    d. the compressor is on level ground

11. Extra precautions must be taken when using air compressors because _____.
    a. they operate under high pressure
    b. they are very expensive and frequently stolen
    c. they fall apart easily, even under ideal conditions
    d. they have been known to disengage without warning

12. The components of an aerial lift include a _____.
    a. base, auger, and bucket
    b. base, platform, and lifting mechanism
    c. platform, boom, and compactor
    d. compactor, control panel, and forks

13. The _____ allows hydraulic fluid to flow and allows the aerial lift to be pushed by hand.
    a. emergency lowering valve
    b. free wheeling valve
    c. stabilizer hardware
    d. steer controller
14. Aerial lifts may be operated _____.
    a. on any type of surface, so long as proper PPE is worn
    b. on sand or gravel, so long as the area is flat
    c. at least 50 yards away from any school zone
    d. on firm surfaces only
15. The inspection and maintenance of aerial lifts ____.
    a. should only be performed by authorized personnel
    b. is only necessary when a problem arises
    c. may be conducted by any laborer
    d. is the same for all types, regardless of manufacturer
16. Most forklifts used in construction are _____.
    a. electric powered
    b. diesel powered
    c. solar powered
    d. hybrid powered
17. The mast of a typical fixed-mast forklift can be tilted _____ degrees forward.
    a. 19
    b. 39
    c. 59
    d. 79
18. The horizontal reach of a fixed-mast forklift is _____.
    a. virtually limitless due to modern technology
    b. determined by the operator, according to manufacturer's instructions
    c. limited by how close the machine can be driven to the pick-up or landing point
    d. proportional to the weight-height ratio of inverse flexibility
19. The most important factor to consider when using a forklift is _____.
    a. whether it has front-wheel or rear-wheel drive
    b. whether it has three wheels or four
    c. its maximum height capabilities
    d. its intended capacity
20. When placing elevated loads, one of the most significant safety concerns is _____.
    a. birds
    b. visibility
    c. ambient noise
    d. engine overheating
21. The mast of a forklift should be titled rearward to carry the load when transporting _____.
    a. long objects
    b. hazardous waste
    c. cylindrical objects
    d. oversized loads
22. Preventive maintenance of a forklift is the responsibility of _____.
    a. the forklift operator
    b. the general foreman
    c. the shipping clerk
    d. the logistics manager
23. When working with a trencher, what determines the maximum depth of the trench?
    a. The skill of the operator
    b. The proximity to natural rivers and estuaries
    c. The size of the boom
    d. The weight and grade of the soil
24. What control panel function must be selected when the speed and directional control are moved from neutral, or when the digging chain control is moved from the stop position?
    a. Boom depth control
    b. Digging chain control
    c. Speed/direction control
    d. Operator presence or deadman switch
25. Which trench control helps start a cold engine?
    a. Digging chain control
    b. Deadman switch
    c. Ignition switch
    d. Choke

26. Before starting up a pedestrian trencher, _____.
    a. the pump clutch must be fully engaged
    b. the pump clutch must be disengaged
    c. the boom depth control should be checked
    d. the digging chain control should be checked

27. With a pedestrian trench, the speed/directional control, boom depth control, and digging chain control should be checked _____.
    a. before starting the engine
    b. every four weeks
    c. when the engine is idling
    d. if the equipment is to be used for more than four continuous hours

28. When mobiling about a job site, the axle lock on a trencher should be engaged or disengaged to _____.
    a. match ground conditions
    b. prevent explosions
    c. avoid contact with water
    d. save fuel

29. Before starting any digging, excavation, or trenchwork, always call _____ to make sure no underground utilities are in the dig path.
    a. 411
    b. 611
    c. 811
    d. 911

30. When first starting a trencher, _____ of clearance should be allowed between the digging chain and any obstacles.
    a. 1'
    b. 2'
    c. 3'
    d. 4'

31. When operating a trencher on a slope, keep the digging boom _____.
    a. low
    b. high
    c. to the right
    d. to the left

32. Both AC and DC capabilities are required on all power tools powered by a _____.
    a. Compactor
    b. Hydraulic sensor
    c. Welding machine
    d. Mud pump

33. Industrial, heavy-duty machines used for manual welding are typically on a _____ percent duty cycle.
    a. 20
    b. 40
    c. 60
    d. 100

34. When it is necessary to remove water from a depth of greater than 20 feet, a _____ should be used.
    a. trash pump
    b. submersible pump
    c. semi-compact pump
    d. hydrophilic pump

35. When working in an area where low, continuous flow is necessary for moving mud, air mixed with water, and other thick fluids, a _____ should be used.
    a. trash pump
    b. submersible pump
    c. diaphragm pump
    d. mud pump

36. The type of pump that would be used with high-viscosity liquids is the _____ pump.
    a. trash
    b. mud
    c. submersible
    d. centrifugal

37. The _____ are major components of a plate-style compactor.
    a. boom and chain
    b. platform and base
    c. vibrator plate and engine bed
    d. fixed-mast and telescoping boom

38. For a backhoe/loader equipped with all-wheel steering, the mode of steering where the operator can choose independent rear-axle operation is called _____ steering.
    a. rear-axle
    b. two-wheel
    c. circle
    d. all-wheel

39. Rough-terrain mobile cranes that are equipped with oversized tires are known as ____.

    a. cherry pickers
    b. boom trucks
    c. backhoes
    d. lattice boom cranes

Figure RQ01

40. The machine shown in Figure RQ01 is a _____.

    a. cherry picker
    b. carry deck
    c. boom truck
    d. lattice boom crane

Trade Terms Quiz

Fill in the blank with the correct term that you learned from your study of this module.

1. Electrical current is measured in _______.
2. Move bulk items around the job site with a(n) _______.
3. A(n) _______ may be part of a circuit or may simply hold a spare fuse.
4. The machine used to supply air to pneumatic tools is called a(n) _______.
5. A machine that moves liquid using rotation is a(n) _______.
6. A machine that moves liquid using hydraulic pressure is a(n) _______.
7. A(n) _______ is used to provide electricity on a job site.
8. The speed or power of an engine can be automatically controlled by a(n) _______.
9. Another term for forklift is _______.
10. To prevent settling of soil after construction, a(n) _______ is used.
11. The rotating part of a centrifugal pump is called the _______.
12. Personnel, tools, and materials can be raised to overhead work areas using a(n) _______.
13. Two types of electrical connectors are _______ and _______.
14. The _______ is designed to protect circuits from overloads.
15. A truck-mounted lift designed for indoor or outdoor use is known as a(n) _______.
16. A liquid is pushed outward from the center of rotation by _______.
17. Electrical potential is measured in _______.
18. A _______ uses compressed air to do work.
19. A machine that uses a boom and chain to dig is called a(n) _______.
20. A(n) _______ is a unit of power equal to one joule per second.
21. In some pumps, the fluid is moved back and forth by a thin wall called a(n) _______.
22. When driving, or _______, a trencher on rough terrain, keep the boom as low as possible.
23. The speed at which fluid is ejected from a pump is called the _______.
24. A control that increases speed according to the movement of the control is a(n) _______.

Trade Terms

Aerial lift
Amperes (amps)
Centrifugal force
Centrifugal pump
Cherry picker
Circuit breaker
Compactor
Compressor
Diaphragm
Diaphragm pump
Forklift
Fuse holder
Generator
Governor
Impeller
Mobiling
Pneumatic
Powered industrial truck
Proportional control
Straight blade duplex connector
Trencher
Twist-lock connector
Velocity
Volt
Watt

Trade Terms Introduced in This Module

Aerial lift: A mobile work platform designed to transport and raise personnel, tools, and materials to overhead work areas.

Ampere (amp): A unit of electrical current.

Centrifugal force: The force that moves a substance outward from a center of rotation.

Centrifugal pump: A machine that moves a liquid by accelerating the liquid radially from the center of an impeller in motion.

Cherry picker: A truck-mounted or trailer-mounted lift designed for both indoor and outdoor use.

Circuit breaker: A device designed to protect circuits from overloads and be reset after tripping.

Compactor: A machine used to compact soil to prevent the settling of soil after construction.

Compressor: A motor-driven machine used to supply compressed air for pneumatic tools.

Diaphragm: A thin, flexible separating wall that flexes back and forth to move a fluid.

Diaphragm pump: A pump that uses a diaphragm to isolate the liquid being pumped from the operating parts in a mechanically driven pump or from the hydraulic fluid in a hydraulically actuated pump.

Forklift: A machine designed to facilitate the movement of bulk items around the job site.

Fuse holder: A device used to hold a fuse. It may be part of a circuit or may simply hold a spare fuse.

Generator: A machine used to generate electricity.

Governor: A device used to provide automatic control of speed or power for an internal combustion engine.

Impeller: A rotating part inside a pump that increases the speed of the fluid moving through the pump.

Mast: The upright member along which forks on a forklift travel.

Mobiling: Term used for driving a trencher around the job site when not digging.

Pneumatic: Run by or using compressed air.

Powered industrial truck: See forklift.

Proportional control: A control that increases speed in proportion to the movement of the control.

Straight blade duplex connector: An electrical connector or style of outlet.

Trencher: A small machine used for digging trenches.

Twist-lock connector: A type of electrical connector.

Velocity: The speed at which a fluid is ejected from a pump.

Volt: A unit of electrical potential and electromotive force.

Watt: A unit of power that is equal to one joule per second.

Additional Resources

This module presents thorough resources for task training. The following reference material is recommended for further study.

Aerial Lifts: Protect Yourself. 2018. US Department of Labor: Occupational Safety and Health Administration. Available at: **https://www.osha.gov/Publications/aerial_lifts_safety.html**

Excavations and Trenching. 2018. Princeton University: Environmental Health and Safety. Available at: **https://ehs.princeton.edu/workplace-construction/construction-safety/excavations-and-trenching**

Hydraulic Systems Safety. Paul D. Ayers. 1992. Colorado State University Cooperative Extension. Available at: **http://nasdonline.org/static_content/documents/1100/d000891.pdfg**

Oregon OSHA Fact Sheet Plus: Compressed Air Piping Systems. Oregon OSHA. 2011. Available at: **https://osha.oregon.gov/OSHAPubs/factsheets/fs44.pdf**

OSHA Fact Sheet: Using Portable Generators Safely. Oregon OSHA. 2011. Available at: **https://www.osha.gov/OshDoc/data_Hurricane_Facts/portable_generator_safety.pdf**

Preventing Injuries when Working with Hydraulic Excavators and Backhoe Loaders. The Centers for National Institute for Occupational Safety and Health. 2003. Available at: **https://www.cdc.gov/niosh/docs/wp-solutions/2004-107/pdfs/2004-107.pdf**

Safe Fueling Procedures. Indiana Constructors. 2018. **Available at:http://indianaconstructors.org/safe-fueling-procedures/?print=pdf**

Soil Compaction Handbook. 2011. Multiquip Inc.® Available at: **http://www.multiquip.com/multiquip/pdfs/Soil_Compaction_Handbook_low_res_0212_DataId_59525_Version_1.pdf**

Trenching and Excavation Safety. The INGAA Foundation, Inc. 2012. Available at: **http://www.ingaa.org/file.aspx?id=18983**

Figure Credits

DitchWitch®, Figure 26(B)

Ingersoll-Rand Co., Figure 7

JLG Industries, Inc., Figures 8, SA02

Link-Belt Construction Equipment Co., Figures 32, 33

Manitou North America, Figure 23

Manitowoc Crane Group, Figures 30A, 31, Review Question Figure 1

Miller Electric Mfg. Co, Figure SA01

Sellick Equipment Ltd., Figure 14

Terex Cranes, Figure 29

Wacker Corporation, Figure 9

Section Review Answer Key

Section 1.0.0

Answer	Section Reference	Objective
1. d	1.0.0	1a
2. c	1.1.2	1a

Section 2.0.0

Answer	Section Reference	Objective
1. c	2.1.0	2a
2. b	2.3.1	2c
3. a	2.1.0	2c
4. b	2.2.2	2b
5. a	2.3.2	2c

Section 3.0.0

Answer	Section Reference	Objective
1. c	3.0.0	3a
2. a	3.1.0	3a
3. b	3.1.0	3a
4. b	3.2.1	3b
5. c	3.3.1	3c
6. b	3.3.2	3c

Section 4.0.0

Answer	Section Reference	Objective
1. b	4.0.0	4a
2. a	4.1.0	4a
3. c	4.2.0	4b
4. b	4.3.1	4c

SECTION 5.0.0

Answer	Section Reference	Objective
1. c	5.0.1	5a
2. b	5.3.1	5c
3. c	5.2.2	5b

SECTION 6.0.0

Answer	Section Reference	Objective
1. c	6.0.1	6a
2. d	6.1.0	6a
3. a	6.2.1	6b
4. c	6.3.1	6c

SECTION 7.0.0

Answer	Section Reference	Objective
1. c	7.2.1	7b
2. c	7.1.0	7a
3. c	7.2.1	7b
4. b	7.3.0	7c

SECTION 8.0.0

Answer	Section Reference	Objective
1. b	8.1.2	8a
2. c	8.2.1	8b

This page is intentionally left blank.

NCCER CURRICULA — USER UPDATE

NCCER makes every effort to keep its textbooks up-to-date and free of technical errors. We appreciate your help in this process. If you find an error, a typographical mistake, or an inaccuracy in NCCER's curricula, please fill out this form (or a photocopy), or complete the online form at **www.nccer.org/olf**. Be sure to include the exact module ID number, page number, a detailed description, and your recommended correction. Your input will be brought to the attention of the Authoring Team. Thank you for your assistance.

Instructors – If you have an idea for improving this textbook, or have found that additional materials were necessary to teach this module effectively, please let us know so that we may present your suggestions to the Authoring Team.

NCCER Product Development and Revision

13614 Progress Blvd., Alachua, FL 32615

Email: curriculum@nccer.org

Online: www.nccer.org/olf

❑ Trainee Guide ❑ Lesson Plans ❑ Exam ❑ PowerPoints Other ______________

Craft / Level: Copyright Date:

Module ID Number / Title:

Section Number(s):

Description:

Recommended Correction:

Your Name:

Address:

Email: Phone:

This page is intentionally left blank.

Glossary

4-to-1 rule: The safety rule for straight ladders and extension ladders, which states that a ladder should be placed so that the distance between the base of the ladder and the supporting wall is one-fourth the working distance of the ladder.

Aerial lift: A mobile work platform designed to transport and raise personnel, tools, and materials to overhead work areas.

Ampere (amp): A unit of electrical current.

Office of Apprenticeship: The U.S. Department of Labor office that sets the minimum standards for training programs across the country.

Assured equipment grounding conductor program: A detailed plan specifying an employer's required equipment inspections and tests and a schedule for conducting those inspections and tests.

Backfire: A loud snap or pop as a torch flame is extinguished.

Base plates: Flat discs or rectangles under scaffold legs that are used to evenly distribute the weight of the scaffold. Base plates come in different sizes for different scaffold heights.

Bevel: A cut made at an angle; an angle cut or ground on the end of a piece of solid material.

Burr: A sharp, ragged edge of metal usually caused by cutting pipe.

Carburizing flame: A flame burning with an excess amount of fuel; also called a reducing flame.

Casters: Wheels that are attached to the bottoms of scaffold legs instead of base plates. Most casters come with brakes.

Centrifugal force: The force that moves a substance outward from a center of rotation.

Centrifugal pump: A machine that moves a liquid by accelerating the liquid radially from the center of an impeller in motion.

Chamfer: An angle cut or ground only on the edge of a piece of material.

Cherry picker: A truck-mounted or trailer-mounted lift designed for both indoor and outdoor use.

Chucks: Parts of a machine that hold a piece of work tightly in the machine. A chuck is normally used only when the pipe or cutter will be rotated.

Circuit breaker: A device designed to protect circuits from overloads and be reset after tripping.

Compactor: A machine used to compact soil to prevent the settling of soil after construction.

Compressor: A motor-driven machine used to supply compressed air for pneumatic tools.

Conduit: A round raceway, similar to pipe, that contains conductors.

Coupling pin: A steel pin used to line up and join sections of a scaffold.

Cross braces: Steel pieces used to connect and support the vertical uprights of a scaffold.

Diaphragm: A thin, flexible separating wall that flexes back and forth to move a fluid.

Diaphragm pump: A pump that uses a diaphragm to isolate the liquid being pumped from the operating parts in a mechanically driven pump or from the hydraulic fluid in a hydraulically actuated pump.

Die: A tool used to make male threads on a pipe or a bolt.

Drag lines: The lines on the edge of the material that result from the travel of the cutting oxygen stream into, through, and out of the metal.

Dross: The material (oxidized and molten metal) that is expelled from the kerf when cutting using a thermal process. It is sometimes called slag.

Duty rating: American National Standards Institute (ANSI) rating assigned to ladders. It indicates the type of use the ladder is designed for (industrial, commercial, or household) and the maximum working load limit (weight capacity) of the ladder. The working load limit is the maximum combined weight of the user, tools, and any materials bearing down on the rungs of a ladder.

Fabrication: The act of putting together component parts to form an assembly.

Female threads: Threads on the inside of a fitting.

Ferrous metals: Metals containing iron.

Flare: A pipe end that has been forced open to make a joint with a fitting.

Flashback: The flame burning back into the tip, torch, hose, or regulator, causing a high-pitched whistling or hissing sound.

Forklift: A machine designed to facilitate the movement of bulk items around the job site.

Fuse holder: A device used to hold a fuse. It may be part of a circuit or may simply hold a spare fuse.

Generator: A machine used to generate electricity.

Gouging: The process of cutting a groove into a surface.

Governor: A device used to provide automatic control of speed or power for an internal combustion engine.

Ground fault circuit interrupter (GFCI): A fast-acting circuit breaker that senses small imbalances in the circuit caused by current leakage to ground and, in a fraction of a second, shuts off the electricity.

Guardrails: Protective rails attached to a scaffold. The top rail is 42 inches above the scaffold floor, and the middle rail is halfway between the top rail and the toeboard.

Harness: A device that straps securely around the body and is connected to a lifeline. It is part of a personal fall arrest system.

Hinge pin: A small pin inserted through a leg and base plate or caster wheel and used to hold these parts together.

Horsepower (hp): A unit of power equal to 745.7 watts or 33,000 foot-pounds per minute.

Impeller: A rotating part inside a pump that increases the speed of the fluid moving through the pump.

Kerf: The gap produced by a cutting process.

Leveling jack: A threaded, adjustable screw located between the legs and the base plates or caster wheels of a scaffold. It is used to raise or lower parts of a scaffold to level it on uneven surfaces.

Male threads: Threads on the outside of a pipe.

Mast: The upright member along which forks on a forklift travel.

Mobiling: Term used for driving a trencher around the job site when not digging.

Neutral flame: A flame burning with correct proportions of fuel gas and oxygen.

Occupational Safety and Health Administration (OSHA): The federal government agency established to ensure a safe and healthy environment in the workplace.

On-the-job learning (OJL): Job-related learning acquired while working.

Outside diameter: A measurement of the outside width of a pipe.

Oxidizing flame: A flame burning with an excess amount of oxygen.

Pierce: To penetrate through metal plate with an oxyfuel cutting torch.

Pipe fitting: A unit attached to a pipe and used to change the direction of fluid flow, connect a branch line to a main line, close off the end of a line, or join two pipes of the same size or of different sizes.

Pneumatic: Run by or using compressed air.

Powered industrial truck: See forklift.

Proportional control: A control that increases speed in proportion to the movement of the control.

Ratchet: A device that allows a tool to rotate in only one direction.

Revolutions per minute (rpm): The number of complete revolutions an object will make in one minute.

Scaffold: An elevated work platform for workers and materials.

Scaffolding: A temporary built-up framework or suspended platform or work area designed to support workers, materials, and equipment at elevated or otherwise inaccessible job sites.

Scaffold floor: A work area platform made of metal or wood.

Soapstone: Soft, white stone used to mark metal.

Straight blade duplex connector: An electrical connector or style of outlet.

Sweat: A method of joining pipe in which solder is applied to the joint and heated until the solder flows into the joint.

Tack welds: Short welds used to hold parts in place until the final weld is made.

Thread gauge: A tool used to determine how many threads per inch are cut in a tap, die, bolt, nut, or pipe. Also called a pitch gauge.

Toeboard: A 4-inch railing attached around the scaffold floor to prevent tools and materials from falling off the scaffold.

Trencher: A small machine used for digging trenches.

Twist-lock connector: A type of electrical connector.

Velocity: The speed at which a fluid is ejected from a pump.

Vertical upright: The end section of a scaffold, also known as a buck, made of welded steel. It supports other vertical uprights or upper end frames and the scaffold floor.

Volt: A unit of electrical potential and electromotive force.

Washing: A term used to describe the process of cutting out bolts, rivets, previously welded pieces, or other projections from the metal surface.

Watt: A unit of power that is equal to one joule per second.

This page is intentionally left blank.

This page is intentionally left blank.

This page is intentionally left blank.